2nd International Telecommunication Conference on Advanced Micro- and Nanoelectronic Systems and Technologies (AMNST 2017)

IOP Conference Series: Materials Science and Engineering
Volume 498

Moscow, Russia
1 – 2 June 2017

ISBN: 978-1-5108-8555-4
ISSN: 1757-8981

Printed from e-media with permission by:

Curran Associates, Inc.
57 Morehouse Lane
Red Hook, NY 12571

Some format issues inherent in the e-media version may also appear in this print version.

Printed by Curran Associates, Inc. (2019)

For permission requests, please contact the Institute of Physics at the address below.

Institute of Physics
Dirac House, Temple Back
Bristol BS1 6BE UK

Phone: 44 1 17 929 7481
Fax: 44 1 17 920 0979

techtracking@iop.org

Additional copies of this publication are available from:

Curran Associates, Inc.
57 Morehouse Lane
Red Hook, NY 12571 USA
Phone: 845-758-0400
Fax: 845-758-2633
Email: curran@proceedings.com
Web: www.proceedings.com

Table of contents

Volume 498

2nd International Telecommunication Conference "Advanced Micro- and Nanoelectronic Systems and Technologies"

1–2 June 2017, Moscow, Russia

Accepted papers received: 7 February 2019
Published online: 16 April 2019

Preface

Papers

IOP Conf. Series: Materials Science and Engineering **498** (2019) 011001 doi:10.1088/1757-899X/498/1/011001

Second International Telecommunication Conference on Advanced Micro- and Nanoelectronic Systems and Technologies

V S Pershenkov, D S Veselov and G I Zebrev*
National Research Nuclear University MEPhI (MEPhI), Kashirskoe shosse 31, Moscow, 115409, Russian Federation

Corresponding author's e-mail address: GIZebrev@mephi.ru

Second International Telecommunication Conference on Advanced Micro- and Nanoelectronic Systems and Technologies took place from June 1st to June 2nd 2017 at National Research Nuclear University MEPhI (Moscow Engineering Physics Institute). It was organized by the Micro- and Nanoelectronic Department of National Research Nuclear University MEPhI (the oldest Microelectronic department in Russia (USSR).

The Conference consists of one plenary and three oral and poster sessions. The plenary session presented the most significant results of Micro- and Nanoelectronic Department MEPHI and other participants. This Conference provides to participants an opportunity to discuss their scientific and technical results in a wide audience of leading industry experts.

The main topics of the Conference were:

- Semiconductor physics;
- Design of micro- and nanosystems;
- Manufacturing technology of micro- and nanosystems.

More than 150 specialists on Micro- and Nanoelectronics took part in the Conference. They represented leading Russian scientific research centres and universities (such as National Research Center "Kurchatov Institute", All-Russia Research Institute of Automatics (VNIIA), Bauman Moscow State Technical University, Moscow Aviation Institute (National Research University), Moscow Power Engineering Institute, Moscow State University of Mechanical Engineering (MAMI), D. Mendeleev University of Chemical Technology of Russia and many others) and universities from Armenia, Belarus, Ukraine, Germany, Colombia, Spain, South Korea, Republic of Uzbekistan.

We would like to thank all of the speakers, participants and organizing committee members for their contribution to the Conference.

IOP Conf. Series: Materials Science and Engineering **498** (2019) 011002 doi:10.1088/1757-899X/498/1/011002

Peer review statement

All papers published in this volume of *IOP Conference Series: Materials Science and Engineering* have been peer reviewed through processes administered by the proceedings Editors. Reviews were conducted by expert referees to the professional and scientific standards expected of a proceedings journal published by IOP Publishing.

MEMS sensors to assess pain sensitivity of human

A A Podorin, V A Shakhnov

Russian Federation, Moscow, Bauman Moscow State Technical University, Department of Design and Technology of Electronic Devices

Corresponding author's e-mail address: alex@podorin.ru

Abstract. This paper considers the possibility of creation of microelectronic systems to assess human pain sensitivity. The focus is devoted to the description of the known methods of pain sensitivity evaluation, the structure of the system to assess human pain sensitivity, as well as the description of the concept of the individual parts of the system with the use of MEMS sensors. A substantiation is presented of the use of the selected field type to create the impact of pain, as well as increased ease of use when using MEMS sensors for removal of photoplethysmogram, which is used in the evaluation of human pain sensitivity. The model of the device for readout photoplethysmogram data with the use of MEMS sensors is presented.

1. Introduction

The development of methods for determining the pain threshold for various people based on medical records (electrocardiogram [1], photoplethysmogram [1], pressure, etc.) is an important tasks. Now, the only individual approaches are known to solve the problems of automatic evaluation of human response to exposure to standard pain. Another problem is the microminiaturization of devices for estimating the impact of pain, which is associated with methods of determining the level of pain.

For example, the problem of the readout of medical records at the same time as the influence of pain in dentistry considered in [2, 3]. It is proposed to assess pain sensitivity using blood tests and of interleukin in blood.

Gorbonosov and Semenov proposed to use photoplethysmogram to assess pain sensitivity in article [4], however, the problem remains unsolved because the source of pain is the object. On the one hand, this creates objectivity in the study because it is difficult to endure the pain, but at the same time complicates the investigation.

In article [5], a variety of options which are used in the evaluation of the impact of pain and pain sensitivity are presented. The main drawback is the use of subjective methods of measurement because in all the given methods, the man assesses the level of pain himself, but for this he uses different scales. For such a measurement method as in [6], consider the dependencies that are associated with sex, age, race and others. The main disadvantage of these options remains the use of people with chronic pain sensations to reduce the error in the subjective measurement of pain, and general measurement subjectivity.

This work presents an approach to the development of MEMS devices which are able to control the effects of pain and evaluate them by means of intelligent software functions for complex medical criteria: the area of the wave of photoplethysmogram, the value of the area of the wave of photoplethysmogram to the impact of pain, during and after the impact of pain, as well as the average

IOP Conf. Series: Materials Science and Engineering **498** (2019) 012001 doi:10.1088/1757-899X/498/1/012001

values for the wave amplitude and also increase time of photoplethysmogram before, during and after the impact of pain.

The purpose is to assess the possibility of design of microelectronic systems for assessing pain sensitivity based on a typical human response to painful effects.

The use of microelectronic systems in this area will reduce the size of known devices for evaluating the impact of pain, simplifying, and speeding up the results for pain sensitivity analysis due to the minimal number of devices with wired connections and use of wireless connection to other devices.

2. The general structure of human pain sensitivity acquisition and evaluation system.

The structure of the human pain sensitivity analysis system includes the following components:

- pain impact device;
- MEMS sensors for the readout of medical conditions and a device for transmission;
- a device for data processing (computer and software).

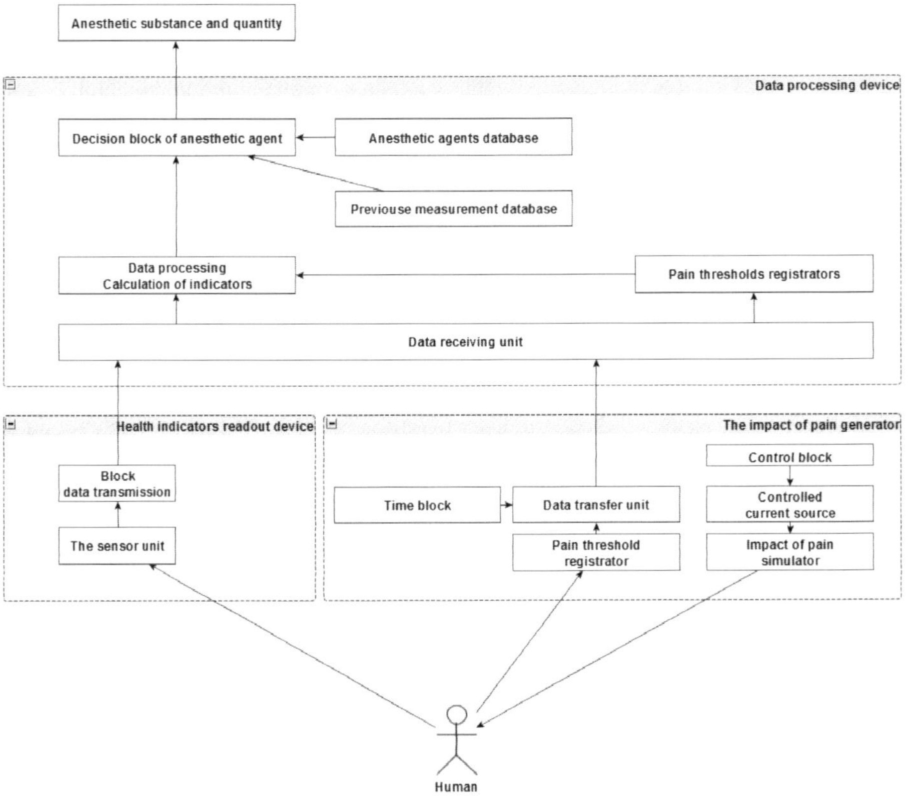

Figure 1. Block diagram of the human pain sensitivity acquisition and analysis system

Anesthetic agent database
Data transmissions block

A person is exposed to the impact of pain from the source of pain impact. Throughout the painful effects, medical records (photoplethysmogram, electrocardiogram, blood pressure and others) are read by a readout device for medical records and transmitted to the processing device (computer with software for data processing). Upon reaching the threshold of pain, a person clicks on the button and press time data is transmitted through the pain impact device to the processing device.

IOP Conf. Series: Materials Science and Engineering **498** (2019) 012001 doi:10.1088/1757-899X/498/1/012001

3. Classification methods of pain impact

When implementing block simulation of pain impact, it is necessary to implement a controlled source of pain. Consider the basic techniques that can be used to create pain impact [7]. In terms of the physical nature of pain, impacts generators are based on the usage of the properties of different physical fields (mechanical, electromagnetic, etc.) to impact human pain.

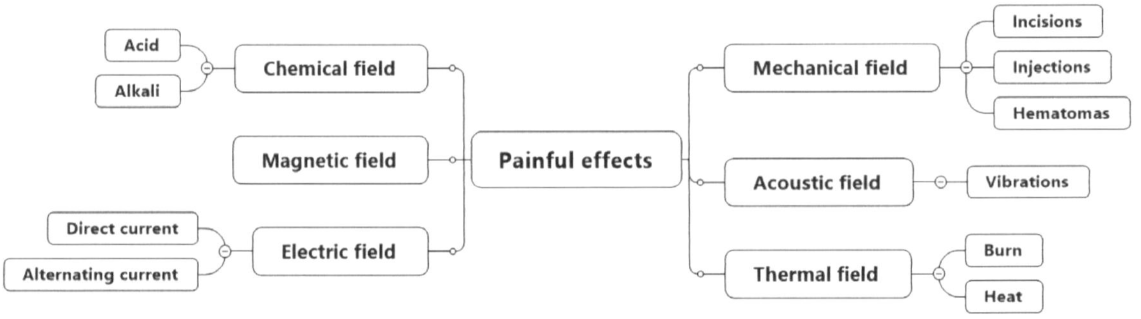

Figure 2. Classification of physical fields by type of painful effects

The mechanical (force) field is presented by various cuts, injections, hematoma, and similar effects. [7] The main disadvantage is the possibility of skin disorders, as well as the inability to accurately manage the impact of pain. It can be used in conjunction with a magnetic field to increase the exposure control, however, this will complicate by several times the system and pain impact procedure (impact tool will be sterilized, or be disposable). This field is difficult to manage in terms of electronic systems, and the use of different tools for applying physical effects requires good fixation, which impedes miniaturization together with the creation of a sufficient level of pain exposure.

The acoustic field is presented by acoustic impact at different frequencies [8]. The nature of the impact is such that it may affect the organs of hearing, as well as on the state of internal organs. At certain frequencies, there may be a resonance with the internal organs, which adversely affects the health of people, and sometimes leads to death. From the standpoint of use in microelectronic systems, this method is interesting because it requires a small amount of energy to create certain frequencies of impacts, nevertheless the possibility of resonance with the internal organs makes this method inapplicable for the purposes at hand.

The thermal field is presented by the impact of temperature. During imitation, it is easily adjustable, however there is the possibility of thermal destruction of the places of contact. In the absence of anesthetic means, the limit of pain sensitivity is likely to be reached before the skin is destroyed, however whilst using anesthesia a subject cannot even feel the burning of the skin and the integrity thereof will be destroyed, which is unacceptable in this case. Thermal effects require a large amount of electricity, which makes it impossible to use wireless systems.

The electric field is presented by the effect of electrical discharge or current. This method allows to create an adjustable painful effect simply, without damaging the object, and makes it easy to produce a level of measurement created by the impact of pain (electric current intensity). It is ideal for use in electronic systems. The main problem is the difficulty in implementing wireless systems, as to assess the needed level of pain it is required to create a voltage of 200V and a current of up to 10mA, which either hinders the creation of wireless systems or requires the use of a sufficiently large battery pack.

The magnetic field - up to a certain point does not affect the object, and after the exposure can be deadly and causes irreparable harm to the object.

The chemical field is presented by the impact of various chemically active substances. It is the main drawback in (замечание: может быть, amidst, «среди») the difficulty of the impact of regulation and material consumption to create these effects. Also, this effect in the majority of the reactions will be destructive, which is unacceptable.

Based on the nature of the discussed effects, it can be concluded that the best way to create controlled impact of pain is the electric field, as electric shock is easy to operate (continuously adjust) (замечание: continuously – это в течении время, может быть, следует использовать слово smoothly), and can be quickly turned off by a fuse in case of emergency.

The source of the impact of pain is a controlled current source. At this stage, the source can be attached to the human skin. Also, this device sends a signal to the processing unit to terminate the painful impact strength of the current at the end of the exposure and end time.

4. The concept of the pain impact device and photoplethysmogram measuring device

These devices are the only ones that are in the known analogues are connected directly to the device using a wired connection, which creates additional inconvenience for the patient and slows downtime measurement response.

The only required information is that of a photoplethysmogram, for which is required a source of infrared radiation and a photoresistor / phototransistor. Also required are a battery pack, the microcontroller unit and wireless information transmission block, for example, Bluetooth.

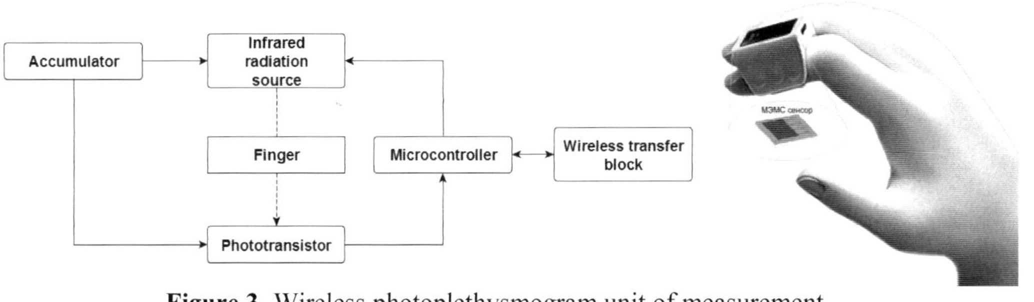

Figure 3. Wireless photoplethysmogram unit of measurement.

This unit may be manufactured in a compact package, analogues of such devices being on the market, however the data transfer protocol is closed, which prevents the use of these devices for third-party projects and restricts their scope to only earmarked values.

The main problem may be the device for creating the pain impact because a sufficiently large amount of energy is needed to create the pain impact, which requires quite a bulky battery pack, as well as the use of the pulsed power supply for 200V output voltage. Based on the calculator from Texas Instruments, for 200V is needed a battery of 12V 0.5A, which takes a moderate amount of space that will allow their use for the wireless measurement.

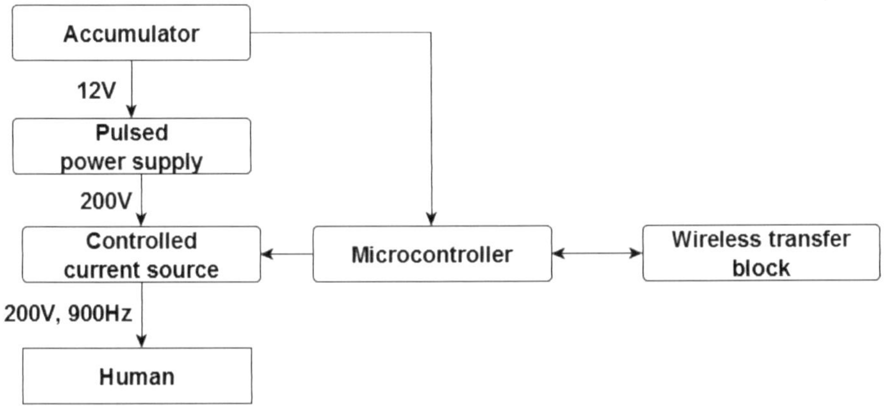

Figure 4. Wireless unit of pain impact

The rest of the device blocks do not interact directly with the patient and their dimensions do not affect the convenience and speed of the procedure.

5. Conclusion

Assessment of pain sensitivity is currently an uncharted area, and known methods of evaluation are subjective and do not give the opportunity to link quantitative indicators with the required number of anesthetic agents to relieve pain in certain medical operations. The difficulty in estimation lies in the use of subjective data, however, the use of large amounts of information reduces the statistical error, and with enough information to move from subjective to objective parameters (photoplethysmogram amplitude, frequency, etc.).

The most simple, accurate and efficient way to create pain is electric current. It is easy to manage and creates a stable pain exposure. The only disadvantage is reduced sensitivity to electrical current for a short time, which is easily eradicated by a change in the measurement site, if re-measurement becomes necessary.

The usage of microelectronic systems makes it possible to reduce the time of measurement, as well as to increase patient comfort by reducing the number of wired connections to the pain sensitivity evaluation unit, nonetheless requires the use of batteries for the wireless data module, which creates some difficulties.

References

[1] Shugailov I A 1984 *Improving the efficiency of anesthesia in the surgical treatment of dental diseases: Ph.D.* (Moscow Dental Institute Semashko)
[2] Melnikova A V and Shugailov I A 2014 *Stomatology* 3 pp 48-52
[3] Shchegoleva V D, Shakhnov V A, Vlasov A I, Boyarkina E S and Kulgashov E V 2008 *Congress of the European Academy of Pediatric Dentistry* 2008. pp. 62-63
[4] Gorbonosov I V and Semenov F V 1994 A method of evaluating the analgesic effect of drugs // Russian patent № 2077261
[5] Mannion A F, Balagué F, Pellisé F and Cedrasch C 2007 Pain measurement in patients with low back pain. *Nature Clinical Practice Rheumatology* **3**.11
[6] Fraenkel L1, Falzer P, Fried T, Kohler M, Peters E, Kerns R and Leventhal H 2012. *J Gen Intern Med.* **27** 5
[7] Ulumbekov E G 2013 *GEOTAR-Media* 2242
[8] Andreev K A, Vlasov A I and Shakhnov V A 2016 *Autom. Remote Control* **77** 7 pp 1281-1285

IOP Conf. Series: Materials Science and Engineering **498** (2019) 012002 doi:10.1088/1757-899X/498/1/012002

Mathematical model of the general problem of state classification in wireless sensor networks

M N Yuldashev and A I Vlasov

Russian Federation, Moscow, Bauman Moscow State Technical University, ul. Baumanskaya 2-ya, 5, Moscow

Corresponding author's e-mail address: yuldashev@gmail.com

Abstract. This paper considers a mathematical model of the general classification problem in wireless sensor networks. We present the fields of application of sensor networks, the existing types of sensors used in data networks are listed, as well as the most typical conditions of sensor networks. The main attention is paid to the analysis of mathematical models of classification tasks, evaluating its parameters. As a result, there has been developed a mathematical model of the problem of the classification of states in wireless sensor networks, taking into account factors such as inaccuracies in the sensor readings, the dynamics of the number of sensors, dynamic network conditions, and their priority. This object allows you to move on to the next stage of the classification network conditions, the development of the classification algorithm.

1. Introduction

The continuous growth of computing power of modern software-controlled devices and the availability of a variety of different sensors has made it possible to build sensory information processing networks that provide the automation of various processes.

Temperature sensors, pressure, sound, motion, and other distance sensors distributed over a controlled area or site gather information about the environment with speed and precision that are ten times higher than when using sensory evaluation techniques. On the basis of the collected data in a short time can be taken the most correct decisions for follow-up, at which the costs of human resources are minimal or nonexistent.

A popular modification of the sensor network is a wireless communication, which expands the range of use by eliminating the need for physical contact between the elements. As a result, sensors can easily be connected to the network, and that serves to be in tight areas or on moving objects. [1]

In the simplest case, the sensor network is a collection of an interconnected wireless network of sensors located in a controlled environment, where the sensor - a module, consisting of the sensor that registers a change in one or the other of the physical parameters of the environment, the data processing unit, and a transceiver battery. Thus, in some systems, the sensor acts as a relay station in a network, thereby achieving a significant coverage area of the network at a low cost.

Wireless sensor networks enable recognition of specific situations in the environment and accordingly respond to them, at which one of the systems can control the behavior of others in algorithms that were made in advance. An important result of combining subsystems is the synergistic effect. Currently, sensor networks are used in industry, agriculture, medicine and other fields. [2]

In recent years, methods of machine learning (Machine Learning) are rapidly gaining popularity. Machine Learning operates with large numbers of structured data and solves the problem of classification, ranking, and regression. They are widely used in various spheres of life, such as medicine, banking, IT, etc. [3]

Machine learning methods focus on the construction of a mathematical model of the behavior of the environment, where the parameters of the system are determined by training to obtain data from the sensors and sample results of the decision of the main module or the operator. The data in the wireless sensor networks are structured and are based on local metadata. This paper proposes the use of a new approach to the application of machine learning to analyze data from the sensor and the subsequent decision of such a problem, as the classification of states of the system and prediction of its behavior in the future. Also proposed are information processing techniques to increase the potential to solve the following tasks: improvement of the noise immunity of the sensor network, adaptation to environmental changes.

The task at hand in using machine learning in sensor networks is the classification of the states of systems as a whole. In other words, the operator does not need to know about the indications of all the sensors (the number may reach hundreds or thousands of units). Nonetheless, the total information is most important to make the correct decisions for future actions. The sensory network system can be described with a limited number of states. In the simplest case, it is sufficient to identify the two states: "correct" and "incorrect."

2. Sensors in wireless sensor networks

The main task of sensor networks is to define the parameters of the environment. For this purpose, we developed a great number of sensors: sensors of temperature, pressure, humidity, light, sound, position, flow and concentration of a substance, vibration frequency, speed, distance, voltage, electric current, and radioactivity.

The development of MEMS technology has reduced the size of the sensors to micro sizes, which expanded the range of applications of sensor networks. For example, it was possible to significantly reduce power consumption, and hence the size of power supplies, control parameters of small objects, place a large number of sensors on the same object, mask sensors.

The main characteristics of the above sensors are sensitivity and accuracy, and for MEMS sensor to also take into account the dimensions of the elements. [4]

3. States of the wireless sensor network

Most wireless sensor networks, of which the main task is to monitor the area or object, have the following standard conditions: violation of temperature conditions, fire, flood, gas leak, explosion, increased vibration, destruction of the object, foreign object detection, lack of controllable objects, etc. Each state of the system is characterized by a certain set of conditions for the environmental parameters and/or object. For example, for fire characterized by a sharp increase in temperature, decrease in air humidity; the destruction of the object - an abrupt change in frequency of the object, change of noise, sharp claps, change in the position of the object, etc. It is worth noting that some of the states of the system cannot be determined only by static reasons, and it is necessary to know their dynamics. [5]

In practice, at one moment a wireless sensor network may have multiple states. For example, fire, explosion or excessive vibration and temperature violation. On the one hand, the operator, first of all, needs to know about the most important state of the system (for example, a fire is more important than the finding of foreign objects), on the other hand, it is necessary to know about the most dynamic states (for example, the consequences of the explosion will come faster than from flood). All the while, different parameters of the sensor networks priority system may differ materially depending on the task.

4. Mathematical model of the general problem of state classification in wireless sensor networks

Let X be a plurality of indications of the sensor network, Y – the set of states of a wireless sensor network.

Each sensor $\overrightarrow{x_k} \in X$, where k = 1..K (K – number of sensors in a sensor network) is a vector in N-dimensional space, where N – the number of unique parameters obtained from the external environment. The value x_{kn} in most cases is a *binary* or *quantitative* trait. [6]

In other words, at a certain point in time the sensor network is represented by a matrix of objects-signs:

$$\begin{pmatrix} x_{11} & x_{21} & \cdots & x_{k1} \\ x_{12} & x_{22} & \cdots & x_{k2} \\ \vdots & \vdots & \ddots & \vdots \\ x_{1n} & x_{2n} & \cdots & x_{kn} \end{pmatrix} \tag{1}$$

Gaps in the data.

Each sensor transmits readings from 1 to N parameters of the environment, i.e., gaps can be present in the data. The most common omissions in the matrix are arranged not by chance, that is, for a long period of time the sensor will not provide data on certain parameters. It turns out that there are many $P \subseteq X$, where $\forall x_{kn} \in P : x_{kn} \notin \mathbb{R}$.

The errors in the data.

As mentioned earlier, the sensors are characterized by error. Consequently, an additional condition to the problem of the classification algorithm is established to ensure noise immunity when errors are permissible.

Sensor tolerance is due to its type, model, and component elements. Also, operating conditions affect the error. It can be considered in order to simplify the problem that one environment parameter is measured by one sensor model: $x_{kn} = x'_{kn} \pm \delta_n$, where n – the measured value of the environment, k = 1..K (K – the number of sensors in a sensor network), δ_n - error for the selected type of sensor. If one environmental parameter is measured by several types of sensors, the number of N can accept a number of different types of sensors, and as a result, the problem is reduced to the above condition.

The dynamics of the sensors and the network conditions.

On the one hand, the number of sensors in the network may constantly change: the defective sensors can be removed from the network, and at any time new can be added. The system should take account of this provision and continue to perform the classification. In other words, the set is not limited, and at some point in time, the system operates with a specific subset $X' \subseteq X$. On the other hand, the set of states of the system, too, has its own dynamics. Some state sensor networks may lose relevance, while the emergence of new types of sensors can be monitored by earlier unknown states. Therefore, the set of states Y is also not limited, however, at some point, the set of parameters of the sensor network $X' \subseteq X$ system can have a plurality of states $Y' \subseteq Y$.

The task of classification.

The problem of classifying the wireless sensor network is to establish a state of dependence (display) $y^{*}\colon X' \to Y'$.

A mathematical model of the general problem of state classification in wireless sensor networks:

Let X - many kinds of indications of the sensor network, Y – set of all possible states of a wireless sensor network. In a certain period of time $t = [T_x; T_{x+1}]$ sensor network described by subsets $X' \subseteq X$ и $Y' \subseteq Y$:

$$X' = \begin{pmatrix} x_{11} & x_{21} & \cdots & x_{k1} \\ x_{12} & x_{22} & \cdots & x_{k2} \\ \vdots & \vdots & \ddots & \vdots \\ x_{1n} & x_{2n} & \cdots & x_{kn} \end{pmatrix}, Y' = \begin{pmatrix} y_1 \\ y_2 \\ \vdots \\ y_L \end{pmatrix}, \quad (2)$$

where k = 1..K (K – the number of sensors in a sensor network at $t = [T_x; T_x)$, n = 1..N (N – number of different sensors at $t = [T_x; T_x)$, l = 1..L (L – the number of wireless sensor network states in $t = [T_x; T_x)$. Wherein $\exists P \subseteq X$, where $\forall x_{kn} \in P : x_{kn} \notin P$ and for $\forall x_{kn} \notin P : x_{kn} = x'_{kn} \pm j$, where j - error for a particular type of sensor.

It is necessary to establish a relationship $y^*: X' \to Y$ taking into account the priorities and the dynamics of network conditions.

To solve this problem, it can be used by machine learning methods. Training on case law implies a finite training sample $X'^M: \{\{x_1, y_1\}, \{x_2, y_2\}, \ldots, \{x_m, y_m\}\}$ [7].

5. Conclusion

A wide variety of sensors allow you to define different states of environment or object by means of a sensor network. If the data from the sensors are displayed to the operator directly, then it must independently analyze the information and make a decision on the system status. Classification function can be passed on to the network. To this end, this paper presents a mathematical model formulation of the general problem of the classification status of the wireless sensor network.

The most popular sensors of sensor networks: a temperature sensor, pressure, humidity, light, sound, position, flow, and concentration of a substance, the frequency of vibration, speed, distance, voltage, electric current, radioactivity. Their main characteristics are the sensitivity, accuracy and overall dimensions.

A wireless network in a certain time interval has a limited number of states, while it is at a particular point can be in one or more states. To classify the important priority of the state and its dynamics.

References:

[1] Yedavalli R K and Belapurkar R K 2011 *J. Control Theory and Appl.* **9** 28-33

[2] Yuldashev M N, Adamov A P and Adamov A A 2016 *Problems of development of advanced micro- and nanoelectronic systems (MES)* **2** pp 248-251

[3] The method for wireless sensor network operation, RF patent for invention №2556423 from 05.07.2013

[4] Michalski R S, Carbonell J G and Mitchell T M 1983 *Machine Learning: An Artificial Intelligence Approach*

[5] Mitchell T. Machine Learning. 1997 *McGraw-HillScience/Engineering Math*

[6] Xu G, Shen W and Wang X 2014 *Sensors* **14** 16932-16954

[7] Yang X and Gandomi A 2012 *Engineering Computations* **29** 464-483

Requirements for radio module of wireless decentralized network

R F Azizov, D A Aminev and Sh M Bakirov

Russian Federation, Moscow, Bauman Moscow State Technical University, Department of Design and Technology of Electronic Devices

Corresponding author's e-mail address: aminev.d.a@ya.ru

Abstract. This paper considers the decentralized network and the features of its functioning. Within it are analyzed the problem of designing radio devices for wireless decentralized networks and the classification of their requirements. The requirements for the software are classified, specifying the structure of the network, protocols, and hardware - the frequency and transmit power, power consumption, dimensions, etc. There are disclosed recommendations for the transceiver design on some of the constituent elements of the radio module.

1. Introduction

The rapid growth of wireless data transmission networks is due to the development of information technologies, reduction of cost, overall dimensions and power consumption of electronic devices in general and small-sized receiver-transmitters including [1-3].

In the decentralized peer ad hoc network, all nodes are equal. Each node on the network can act as a repeater. Lack of leading allows network devices to operate independently of the composition and amount of the radio network participants [4, 5]. A generalized scheme of a decentralized network structure and radio module is shown in Figure 1.

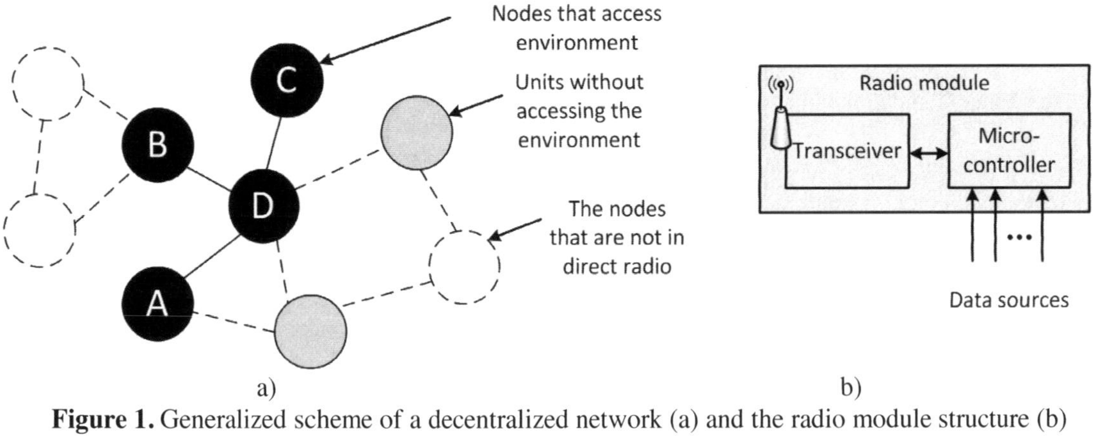

Figure 1. Generalized scheme of a decentralized network (a) and the radio module structure (b)

This structure of the radio module is due to in part to the fact that it became possible to develop complex routing algorithms, compression and data integrity checks directly on the CPU of mobile devices using the operating system tools. This simplifies the process of development of special software and integration into the existing software environment and provides a broad set of software development tools. The receiver-transmitter is necessary to implement only the basic algorithms for broadcast [6].

Nevertheless, the construction of a wireless system in a decentralized network of data exchange is a complex scientific and technical problem in which is placed the radio module design task, taking into account the set of requirements of diverse requirements.

2. Requirements for the radio module of the wireless decentralized network

Basic requirements for the radio module, similar to the requirements of the transceiver, part of the radio module and meet the requirements for transmitters which are not subject to licensing in accordance with the decision of The State Commission for Radio Frequencies [7]. These include requirements to: the dimensional characteristics; value; the frequency at which the transceiver operates; autonomous work; computing power.

In turn, the requirements for the transceiver are formed under the influence of requirements, applicable to:

– the structure of the network (depending on the scenario of application, the volume of data transmitted and the reliability requirements);
– frequency (depending on the desired speed and the allowable size of the desired distance data);
– power (depending on the communication range, the allowable energy and allowable size);
– nutrition (depending on the size characteristics and the nature of the data exchange requirements for power supply);
– antenna (depending on the transmission distance, signal direction and allowable size of the device).

The classification of requirements for radio module is shown in Fig. 2.

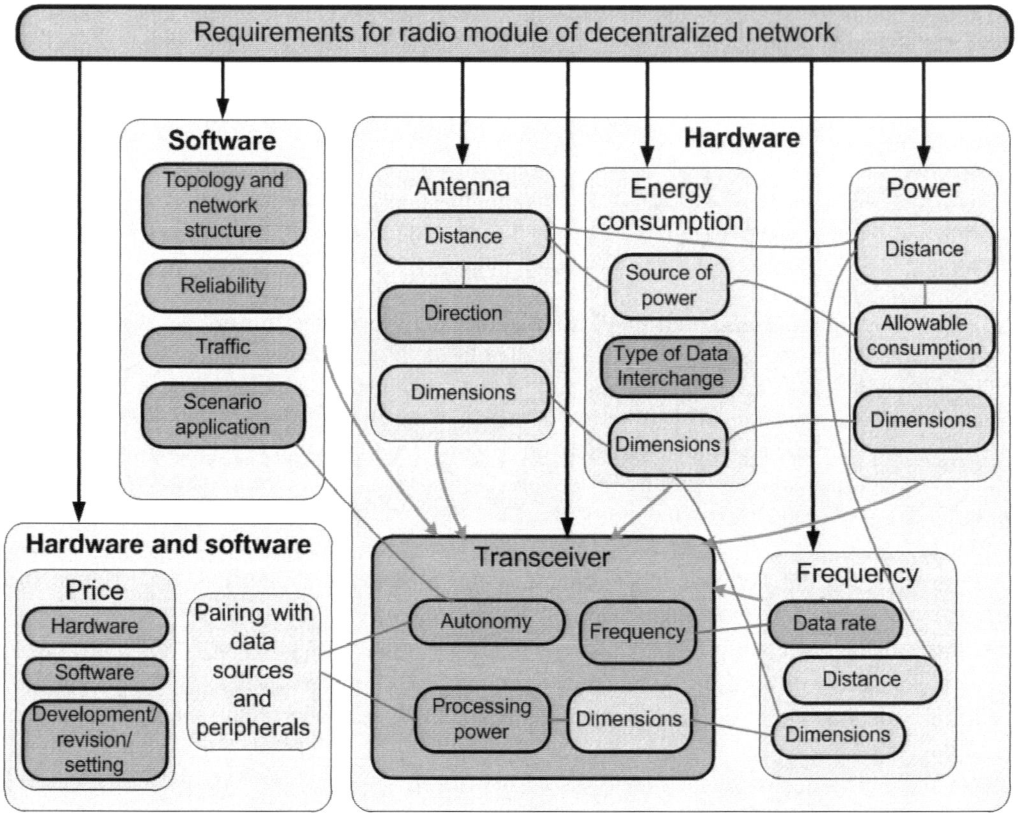

Figure 2. Classification of requirements for radio module

The network structure is determined by the application scenario: a set of requirements to the logic of building the information exchange specified in the terms of reference; network with one rank or a lot of rank. The key requirements are necessary degree of mobility of the nodes, the need to build a dynamic routing information exchange between members of the network, the relationship (rank) sites, the possibility of using the network coordinator, the requirements for the volume of transmitted data, the reliability of data delivery requirements.

The reliability of the network can be achieved using the following options:

- packages of delivery-confirmation of data (acknowledgment);
- checksum;
- error correction codes;
- high-level protocol with delivery control and data integrity.

The required data rate determines the carrier frequency and method of encoding (OOK, QPSK, GFSK, 2FSK, 4FSK) [8-10]. Great influence on the choice of the frequency used is provided by the limits of overall dimensions of the antenna since in such embedded solutions the antenna can be the dimensional unit.

The power increase leads to increase in the communication range, but likewise increases energy consumption, size, and complexity of the scheme. It should also be noted that high-power transmitters may interfere with other members of the network, in other words, significantly increase the load on the network.

The power supply system depends on the scenario of using the transmitters, which is influenced by the degree of mobility of devices; maximum permissible overall dimensions; maximum weight; requirements for tolerability (ammunition).

The transmitter must meet the requirements for frequency, output power, time- and speed-related characteristics of the network, have the computing power to ensure the functioning of the selected network type.

The radio transmitter may be a separate device, which is used to control a single processing core, such as a microcontroller, as well as system-on-chip (SoC), where the transceiver with the nucleus are combined in a single device. The choice depends on priorities: smaller dimensions and less processing power or more computing capabilities with a separate control unit. It is proposed to consider three options for the implementation of the transceiver:

– simple logic in SoC for small networks;
– complex logic in a single powerful computer core;
– full or partial support of the network stack by means of the operating system kernel [11-14], a simplified transceiver control.

For example, for a simple radio network can be implemented on a microcontroller STM32F031P4F6 (16/4) ST Microelectronics Transceiver CC1101 by the company Texas Instruments, and for a complex network used a STM32L030 microcontroller (64/8). In the case of software integration using the system-on-chip SoC 430 from Texas Instruments with the installation of the Linux networking stack.

The antenna is one of the key elements that determine the quality of the wireless network performance and moreover has a significant impact on the final size of the device. Its main characteristics are: length, gain, directivity, overall performance, and price of production. The most common types of antennas used in the mobile decentralized network nodes are whip, chip, and PCB antenna [10]. Whip antennas are high gain (2-10 dBi), with a wide acceptance angle and relatively high cost. Nonetheless, their use is rather difficult in compact devices. Chip antennas have average characteristics (0-5dBi) and small dimensions, characterized by the simplicity of installation and average cost. PCB antennas have good characteristics at high frequencies and low cost. The market has a large selection of ready-made solutions.

A distinctive feature of the receiver-transmitter decentralized network is the inability to predict the direction and distance to the site it is desired to link to. Therefore, there is a need for antennas with high radiation pattern and high gain. The ideal antenna must have a low cost, small size and still have good performance. However, during the design of the antenna it is usually necessary to reach a compromise between these characteristics.

Selecting an antenna for a receiving-transmitter self-organizing decentralized network is a challenging task and its solution is offered by a number of recommendations:

– recommendations for the definition of requirements for mobility network participants;
– recommendations for the definition of requirements for the mass-dimensional parameters of network nodes;
– recommendations on the choice of the carrier frequency used.

For non-mobile nodes can be topical use of directional antennas, however, for dynamically changing networks should be chosen antennas with maximum radiation pattern.

For devices with hard limits on the size of the structural elements may be most applicable PCB and chip antennas. Antenna size is usually determined by the wavelength, and hence the carrier frequency. The higher the frequency used, the smaller the overall size of the antenna. Therefore, due to strict requirements for mass-dimensional indicators of receiving and transmitting devices, high frequency transceivers may be a higher priority.

There transceivers, operating in the unlicensed zones 868 MHz and 2.4 GHz, have attained widespread use [10]. The use of transmitters at high frequencies can significantly reduce weight and size parameters of the antenna and the decentralized network nodes in the whole. Thus, the proposed recommendations take into account the basic requirements for the development of decentralized networks of nodes and allow to choose the best variant of the antenna.

Power supply of the radio module can be implemented based on the following elements [10, 11]:

- not replaceable (solar panels, vibration, radio or other energy sources);
- rarely replaceable elements (batteries);
- often replaced elements (batteries or accumulators);
- stationary power supply (in the case of a limited autonomy).

3. Conclusions

Thus, for the design of the radio module for the decentralized network, the basic requirements are: dimensional characteristics; value; the frequency at which the transceiver operates; autonomous work; computing power.

Taking into account the considered requirements are implemented the following series of receiving and transmitting radio modules at frequencies of 433/868 MHz in the company "Open development" [15]:

– compact and low-power module OD-RF1101 with microcontroller, with data rate of up to 500 kbps;

– ISM module for loaded networks with short-range data transfer, with data rate of up to 500 kbps, range up to 1.5 km;

– OD-LORA module for data transmission over long distances (up to 15 km), with data rate of 37 kbps.

References

[1] Vishnevsky V M, Portnoi S L and Shakhnovich I V: *WiMAX Encyclopaedia. Way to 4G* pp 470 (Tekhnosfera, Moscow (2010))
[2] Basharin G, Gaidamaka Y and Samouylov K: 2013 *Autom. Control Comput. Sci.* **47**(2) 62–69
[3] Stasiak M, Glabowski M, Wisniewski A and Zwierzykowski P *Modelling and Dimensioning of Mobile Wireless Networks: From GSM to LTE* (Willey, Chichester (2010))
[4] Vishnevsky V M and Semenova O V 2012 *Academic Publishing* (London) p 317
[5] Iversen V: *Teletraffic Engineering and Network Planning* (Technical University of Denmark, Lyngby (2011))
[6] *Information Technology - Telecommunications and Information Exchange Between Systems - Local and Metropolitan Area Networks* - Specific Requirements Part 15.4: Wireless Medium Access Control (MAC) and Physical Layer (PHY) Specifications for Low-Rate Wireless Personal Area Networks (WPANs) - Amendment: Alternative Physical Layer Extension to Support the Japanese 950MHz Band / Note: supplement to ANSI/IEEE 802.15.4-2006*Approved 2009-07-28
[7] The decision a meeting of GKRCh of the Ministry of Telecom and Mass Communications of the Russian Federation of July 1, 2016 (protocol No. 16-37). El. resource. URL: http://minsvyaz.ru/ru/documents/5138/. Date of the address 09.01.2017
[8] Huang T, Chen Z, Xia F, Jin C and Liang L 2010 *Conf. on Green Computing and Communications (GreenCom) & Cyber, Physical and Social Computing (CPSCom)* pp 50–54
[9] Casas P, Fillatre L, Vaton S and Nikiforov I. 2010, *J. Netw. Syst. Manage* **19**(1) 58–83
[10] Aminev D A and Azizov R F 2015, *Distributed computer and communication networks: control, computation, communications (DCCN-2015)* ISBN 978-5-91450-170-6, pp. 394-399
[11] Schneps-Schneppe M and Namiot D 2012 *8th Advanced International Conference on Telecommunications, AICT* pp 18–23
[12] Chin E, Felt A P, Greenwood K and Wagner D 2011 *Proc. 9th Int. Conf. on Mob. Syst. Appl. Services* pp. 239–252 http://dx.doi.org/10.1145/1999995.2000018
[13] Shakhnov V A, Vlasov A I, et al. Method of functioning of wireless touch network, Patent for the invention of RUS 2556423 05.07.2013
[14] Vlasov A, Yudin A 2011, *Communications in Computer and Information Science* **156** 180-192 the address 09.01.2017

Features of the model of main functional failures of digital CMOS VLSIs under the action of ionizing radiation

V M Barbashov and N S Trushkin

National Research Nuclear University MEPhI (Moscow Engineering Physics Institute), Kashirskoe shosse 31, 115409 Moscow, Russian Federation

Corresponding author's e-mail address: VMBarbashov@mephi.ru

Annotation. The methods of functional-logical modeling of very large digital integrated circuits (VLSI) under the influence of ionizing radiation are considered. It is shown that the multiplicity of the node and the power of the spectrum are more accurately described when using the concept of fuzzy multiplicity. Methods are proposed for predicting LSI failures when exposed to ionizing radiation, which are based on the model of fuzzy digital and probabilistic reliability automata.

1. Introduction

As a rule, under the influence of radiation, the model of a digital automaton using the apparatus of probability theory is used. In the case when it is necessary to take into account the physical mechanisms of VLSI failure, the construction of a functional-logic model of this class assumes the use of a fuzzy digital Brauer automaton and a transition from the axiomatics of the Boolean lattice to the axiomatics of the vector lattice for each $x \in X$ [1]. The real nature of VLSI radiation behavior is determined by the ratio of the radiation-sensitive parameters of its elements, taking into account the multiplicity of the node and the power of the VLSI spectrum. In this case, a more accurate estimate is possible when using the notion of fuzzy multiplicity. The relationship between the probability density distribution function and the criterial membership function (CFR) determines, in the final analysis, the appropriateness of using functional-logical models of VLSI radiation behavior for each specific case. Such a comparison is a necessary step in the analysis of the VLSI radiation resistance.

2. Topology tree of the modeling environment

In paper [1], the multiplicity of the node and the power of the spectrum of the VLSI tree were determined to estimate the power of the tree spectrum of the CMOS VLSI topology with the use of fuzzy numbers under the radiation effect. In this case, a more accurate assessment of the LSI stability is possible when using the notion of fuzzy multiplicity.

As a rule, with increasing the absorbed dose of ionizing radiation, the amplitude of the pulse of the logical element $l \in [0,1]$ decreases [2], and the intrinsic probability of element switching also decreases. This decrease in probability can be interpreted as an increase in the threshold of operation $l_{\Pi} \in [0,1]$.

As an example, Fig. 1 shows the experimental results of the logical zero output voltage spread for CMOS VSLI chips of RAM 1617RU6 and their transistor structures that correspond to the normal

Gaussian distribution. Since the dispersion σ^2 depends on the absorbed radiation dose, the maximum dose (6) can also be used to determine the maximum radiation dose.

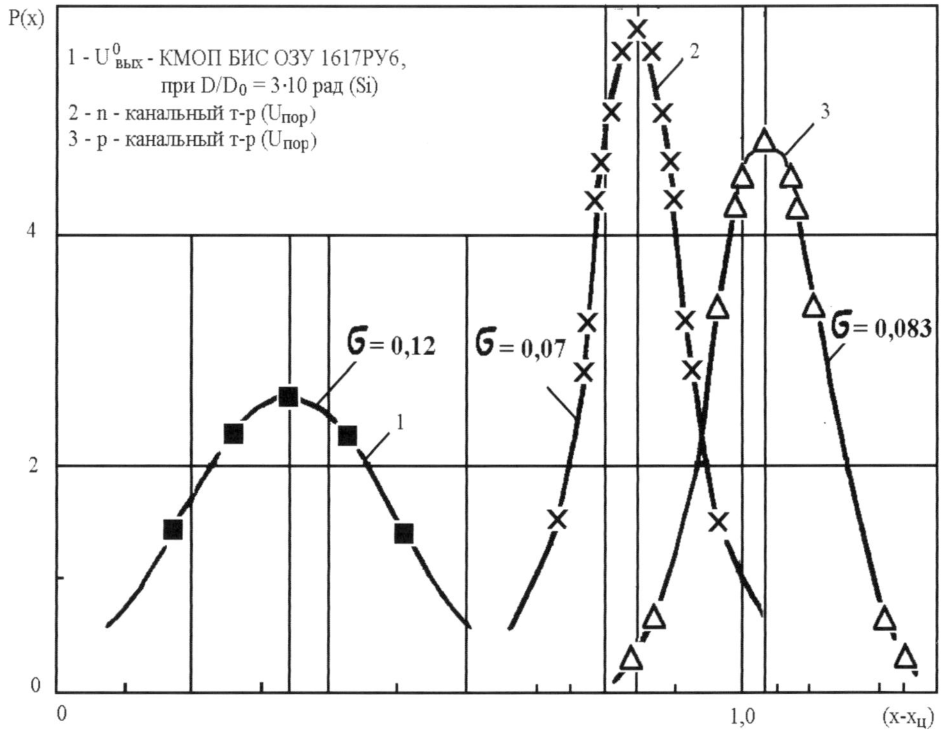

Figure 1. Distributions of averaged voltages over the chip logical zero output of CMOS VSLI RAM 1617RU6 from the dose (1), threshold voltage of n-channel and p-channel MOS transistors (2, 3) obtained by an electron accelerator with Ee = 130 keV

Let n branches leave the node in logical element connections. The node is selected with the maximum number of links (critical node). The multiplicity of the node $m = n$ is realized during normal operation. At the same time, if $l_П$ starts to approach l, an instability zone appears in which the elements switch randomly and the multiplicity of the node becomes undefined. Its magnitude, as is known [3], can be expressed by an indistinct number. To simplify the estimates, triangular $(L - R)$ fuzzy numbers are used in this paper.

At the same time, the thresholds $l_П$ are distributed according to some $f(x)$ law such that $f(x) \underset{\substack{x\to 0\\ x\to l}}{\to} 0$ (Fig. 2). Since the mathematical expectation $\xi = l_П$, the increase in the radiation dose $\xi \to l$ and the distribution is shifted to the right (dotted line in Fig. 2).

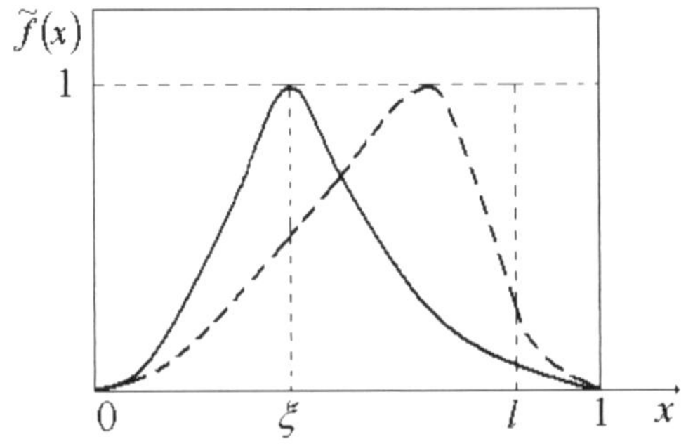

Figure 2. Logic threshold switching threshold function - $\tilde{f}(x)=\dfrac{f(x)}{f(\xi)}$

The fuzzy number mode obviously depends on the magnitude

$$S=\int_0^l f(x)dx,$$

and the blurring is expressed by the relation [4]:

$$\min\{(S(\xi)-S(l)),\ (S_m(\xi)-S(l))\},$$

where S_m is the maximal value of S.

In this case, we consider fuzzy numbers of the multiplicity of nodes for two distribution functions of l_{II} - Gauss and Laplace.

For the normal Gaussian distribution law, the following equality holds:

$$S_G=\frac{1}{\sqrt{2\pi}\cdot\sigma}\int_0^l e^{-\frac{(x-\xi)^2}{2\sigma^2}}dx=\Phi\left(\frac{\xi}{\sigma}\right)+\Phi\left(\frac{l-\xi}{\sigma}\right), \quad (1)$$

$\Phi(z)$ is the probability integral.

On the right in (1), the function is symmetric with respect to $\xi=\frac{l}{2}$ and there is a weak maximum equal to $2\Phi\frac{l}{2\sigma}$ while the ratio $\frac{l}{\sigma}>4$.

This feature of the function makes it possible to use a simple trapezoidal approximation (Figure 3) [5], where $\alpha=\frac{l}{\sigma}$, $x=\frac{\xi}{\sigma}$, $\alpha>4$, $0\le x\le\alpha$.

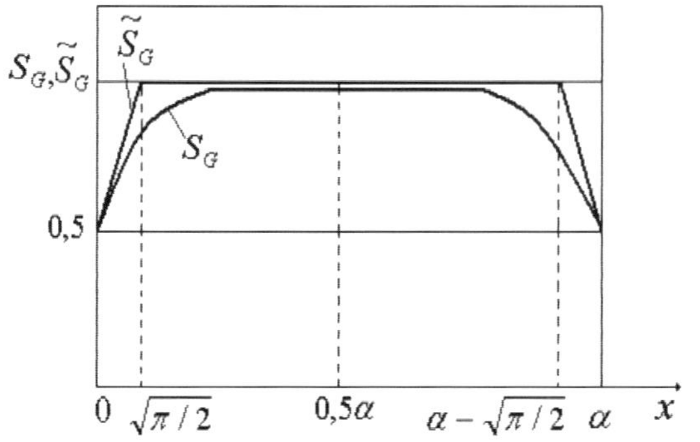

Figure 3. Trapezoidal approximation of S_G for the normal Gaussian distribution law

In this case, the fuzzy-number mode can be written in the form of the expression:

$$M_G = [n \cdot \widetilde{S}_G(x)], \tag{2}$$

The blur takes the following form:

$$\Delta_G = 2[n \cdot \min\{(\widetilde{S}_G(x) - 0{,}5), (1 - \widetilde{S}_G(x))\}], \tag{3}$$

where $\alpha - \sqrt{\pi/2} < x < \alpha$.

In the expressions (2) and (3), [Z] is the whole part of Z.

Then from formulas (2) and (3) we obtain a $(L - R)$ fuzzy number, which is represented as an isosceles triangle with a base Δ_G (Fig. 4).

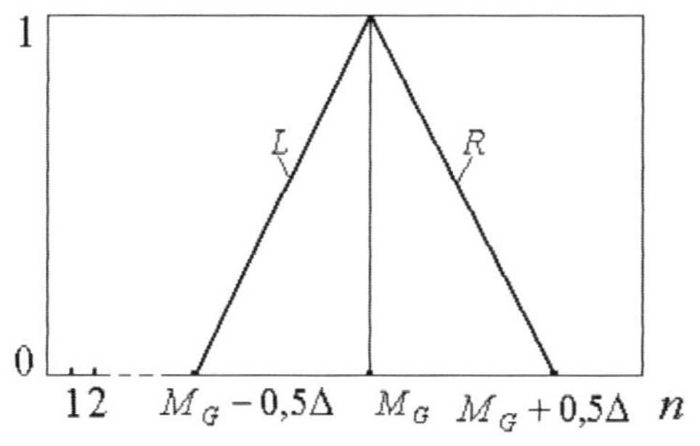

Figure 4. Multiplicity fuzzy number of the node connecting the logical elements

If $x < \alpha - \frac{1}{2}\sqrt{\frac{\pi}{2}}$, then the blurring does not include n and corresponding absorbed dose of ionizing radiation is considered catastrophic. In this case, the limiting absorbed dose is determined by the following equalities:

$$x_{\max} = \alpha - \frac{1}{2}\sqrt{\frac{\pi}{2}} \text{ или } \xi_{\max} = l - \frac{\sigma}{2}\sqrt{\frac{\pi}{2}} \approx l - 063\sigma, \tag{4}$$

The obtained expressions (4) are valid for one node.

For a real system, there are tree intersections, which greatly complicates the analysis. For the Laplace distribution we have:

$$S_L = \frac{1}{2\beta}\int_0^l e^{-\frac{|t-\xi|}{\beta}} dt = 1 - \frac{e^{x-\alpha} + e^{-x}}{2}, \tag{5}$$

where $x = \frac{\xi}{\beta}$, $\alpha = \frac{l}{\beta}$, $\alpha > x, \alpha > 4$.

The Laplace distribution function (5) is analogous to the Gaussian distribution (1) and can also be approximated by a trapezoidal function $\widetilde{S}_L$ (Fig. 5).

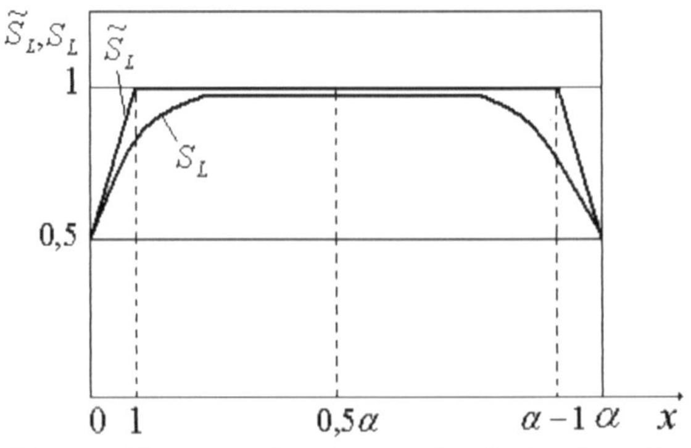

Figure 5. Trapezoidal approximation S_L for the Laplace distribution.

In this case, the mode and the blurring of the fuzzy number are correspondingly equal to the relations [6]:

$$M_L = [n \cdot \widetilde{S}_L(x)], \tag{6}$$

$$\Delta_L = 2[n \cdot \min\{(\widetilde{S}_L(x) - 0{,}5), (1 - \widetilde{S}_L(x))\}], \tag{7}$$

where $[Z]$ is the integer part of Z, $\alpha - 1 < x < \alpha$.

The use of the Laplace distribution is preferable, since it has a much smaller spread at the area of mathematical expectation.

For asymmetric distribution laws, formulas (2) - (7) remain valid, although with different limits of variation of the generalized parameter x. The normalization of x is effected by a certain characteristic parameter of the distribution spreading [7].

To estimate the power of a tree, it is necessary to sum up the fuzzy numbers, however the mean values, as determined in [2], are calculated through the mean value of the modes and the blurrings.

In this case, the average value of the mode is:

$$M_\Sigma = \frac{\sum_{i=1}^{K} M_i}{K}, \tag{8}$$

where K is the number of nodes in the tree, M_i is the mode of the fuzzy number of the node.

The average blur is determined from the ratio:

$$\Delta_\Sigma = \frac{\sum_{i}^{K} \Delta_i}{K}, \tag{9}$$

where Δ_i is the fuzzy number blur for the i-th node.

Thus, the power of the tree is also expressed by an indistinct number, though the mode and fuzziness of which, determined by formulas (8) and (9), are not necessarily integer. If the blur does not capture the maximum power values given in [2], then the corresponding absorbed radiation dose is considered catastrophic, and less than that which is had for one node [8].

IOP Conf. Series: Materials Science and Engineering **498** (2019) 012004 doi:10.1088/1757-899X/498/1/012004

However, this assessment remains overestimated, since it does not take into account the fact that for nodes of the intermediate level of the tree, if the element is not switched on, the multiplicity of the corresponding node is considered equal to 0.

3. Conclusion

The real nature of CMOS VLSI performance under ionizing radiation is determined by the specific spectrum of the VLSI topology tree and the ratio of radiation-sensitive parameters: the multiplicity of nodes, the power of the tree spectrum, and the fuzzy-number mode. In this case, the relationship between the distribution function of the spread and the fuzzy number mode determines, ultimately, the appropriateness of the use of functional-logical models of VLSI radiation behavior for each specific case. Providing such a comparison is a necessary stage in the general procedure of the VLSI radiation resistance analysis. It should be noted that in different ranges of levels or intensity of impact, the power assessment of the tree spectrum of the VLSI topology can be either fuzzy or probabilistic. In this case, fuzzy multiplicity can be specified in n-dimensional space with preservation of the basic model parameters. Radiation effects arising in VLSI at different irradiation levels are estimated in the model by changing the same parameter of the system. Such a structure of the model is probabilistic-fuzzy with operators of probability type, and the criterion membership function is a superposition of the statistical and deterministic criterion membership function. At the same time, the interdependence of fuzzy multiplicity and probabilistic logic is determined, which allows the most accurate estimation of the VLSI functioning under the influence of radiation.

References

[1] Barbashov V M and Trushkin N S 2016 *IOP Conf. Ser .: Mater. Sci. Eng.* **151** 012011
[2] Barbashov V M, Podlepetsky B I and Trushkin N S 2016, *Sensors and Systems* **4** pp 54-57
[3] *Fuzzy sets and the theory of possibilities* 1986, Ed. R.R. Yager. M . (Radio and Communication) pp 408
[4] Barbashov V M and Trushkin N S 2009, *Microelectronics* **38** 1 pp 34-47
[5] Sogoyan A V, Chumakov A I and Nikiforov A Yu 1999 *Russian Microelectronics.* **28** 4 pp 224-235
[6] Barbashov V M, Trushkin N S and Kalashnikov O A 2015 *Russian Microelectronics* **44** 5 pp 312-315
[7] Sogoyan A V and Polunin V A 2015 *Russian Microelectronics* **44** 1 pp. 54-59
[8] Barbashov V M 2011 *Sensors and systems* **6** pp. 29-34

Impact of switches resistance on successive approximation of ADC dynamic performance

Y Bocharov[1*], V Butuzov[1,3], V Shunkov[1], O Kus[2] and V Prokopyev[2]

[1] National Research Nuclear University MEPhI (Moscow Engineering Physics Institute), Kashirskoe shosse 31, 115409 Moscow, Russian Federation
[2] OKB Fifth Generation, LLC, Nikolaeva street 11, 630090 Novosibirsk, Russian Federation
[3] Megarad, LLC, Nikolaeva street 11, 630090 Novosibirsk, Russian Federation

[*]Corresponding author's e-mail address: yibocharov@mephi.ru

Abstract. This paper discusses the degradation of the dynamic performance in switched capacitors based on successive approximation analog-to-digital converters (SAR ADCs), caused by the resistance of the switch, which connects a voltage reference to a common bus of switched capacitor array. Based on the analysis of the phenomenon, the techniques for reducing its influence are shown. They were used in the design of micro-power SAR ADC, which is presented in the paper.

1. Introduction

A successive approximation analog-to-digital converters (SAR ADCs) are widely used due to the low power consumption and relatively simple structure. Modern SAR ADC consists of the following blocks: a digital-to-analog converter (DAC) based on capacitor array, a comparator and a control logic that provide the implementation of successive approximation algorithm and general synchronization.

Figure 1 shows the simplified structure of an 8-bit SAR ADC with input signals in the range from 0 to a full-scale voltage V_{FS}. For clarity, this figure shows an ADC having single-ended input and a single array of switched capacitors in the DAC. It forms the sequence of voltages required for binary search and serves as a sample-and-hold circuit.

In the sampling phase, all capacitors are connected in parallel. Their bottom plates are connected to the source of the input signal, and their top plates are connected to a reference source with a voltage equal to half of a full scale. In the phase of determining the bit values, the bus of the top plates of the capacitors is disconnected from the reference voltage. The bottom plate of each capacitor, starting with the most significant bit (MSB), is connected to the reference source of the full-scale voltage V_{FS}, while the bottom plates of the other capacitors are connected to the ground bus. The comparator determines the values of the bits, comparing the voltage on the top plate bus with a voltage that is equal to half of the scale.

A great number of papers are devoted to the analysis of SAR ADC errors and the methods for reducing them [1–15]. There are many papers, which propose the methods for improving the dynamic performance of these ADCs [16–20]. Despite the large number of publications on this issue, one point concerning the impact of switches resistance on SAR ADC dynamic performance, has not been considered in previous papers. The purpose of this article is an explanation of this phenomenon, its analysis and discussion of techniques for reducing its impact on the performance of ADCs.

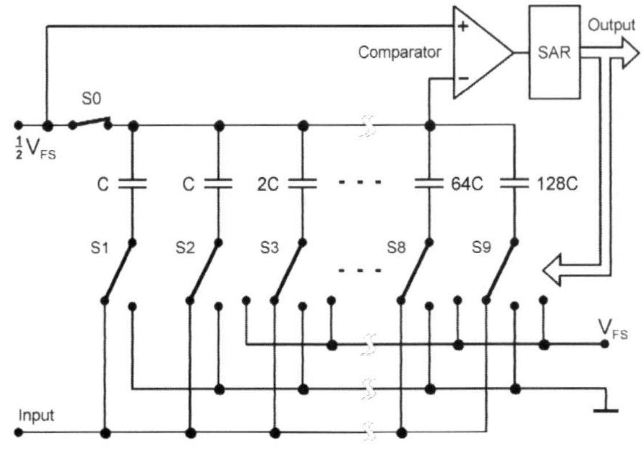

Figure 1. Exemplary architecture of 8-bit SAR ADC.

2. Analysis of the phenomenon and identifying ways to reduce its impact on ADC dynamic performance

2.1. Summary of phenomenon

Figure 2 demonstrates the effect in question. The essence of this effect is that, because of the resistance between the reference source and the bus of the top capacitor plates, the potential of this bus during the sampling phase does not remain constant. As shown in Fig. 2, the voltage on the bus of the top plates does not accurately match to the voltage of the reference source. It also contains a component proportional to the derivative of the input signal. The attenuated and phase-shifted input signal is applied to the bus through the capacitors of the DAC matrix that are connected in parallel during the sampling phase.

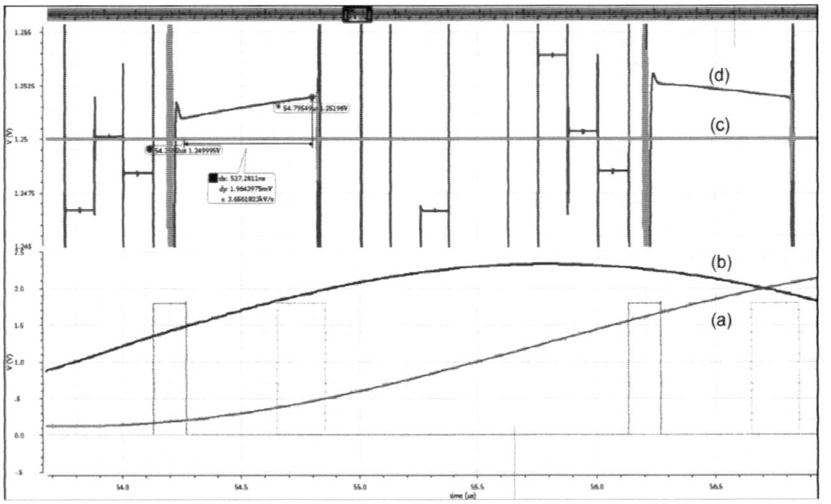

Figure 2. Simulated waveforms during the sampling and conversion: (a) input signal; (b) derivative of input signal; (c) output of voltage reference source; (d) voltage at the top-plate bus of capacitor array.

This effect appears as a reference source of output voltage error. This causes a degradation of the parameters of the SAR ADC, including the effective number of bits (ENOB), as the frequency of the input signal increases, as shown in Figure 3. This effect most affects the dynamic performance of ADC, which features a resolution of 12 bits and more.

IOP Conf. Series: Materials Science and Engineering **498** (2019) 012005 doi:10.1088/1757-899X/498/1/012005

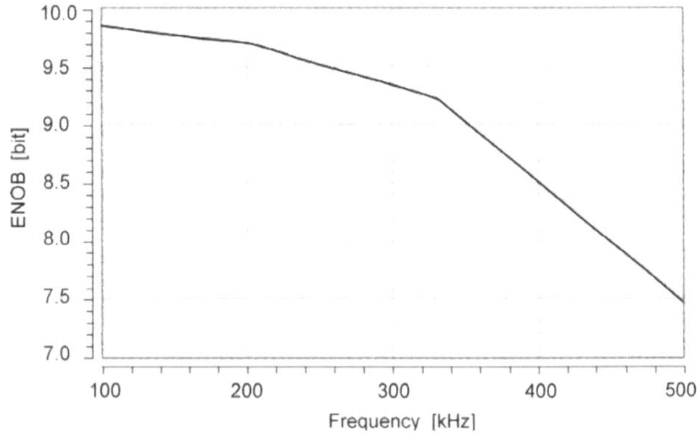

Figure 3. Simulated plot of ADC effective resolution degradation with increasing the frequency of the input signal.

2.2. Detailed analysis

The equivalent circuit of the capacitive DAC is shown in Figure 4. A capacitor array of the DAC consists of higher-bits and lower-bits sub-arrays, which are connected via an attenuating capacitor C_a.

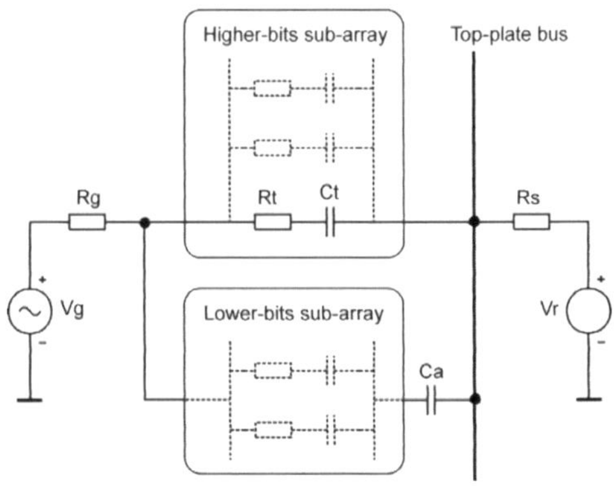

Figure 4. Simplified AC equivalent circuit of capacitive array and switches during the sampling phase.

During a sampling phase, all capacitors are connected in parallel. They are connected between the input source V_g and the reference source V_r through closed switches with resistance $R_g + R_t$. The reference source is connected through a switch with resistance R_s to a top-plate bus of capacitors. The total capacitance of a higher-bits sub-array is $C_t = C_0 (2^n - 1)$, where C_0 is the unit capacitance and n is resolution of a higher-bits sub-array. The total capacity of a lower-bits sub-array with a series capacitor C_a is equal to C_0 and the total capacity of DAC capacitor array is $C_s = C_t + C_0 = 2^n C_0$.

A voltage transfer ratio from the input signal source to the top-plate bus of the DAC:

$$K_v(f) = K_s \frac{f}{f_0}\left(1+\left(\frac{f}{f_0}\right)^2\right)^{-\frac{1}{2}}, \quad K_s = R_s\left(R_s + R_g + R_t\right)^{-1}, \qquad (1)$$

where f – input frequency and $f_0 = [2\pi C_s(R_g + R_t + R_s)]^{-1}$. For the input signals frequencies much lower than f_0, the transfer ratio is $K_v(f) \approx K_s f/f_0 = 2\pi f C_s R_s$. The total capacity of the DAC array and the resistance of the switch that connects the reference source to the top-plate bus determine it. In this approximation, it is independent of the signal source resistance and resistance of DAC switches.

IOP Conf. Series: Materials Science and Engineering **498** (2019) 012005 doi:10.1088/1757-899X/498/1/012005

Deviation of the top-plate voltage V_{tp} from the reference voltage $\Delta V = (V_{tp} - V_r)$ causes the relative error in decibels relative to the carrier that is equal to $\delta(f) = 20\ \lg(\Delta V/V_g) = 20\ \lg[K_v(f)]$.

2.3. Methods for improving the dynamic performance of SAR ADC

According to the analysis performed, the effect of the reference switch resistance on the dynamic performance of the ADC can be mitigated both by using a bootstrapped switch, which has a low and stable resistance, and by minimization of the unit capacitor value.

The refined version of the 10-bit SAR ADC utilizes bootstrapped switches to improve dynamic performance. Fig.5 shows the schematic diagram of the bootstrapped switch. The circuit consists of two switches, which are based on a modified known scheme [21–23]. The second dummy switch performs the function of charge injection compensation.

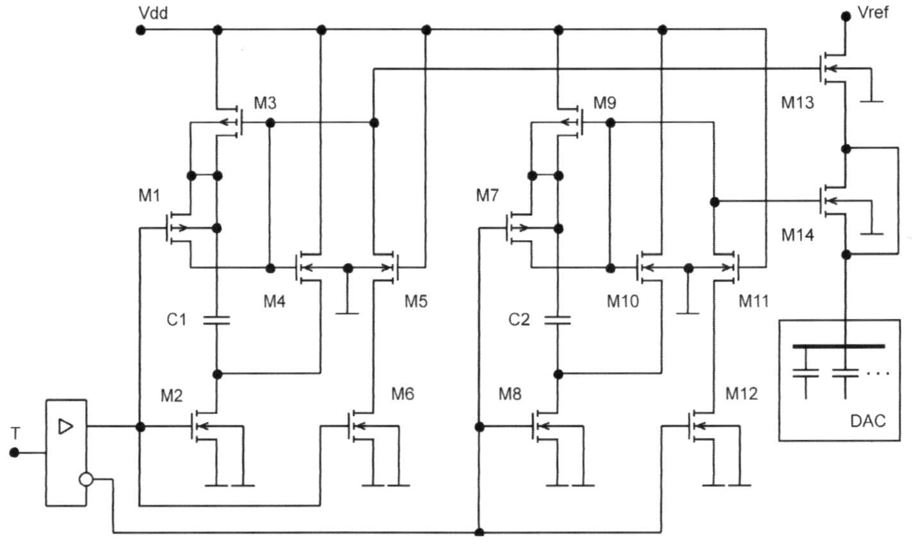

Figure 5. Schematic diagram of bootstrapped switch with charge injection compensation by a dummy switch.

In addition, the capacitive array of the DAC in a new version of ADC uses a unit capacitor with capacitance reduced to 60 fF. This allowed not only improving the dynamic parameters, but also reducing the ADC power consumption with the maintenance of acceptable linearity and noise performance. Fig. 6 shows implementation details of the finger-type unit capacitor used in the ADC.

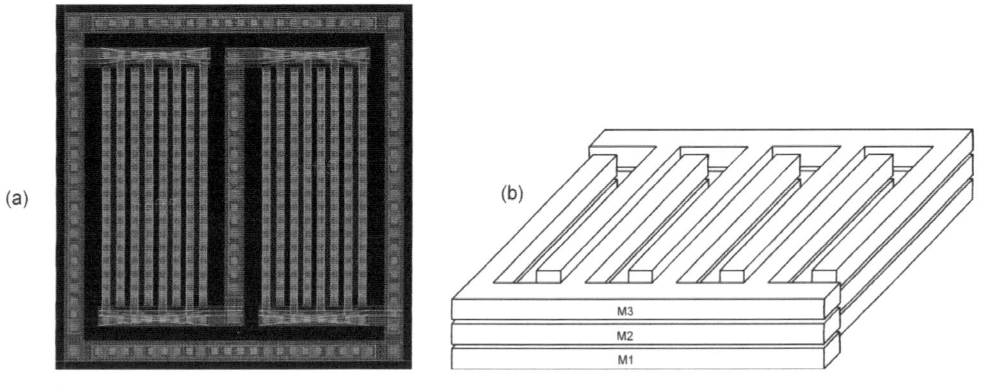

Figure 6. Implementation of the unit capacitor: (a) layout, (b) stack of metal layers.

Fig. 7 illustrates enhancement of ADC dynamic performance achieved by using the proposed techniques. At a sampling rate of 1 Msps and an input signal frequency of 333 kHz, the effective resolution of ADC has been improved from 9.32 bits to 9.85 bits.

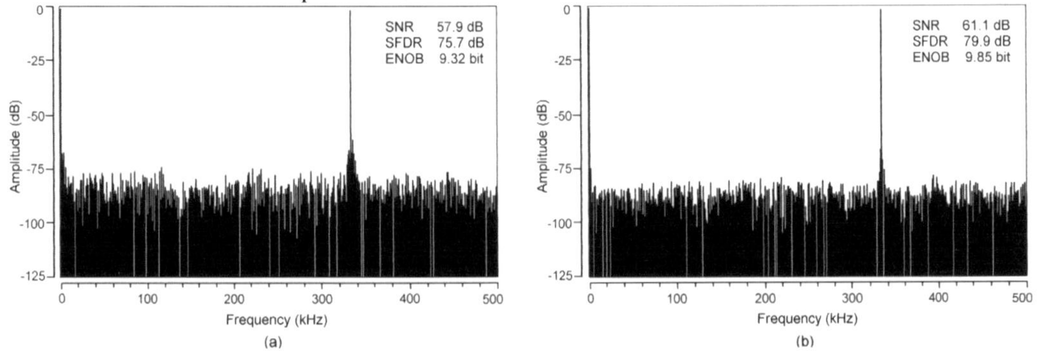

Figure 7. Output spectrum and dynamic performance of (a) prototype ADC and (b) advanced version of ADC with bootstrapped reference voltage switch and smaller capacitances.

A 10-bit SAR ADC using the methods proposed has been implemented in a commercially available 180 nm CMOS process. ADC has low power consumption, which is less than 0.4 mW at a sampling rate of 1 Msps. Fig. 8 shows the layout of the chip core that is inclusive of the capacitor DAC and switch circuitry.

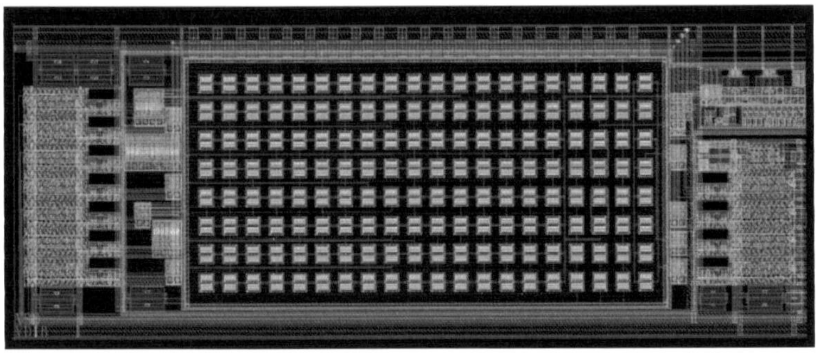

Figure 8. Layout of capacitor arrays and switches of designed ADC.

3. Conclusion

We considered the phenomenon of degradation of the dynamic parameters of the SAR ADC with capacitor DAC due to the influence of the impedance of the reference source, as well as methods for reducing the impact of this effect. A low-power 10-bit ADC with improved dynamic performance was developed using the proposed methods.

Acknowledgments
The Russian Science Foundation (Grant 16-19-00167) supported this work.

References

[1] Ginsburg B P and Chandrakasan A P 2007 *IEEE J. Solid-State Circ.* **42** pp 739–747
[2] Saberi M, Lotfi R, Mafinezhad K and Serdijn W A 2011 *IEEE Trans. Circuits Syst.– I* **58** pp 1736–1748
[3] Jun M, Yawei G, Yue W, Xu C and Xiaoyang Z 2013 *J. Semicond.* **34** pp 085014
[4] Zhang C and Wang H 2012 *IEEE Trans. Instrum. Meas.* **61** pp 587–594
[5] Weiru G, Fan Y and Junyan R 2014 *J. Semicond.* **35** pp 085006

[6] Wagdy M F 1990 *IEEE Trans. Circuits Syst.* **37** pp 959–962
[7] Cho S H, Lee C K, Kwon J K and Ryu S T 2011 *IEEE J. Solid-State Circ.* **46** pp 1881–1892
[8] Liu W, Huang P and Chiu Y 2011 *IEEE J. Solid-State Circ.* **46** pp 2661–2672
[9] Yan L, Siliang H, Donghui W and Chaohuan H 2012 *J. Semicond.* **33** pp. 125009
[10] Du L, Ning N, Zhang J, Yu O and Liu Y 2013 *IEEJ Trans. Electrical and Electronic Eng.* **8** pp 408–414
[11] Fan H, Han X, Wei O and H. Yang H 2013 *Analog Integr. Circ. Sig. Process.* **74** pp 239–254
[12] Park H and Ghovanloo M 2013 *Analog Integr. Circ. Sig. Process.* **75** pp 459–465
[13] Yingying C and Dongmei L 2013 *J. Semicond.* **34** pp 045007
[14] Zhang D and Alvandpour A 2014 *IEEE Trans. Circuits Syst. – II* **61** pp. 666–670.
[15] Ke L, Zhankun D, Li S and Xiao M 2015 *J. Semicond.* **36** pp. 125004
[16] Inanlou R, Shahghasemi M and Yavari M 2013 *Analog Integr. Circ. Sig. Process.* **77** pp. 257–269
[17] Sedighi B 2013 *Analog Integr. Circ. Sig. Process.* **77** pp. 459–469
[18] Harpe P 2016 *IEEE Solid-State Circuits Mag.* **8** pp. 64–73
[19] Zhu Y, Chan C-H, Chio U-F, Sin S-W, U S-P, Martins R P and Maloberti F 2010 *IEEE J. Solid-State Circ.* **45** pp. 1111–1121
[20] Kull L, Toifl T, Schmatz M, Francese P A, Menolfi C, Brandli M, Kossel M, Morf T, Andersen T M and Leblebici Y 2013 *IEEE J. Solid-State Circ.* **48** pp. 3049–3058
[21] Abo A M and Gray P R 1999 *IEEE J. Solid-State Circ.* **34** pp. 599-606
[22] Dessouky M and Kaiser A 2001 *IEEE J. Solid-State Circ.* **36** pp. 349–355
[23] Razavi B 2015 *IEEE Solid-State Circuits Mag.* **7** pp. 12–15

IOP Conf. Series: Materials Science and Engineering **498** (2019) 012006 doi:10.1088/1757-899X/498/1/012006

«On-line» neutron fluence registration by silicon bipolar transistor

V I Butin[1,2,*], I V Butin[3], A V Butina[1]

[1] Dukhov Research Institute of Automatics (VNIIA), Moscow, Russia
[2] National Research Nuclear University MEPhI, Moscow, Russia
[3] Bauman Moscow Technical University (BMSTU)

*Corresponding author's e-mail address: butin.valentin@gmail.com

Abstract. An efficient practical method of bipolar transistor application for neutron fluence registration is presented. The method is focused on the neutron fluences damage effect correlation for various sources to reference source. The bipolar transistor connection diagram and an example of device realization based on STM32 microcontroller are described. STM32 is used for providing test signals and parameter registration used for gain factor calculation of the bipolar transistors monitor (BTM). The proposed approach allows to consider the features of BTM that is pre-irradiated at reference source and their application as the neutron fluence monitors. These connection diagram and constructive solutions provide "on-line" neutron fluence registration for different sources in terms of the reference source as required for the verification of radiation test results.

1. Introduction and overview

The neutron irradiation test is performed to determine the susceptibility of semiconductor devices to degradation in the neutron environment [1]. Devices under test (DUT) are to detect and measure the degradation of critical semiconductor device parameters as a function of neutron fluence and to determine whether the specified semiconductor device parameters are within the specified limits after exposure to a specified level of neutron fluence. Typically, a specific parameter, such as minority-carrier lifetime or current gain factor is measured as a function of particle fluence, and the rate of parameter degradation is determined for specific measurement conditions. The degradation rate has been referred to in literature as the damage factor.

The neutron fluence used for device irradiation can be obtained by measuring the amount of radioactivity induced in a fast-neutron threshold activation foil such as ^{32}S, ^{54}Fe, or ^{58}Ni, irradiated simultaneously with the device. The conversion of the foil radioactivity into neutron fluence requires knowledge of the neutron spectrum incident on the foil. If the spectrum is undefined, it can be determined by using the following E720-E722 DoD adopted ASTM standards or their equivalent.

In ASTM standard E722 [2], the neutron fluence is described in terms of the monitor response to the equivalent monoenergetic neutron fluence per unit. The monitor foil is used to predict the equivalent monoenergetic neutron fluence, which is only valid in case of a constant energy spectrum.

Neutron exposure of a sample is commonly given in terms of the neutron fluence, n/cm^2. However, the amount of displacement damage from neutrons in a given material depends on neutron energy. Therefore, in order to allow meaningful comparisons between experiments using different neutron energy spectra, neutron fluences are often determined in terms of ("normalized to") an equivalent 1-MeV neutron fluence, which is the fluence of 1-MeV neutrons that produces the same electronic effect as the neutron spectrum used in a particular study [3].

The radiation response of many types of devices (e.g., bipolar transistors, solar cells, focal-plane arrays, and other detectors) can be predicted reasonably well based on calculations of the amount of displacement damage energy imparted to the primary knock-on atoms [4, 5].

The Messenger-Spratt equation [6] was used to describe the radiation response of the common emitter dc current gain factor of a bipolar transistor:

$$\frac{1}{H_{21E}(F_n)} = \frac{1}{H_{21E}(0)} + K_N \cdot F_n, \qquad (1)$$

where $H_{21E}(0)$ and $H_{21E}(F_n)$ are the initial gain and final gain respectively, K_N is the particle- and energy-dependent displacement damage factor, and F_n is the particle fluence. The damage factor is determined experimentally by gain factor measurements of the DUT (for a particular set of device operating conditions) after incremental exposures at a given particle energy.

To compare the results of the radiation test for different neutron energy spectra, one uses a comparison method for neutron fluences F_x in terms of the equivalent to the reference reactor, neutron fluence F_{ref}, which produces the same damage effect [7-9].

The connection diagram of the bipolar transistor and the device based on the STM32 microcontroller are proposed by the authors to allow carrying out the active mode measurements with a minimum quantity of additional lines.

2. Background information

There is a well-known function of δ(E) dependence for Si [2, 5, 6], which can be used to estimate the damage effect Dn of neutron radiation with a spectral distribution φ(E):

$$D_n = \int \varphi(E) \cdot \delta(E) \cdot dE \qquad (2)$$

An alternative method of determining the effectiveness of neutron radiation is a method to compare the results of the irradiation of silicon structures to the results for the reference source by using the bipolar transistor monitors (BTM). The gain factor dependence (H_{21E}) on the neutron fluence can be described by the Messenger-Spratt equation.

Types of bipolar transistors (planned to be used as BTM) are determined by the range of neutron fluence values.

Using the results of gain factor H_{21E} at a fixed value of emitter current (the stage we called “calibration”), one can calculate values of K_N.

3. The algorithm of neutron fluence measurement using BTM

We propose using the following sequence of operations for radiation tests:

1. The voltage V_{cc} should be set at the programmable power supply output.

2. The Data acquisition/switch unit measures the current consumption for each of the BTMs (as a sum of collector and base current for each transistor).

3. The gain factor value H_{21E} for BTM is calculated using equation:

$$H_{21E} = \frac{I_{Collector}}{I_{Base}} = \frac{I_{Emitter} \cdot R_{Base}}{V_{cc} - V_{BE}} - 1 \qquad (3)$$

where V_{BE} – voltage decrease at emitter junction, V;
V_{cc} – voltage value at output of programmable power supply, V;
R_{Base} – resistance in transistor base line, Ohm.

4. The neutron fluence value corresponding to the neutron fluence of the reference reactor is calculated in the final step.

IOP Conf. Series: Materials Science and Engineering **498** (2019) 012006 doi:10.1088/1757-899X/498/1/012006

Base current must be kept constant during the radiation tests. Thus, it is needed to take into account the coefficient correlating the dependence of the damage factor of the transistors and the injection level of the minority carriers in the base of the BTM.

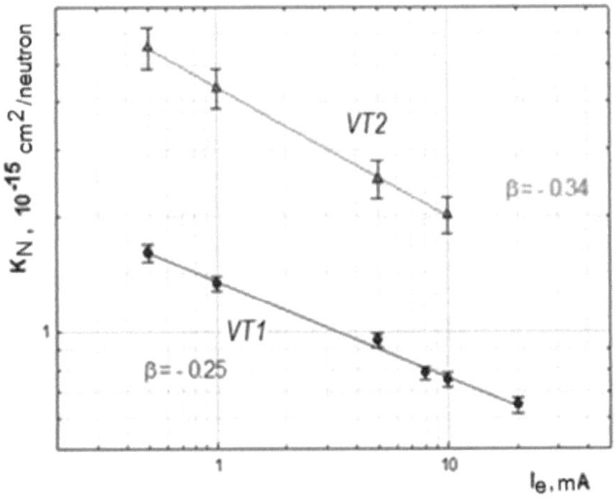

Figure 1. Dependence of damage constant (K_N) on emitter current (I_e) for high frequency (VT1) and low frequency (VT2) BTMs.

The executed research resulted in the following dependence for the constant K_N (see Fig.1):

$$K_{N\,(I\_x)} = K_{N\,(I\_emitter)} \cdot \left(\frac{I_x}{I_{emitter}}\right)^{\beta} \quad (4)$$

where $K_{N\,(I\_emitter)}$ and $K_{N\,(I\_x)}$ are values of the damage constant for emitter current of BTM at the "calibration" ($I_{emitter}$) stage and at the "measurements" (I_x) stage, respectively, and β is the exponential function index.

The values of the constants of damage K_N for several values of I and β are determined during the calibration process.

4. The bipolar transistors connection

Fig.2 shows a connection diagram of the BTM: for each BTM there is only one additional line of communication, which is essential for multi-point measurements. The attenuator at the output of BTM is used for reducing BTM output voltage to the ADC range measurement.

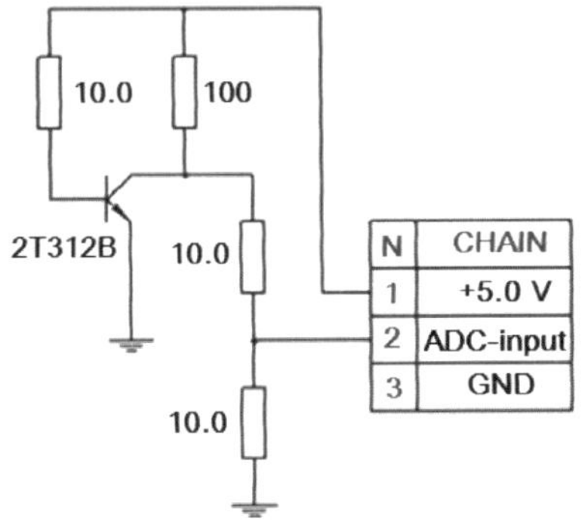

Figure 2. The BTM connection diagram for «on-line» registration.

The irradiation of BTM is carried out in active mode. The values of I_C and I_B are measured remotely.

The value of the coefficient H_{21E} of BTM can be determined using the results of measurements as in (3).

5. The device for results processing

The multi-channel registration of BTM parameters can be developed using modern microcontroller devices [10]. It allows simplifying the processing of measurement results. Besides, the autonomy of the portable microcontroller-based devices is their certain advantage.

The key item of a developed device is microcontroller STM32 [10, 11]. It digitalizes the analog signal coming from BTM1 to BTM4, records the digital signals in the microcontroller data register, calculates the current values of BTM informative parameter (H_{21E}), and transforms the data in a code combination. This code comes to LED indicator through the parallel input/output port of the microcontroller.

Fig.3 presents the "calibration" stage: the linearly increasing voltage is formed at DAC-output of the STM32 microcontroller and comes to transistor base of BTM.

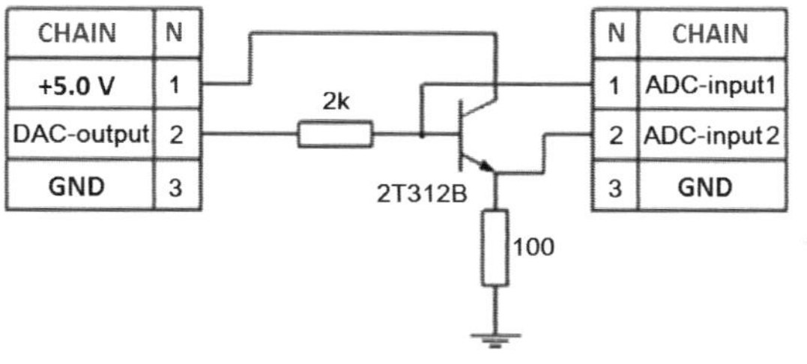

Figure 3. Schematic diagram for BTM at "calibration" stage

BTM is opened by DAC output voltage. When the value of this voltage gets to a previously determined level (for example, 100 mV or 1.0 V for emitter current properly 1.0 mA or 10 mA), the device fixes the voltage value at DAC-output, measures the BTM-transistor base voltage, calculates BTM gain factor (H_{21E}) and indicates its value. Each of the transistors is connected separately at the "calibration" stage for the measurement of parameters.

At the "measurements" stage, the code signals are formed at ADC-inputs ADC1 to ADC4, each of which corresponds to voltage value at resistor loads BTM1 to BTM4. Afterwards, gain factor (H_{21E}) for each of the BTM1 to BTM4 transistors is calculated. The result of calculation of BTM value of H_{21E} is indicated on the LED display.

6. The proposed device architecture

The developed device for neutron fluence effective value registration is based on the 32-bit STM32F407VGT6 microcontroller with ARM Cortex-M4F core, consisting of:
- 12-bit DAC;
- 2x12-bit ADC, which can be expanded to 16 channel;
- interface for parallel data output on the indicator;
- open-source hardware and software.

The proposed architecture of STM32 allows organizing the process of multichannel BTM-parameter measurements with ±1.0 mV precision, primary information processing and indicating the measurement results.

For software development, we used STM32Cube [12, 13] ("STMicroelectronics"). The STM32Cube HAL library consists of in-line HAL-abstraction software for STM32, providing the maximum of code translation inside the STM32 series. The set of built-in software components (RTOS, USB, FatFS, TCP/IP, Graphics) was configured for their combined action.

We have developed the driver program for the STM32 microcontroller, which realizes the previously described algorithm. The program was debugged and the microcontroller was programmed for the device prototype.

IOP Conf. Series: Materials Science and Engineering **498** (2019) 012006 doi:10.1088/1757-899X/498/1/012006

7. Experimental setup for device prototype testing

The device prototype during the experiments of 2T312B bipolar transistors was irradiated under the fields of two types of sources (RISI and IPPE, Russia).

The set of 8 transistors was pre-irradiated at the "calibration" stage at reference source BARS-4 (RISI). The measurement results of gain factor H_{21E} for each of the transistors were used for further calculation of the K_N and β parameters (for values of emitter current equal to 1.0 and 10.0 mA before and after irradiation).

Then we repeated the irradiation of four transistors of the set at the same fluence and measured the gain factor H_{21E} for each the 2T312B transistors. The uncertainty of the calculated values of neutron fluence from the standard detectors registrations was less than 10%.

The remaining set of four transistors was placed in a FS-1M (IPPE) source, the spectral-energy characteristics of the neutron radiation of which differ significantly from the reference reactor. Comparing the results of the irradiation, we see the differences between the neutron fluence of more than 0.1 MeV energy (that were obtained from the results of "on-line" measurements of the gain factor H_{21E} for each of the four transistors) and standard detectors registrations is less than 10%.

A novel device architecture has been proposed for the processing of the results of research experiments. Thus, we used two different algorithm variants for device realization – both for calibration and measurement – each of which was implemented as a separate program stored in the microcontroller before testing.

8. Conclusion

The practical method for bipolar transistor application for "on-line" neutron fluence registration is described. This method is focused on the neutron fluence damage effect correlation of various sources to reference source.

The proposed method is based on the application of pre-irradiated bipolar transistors monitors (BTM) and the SMT32 device as a self-contained unit.

The developed device prototype was used in radiation tests of two types of neutron sources (RISI and IPPE, Russia). Comparing the results of the irradiation, the uncertainties of the neutron fluence of more than 0.1 MeV energy from the standard detectors registrations was determined to be less than 10%.

References

[1] MIL-STD-883E. Test method standard. Microcircuits. Dec.1996
[2] ASTM E722-85. Standard practice for charactering neutrons fluence spectra in terms of an equivalent monoenergetic neutron fluence for radiation hardness testing of electronics.
[3] Srour J R, Marshall C J and Marshall P W, 2003, *IEEE Trans. Nucl. Sci.* **50** pp 653–668
[4] Summers G P, Dale C J, Burke E A, Wolicki E A, Marshall P W, and Gehlhausen M A, 1987 *IEEE Trans. Nucl. Sci.* **34** pp. 1134–1139
[5] Dale C J, Marshall P W, Cummings B, Shamey L, and Holland A 1993 *IEEE Trans. Nucl. Sci.* **40** pp. 1628–1637
[6] Messenger G C and Spratt J P 1958 *Proc. IRE* **46** pp. 1038–1044
[7] Butin V I, Zinchenko V F and Romanenko A A 2003 «The system of radiation tests of electronic devices». (Vladimir)
[8] Andreyev S A, Litvin V I, Snopkov A A and Cherashev V I 2014 *Voprosy atomnoy nauki i tehniki. Series: Fizika yadernyh reaktorov (in Russian)* **3** pp. 3-10
[9] Vizkelethy G, Bielejec E S, Doyle B L and Buller D L 2007 *SANDIA REPORT* SAN 02006-7746
[10] Vlasov A and Yudin A 2011 *Communications in Computer and Inf. Science* **156** pp 180-192
[11] Butin I V 2016 *Science, technic and education* **7 (25)** pp. 66-73
[12] RM0090. Reference manual. Available: http://www.st.com
[13] UM. User manual. Description of STM32F4xx HAL Drivers. Available at: http://www.st.com

IOP Conf. Series: Materials Science and Engineering **498** (2019) 012007 doi:10.1088/1757-899X/498/1/012007

Total dose measurements by p-channel transistors of ICs

V I Butin[1,2,*], I V Butin [3] and A V Butina[1]

[1] Dukhov Research Institute of Automatics (VNIIA), Moscow, Russia
[2] National Research Nuclear University MEPhI, Moscow, Russia
[3] Bauman Moscow Technical University (BMSTU), Moscow, Russia

*Corresponding author's e-mail address: butin.valentin@gmail.com

Abstract. In this paper, a new approach of p-channel MOS-transistors used for total dose monitoring at semiconductor components under radiation is presented. The calibration results enable to determine the numerical parameters for the MOSFETs electro-physical model of dose effects. We propose to calculate the total dose of ionization flux using the results of MOS-transistors drain current measurements, under a fixed gate and drain voltage.

1. Introduction

The irradiation of large-size objects (such as spacecraft) that consist of heavy solid elements is connected to some difficulties, namely, the accurate estimation of total dose for radiation-sensitive elements. The actual value of total dose accumulated in radiation-sensitive regions of a semiconductor component usually differs from values for regular dosimeters that are placed at the surface of research objects. The reason for such a difference is the transformation of the radiation spectrum during its transmission through construction materials of complex objects and systems.

MOS integrated circuits are nowadays in general use in electronics. However, such devices are rather sensitive to total dose so it is possible to use semiconductor devices as total dose monitors. They respond to total ionizing dose (TID) and their design is similar to that of critical units of electronics.

The purpose of this paper is to introduce the technique of application of p-channel MOSFETs within the digital IC as detectors for quick and low-cost measurements of TID accumulated in the semiconductor components in the volume of large-size research objects. We first provide the basic principles and a new scheme for on-line TID measurements, and also discuss the main advantages of such scheme. Then, we propose and describe the research algorithm. Finally, we consider the results of the irradiation tests with on-line TID measurements for 1594TL2T under static gamma-irradiation. The calculated values of TID determined for the experimental detector are in good accordance with the values of regular dosimeters of the simulation facility.

2. Theoretical basis

Recent reports [1-6] have shown that radiation impact of MOS-structures produces the positive charge accumulation at the gate insulator and increases the surface state density at the semiconductor-dielectric interface. In our previous papers [7], experimental data was reported and the electro-physical model was proposed, which described dose effects kinetics for n-channel MOSFETs sensitive to irradiation conditions (environment temperature, radiation intensity and electrical mode). The practical realization of the proposed electro-physical model allows us to describe mathematically the behavior of MOS-transistors under irradiation and predetermines the use of MOS-devices for TID detectors [8-10]. We propose to use p-channel MOS-transistors within the digital IC for TID monitoring.

It is necessary to ensure the compliance with accumulated dose in the sensitive volumes of the detector and MOS-devices in research electronics. Thus, the p-channel MOS-transistors within the simple digital IC (i.e. inverters, NAND circuits, NOR circuits, Schmitt triggers et al.) can be effectively used as sensitive elements, as they are in the same design as dose-sensitive electronics components.

We measured on-line the TID using the test setup, shown in Fig. 1. During the experiments, the application of p-channel MOS-transistors within the CMOS integrated circuit 1594TL2T (device under test, DUT) was researched for on-line TID measurement. The DUT was placed under a Co^{60} gamma ray source. The desktop PC for accessing the measurement equipment was positioned near the irradiation room. It was assembled from "PXI" standard test devices and was managed with the special LabVIEW program for remote control of the generator and multimeter.

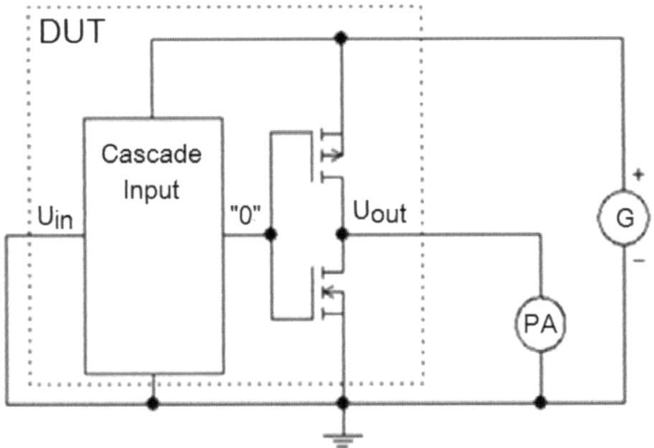

Figure 1. Schematic diagram of the test setup. G – generator, PA – multimeter

The main advantage of the proposed scheme is that the researcher is supposed to measure only drain current of the proper MOS-transistor in over-threshold mode during multichannel registration. Besides, one needs N+1 cables for N detectors. It is especially important for typical automated measurement systems, when there is difficulty in the registration of the total transfer characteristics. However, it is possible to take point-by-point measurements for a fixed power supply value.

At the stage of pre-calibration before and after irradiation, it is propose to register transfer characteristics in the over-threshold mode of p-channel MOSFETs within the IC.

The irradiation of the p-channel MOS-transistor with dose $D(t) = P \cdot t$ changes its transfer characteristics $I_{DS}(D)$ [11,12]:

$$I_{DS(D)} = \frac{W}{2L} \frac{\mu_0 \cdot C_{ox}}{(1 + K_\mu \cdot \mu_0 \cdot D)} \left(U_{GS} - U_{T0} - K_{\Delta V_T} \cdot D\right)^2 \quad (1)$$

where W is channel width, L is channel length (these parameters are typical and dependent from technology of MOS-devices);

μ_0 is the initial value of carriers mobility in the channel;

C_{OX} is specific capacity of the gate insulator;

K_μ is dose sensitivity coefficient of electrons mobility in the channel; U_{GS} is gate-source voltage (in our case $U_{GS} = U_{CC}$, U_{T0} is threshold voltage);

$K_{\Delta V_T}$ is dose sensitivity coefficient.

Using the notations $a = K_{\mu} \cdot \mu_0$ and $b = \frac{K_{\Delta V_T}}{U_{GS} - U_{T0}}$, the simplified expression for p-channel MOS-transistor current by fixed power supply value may be expressed as follows:

$$I_{DS(D)} = I_{DS(0)} \frac{1}{(1 + a \cdot D)} (1 - b \cdot D)^2 \quad (2)$$

where $I_{DS(0)}$ is the initial current value (before irradiation).

The values of coefficients a and b can be calculated through calibration data of the radiation facility with determined spectrum and intensity – for instance, a Co^{60} source.

3. Details of the model

We propose that p-channel MOS-transistors within the digital IC should be used as detectors for measurements of TID as following:

1) *Calibration.*

The drain current I_{DS} change of p-channel MOS-transistors within the digital IC is experimentally measured under radiation by a facility with determined spectrum and dose rate. During this experiment, the power supply value is fixed. The irradiation dose is about 20-50 krad (Si).

2) *Extraction of model numerical parameters.*

The numerical values of a and b are calculated for equation (2) through the calibration data.

3) *Use of TID detectors*

The drain current I_{DS} change of p-channel transistors is measured under conditions with unknown spectrum and composition of irradiation. The fixed power supply value provides the over-threshold mode. Using the parameters a and b that were determined at the stage of calibration, one can calculate the corresponding value of TID for a given irradiation time.

The time diagram of source power of the relative change and the initial measurements data are used to determine the real value of TID for over-radiated detectors in experimental research under static and quasi-static irradiation.

4. Results

Let us discuss the experimental research of ICs 1594TL2T for radiation hardness under static gamma-irradiation by a facility with determined radiation spectrum and dose rate. The digital IC consists of six inverted Schmitt triggers that were used as TID monitors.

The automated measuring system was developed on the basis of modular equipment National Instruments (chassis NI PXI-1042, DC Power Supply PXI-4110, Digital Multimeter PXI-4072) and software LabVIEW2014. This system provided the power supply U_{DS} with proper accuracy (up to several mV) and it enabled to take the current I_{DS} measurement under condition of changing voltage U_{DS} within determined time intervals.

1) *Calibration*

The detectors were irradiated at the voltage U_{DS} = 1 V. The certain voltage value at the inputs of the transistors was 0 V. The value of TID at the end of irradiation was equal to 30 krad (Si). Fig.2 represents the experimental curve I_{DS} (D). The numerical values of coefficients of equation (2) were determined using this curve:

$$a = 0.0132;\ b = -\ 0.0086;\ I_{DS}(0) = 3.305\ mA.$$

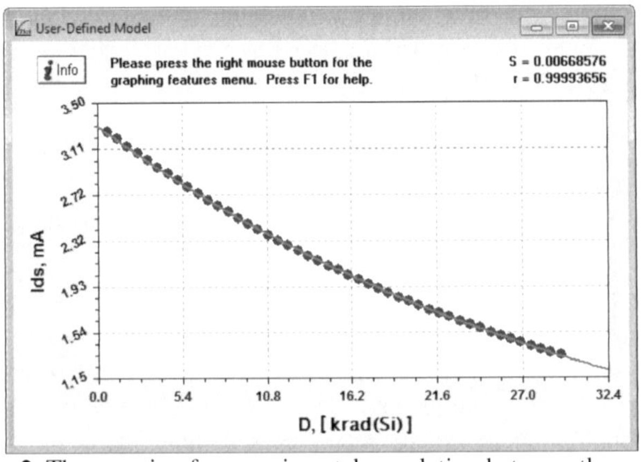

Figure 2. The equation for experimental correlation between the current I_{DS} and total dose D and its coefficient values:
$y=3.305(1/(1+a{\cdot}x))(1+b{\cdot}x)$

2) Application of detectors for research of electronics radiation hardness

The detectors were mounted at the experimental board closely to research objects during the experiments. The irradiation was carried out at the facility with determined spectrum and dose rate. The current I_{DS} was measured during the experiment at intervals of 10 minutes.

The detectors functioned in the proper electrical mode of the facility before, during, and after the irradiation and the control parameter annealing was observed.

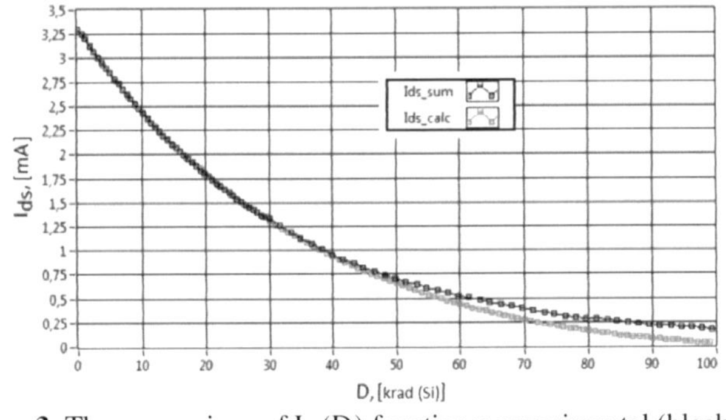

Figure 3. The comparison of I_{DS} (D) functions: experimental (black) and calculated after parallel translation (red)

Finally, Fig.3 illustrates the difference between values of TID determined for the experimental detector and values of regular dosimeters for the simulation facility (imprecision is less than 10%).

Thus, the irradiation experiments revealed the detector tolerance within TID ranging from 0 to 35 krad (Si) (without the TID value that was accumulated during calibration) is less than 10%.

3) Stability of calibration results

It is important to point out that the application of p-channel transistors within CMOS integrated circuit as TID detectors is possible only for stable values of parameters that were determined during the calibration. The experimental research revealed that the degradation of function I_{DS}(D) remained quasi-stable after 1000 hours post irradiation.

5. Conclusion

In this article, there was considered the application features of p-channel MOS-transistors within theCMOS integrated circuit as a TID detector for total ionizing dose monitoring in DUT under irradiation.

This work presented one of the applications of electro-physical modeling for the forecasting of the degradation of the MOS-transistors TID characteristics.

In the proposed method, the TID value of the detector is in good accordance with that of ICs for a wide spectrum of radiation energy.

References

[1] Oldham T R and McLean F B. 1987 *Basic mechanisms of radiation effects in electronic materials and devices (Harry Diamond Laboratory Technical Report)* HDL-TR-2129
[2] McWhorter P J 1986, *Simple technique for separating the effects of interface traps and trapped-oxide charge in metal-oxide-semiconductor transistors* ed. P. J. McWhorter, P.S. Winokur *(Applied Physics Letters.)* **48** 2 pp 133-134
[3] Oldham T R *Basic Mechanisms of TID and DDD Response in MOS and Bipolar Microelectronics* (DSFG, Inc. NASA Goddard, Radiation Effects and Analysis Group)
[4] Oldham T R, 1999 *Ionizing Radiation Effects in MOS Oxides*, Advances in Solid State Electronics and Technology (ASSET) Series, World Scientific Publishing, Singapore
[5] Oldham T R and McLean F B 2003, *IEEE transactions on nuclear science* **50** 3
[6] Mazur J 2002 *Radiation Effects – From Particles to Payloads* (IEEE NSREC Short Course) pp II-1 – II-69
[7] BarmakovYu N, Butin V I and Butina A V 2013 *Questions of atomic science and technics. Series: Physics of radiation effects on radio-electronic equipment* **3** pp 5-9
[8] Winokur P S, Schwank J R, McWhorter P J, Turpin D C and Dressendorfer P V 1984 *IEEE Transactions on Nuclear Science* **31** (6) pp 1453-1460 DOI: 10.1109/TNS.1984.4333529
[9] Ma T P, Dressendorfer P V, 1989, *Ionizing Radiation Effects in MOS Devices and Circuits* 1st edition (Wiley New York)
[10] Schwank J 2002, *Total Dose Effects in MOS Devices* (*Radiation Effects – From Particles to Payloads*) IEEE NSREC Short Course
[11] Muller R S and Kamins T I 1989 *Device Electronics for Integrated Circuits* 2nd Edition, (Russian Translation published in Moscow, USSR 1989)
[12] Schwank J R, Shaneyfelt M R and Dodd P E *Radiation Hardness Assurance Testing of Microelectronic Devices and Integrated Circuits: Radiation Environments, Physical Mechanisms, and Foundations for Hardness Assurance* (Sandia National Laboratories Document Sand) 2008-6851P

The use of quantum-kinetic models in system design synthesis of carbon nanostructures

G V Abramov[1], A N Gavrilov[2], O Ja Kravets[3] and N V Varlamov[4*]

[1]Voronezh State University, Universitetskaya Ploshad 1, 394018 Voronezh, Russian Federation
[2]Voronezh State University of Engineering Technologies, Prospekt Revolutsii 19, 394036, Voronezh, Russian Federation
[3]Voronezh State Technical University, Moskovsky Prospekt 14, 394026 Voronezh, Russian Federation
[4]National Research Nuclear University (MEPhI), Kashirskoe shosse 31, 115409 Moscow, Russian Federation

*Corresponding author's e-mail address: * nvvarlamov@mephi.ru

Abstract. In the process of designing a carbon nanostructure (CNS) synthesis system, the main task is to determine the technological and functional parameters of synthesis. We offered a mathematical model of the process for the purpose of studying the synthesis processes using the graphite thermal evaporation, based on the Boltzmann kinetic equation, which takes into account elastic and non-elastic particle interactions. We offer a numerical solving method using the parallel computing technologies. We conducted the study of the effect of the main synthesis parameters on the carbon cluster formation and the amount of obtained material containing CNS.

1. Introduction

An intense interest in the study of obtaining methods, structure and properties of nano-scale systems is due to the diversity and uniqueness of variants of their practical use. Small sizes of the structural components, which are under 100 nm, are the reason why the properties of the obtained nanomaterials differ from those of bulky analogues. For example, the application of carbon nanostructures (CNS) as polymer fillers allows creating composite systems with new and often unique performance features as compared with traditional materials. This allows using such materials, in the long view, in various fields of modern industry such as: automotive and aircraft engineering, electronics, powder-based metallurgy, paint and coating production [1].

The wide use of CNS (fullerenes, nanotubes, nanofibers) in new high-technology solutions is hindered by their high cost and low performance of the existing synthesis methods [2]. First of all, this is due to the lack of fundamental study of the CNS formation processes.

The known CNS synthesis technologies mainly imply the use of various methods of thermal evaporation of atomic carbon, or disintegration of carbon-containing gases with farther sedimentation of the formed structures on a cooled surface. One of the most spread methods in this group of synthesis methods is the graphite thermal sublimation of highly purified graphite with an arc discharge plasma in the buffer gas media (usually He or Ar) [3]. The method is distinguished by the fact that in various technological conditions of synthesis one can get multi-layer, one-layer nanotubes (with the use of catalyzer) or fullerenes of the C_{50}-C_{92} series with various outputs and quality [4-5].

The study of the processes that take place at the graphite thermal evaporation synthesis method is a complicated issue. The process goes fast; it is performed in plasma in quite a small amount. Therefore, a mathermatical model is used for the purpose of studying the general regularities and features of the process. However, building an adequate nanostructure synthesis model even for the trivial case is a nontrivial task. Multiple mutually interacting phenomena taking place in nanostructures at phase and structural transformations, which must be taken into account, significantly complicate the designing process [6]. The use of more accurate models of nanosystems results in complicated solving algorithms and extensive computational costs. The use of simple models can result in non-efficient, narrow-bound and even incorrect design solutions.

2. Problem setting

For the study of the carbon nanostructure synthesis technology based on graphite thermal sputtering, a mathematical model of the process that takes into account its specific character is needed. It must enable accounting for the whole complicated mechanism of spatial CNS formation in low-temperature plasma, with multiple influencing factors such as:

– temperature fields in reaction chamber;

– moving boundaries of electrodes that form the flows of carbon ions evaporated from the anode, and electron emission from the cathode;

– parameters of the buffer environment and electromagnetic field that influence the distribution and travel of charged particles in the multicomponent plasma with account for their interactions;

– energetical and spatial conditions required for forming stable covalent bonds in carbon.

The model must also allow the study of the process with various degrees of detalization (both at the level of particle flow, and the interactions of separate particles with each other).

3. Development of a mathematical model

The models with various levels of hierarchical detalization can be used for mathematical modelling: classical, semi-classical and quantum-mechanical ones, which differ from each other by the level of accuracy of the object description, possibility to take into account the specific features of the process and the complexity of numerical solution. Classical models (molecular dynamics method, classical potential method, and single-particle approximation) allow nanosystem modelling with low computational costs; however, they show low accuracy of the modelling results and do not let describe the processes of nanostructures interactions and formation [7]. The use of semi-classical models (hydrodynamic, diffusion-drifting, and Monte Carlo models) secure acceptable compromise between the computational costs and accuracy of the physical phenomena descriptions from the point of view of the application for the practical engineering; however, they do not allow a full-scale study of the processes of nanostructures formation, growth and interactions [8]. The models based on the quantum-mechanical approaches (Schrodinger, and quantum-kinetic approaches) have been hardly used in the practical engineering up to the present, though they provide high accuracy of the applied tool for describing physical processes [9]. This is due to extensive computational costs when the models of this hierarchical level are used. In the case of the large-dimension nanosystem modelling, the physical limits of a classical computer are the factor that hinders the wide use of such models. Basing on the quantum-kinetic models (Boltzmann kinetic equation, and the quantum-mechanical version of the Liouville equation), one can obtain more simple models, as compared with the Schrodinger ones, for describing nanostructures in a wide range of linear dimensions with account for various interactions. The use of the algorithms of parallel and distributed calculations organization allows the extension of the above stated problem.

This paper considers the quantum-kinetic approach to the CNS synthesis process modelling using the graphite thermal sublimation with account for elastic and non-elastic collisions in the arc discharge plasma. We used the system of Boltzmann equations (1) for each plasma component complemented with the system of Maxwell equations (2) that describes electromagnetic field, to describe the processes of the CNS arc discharge synthesis in the multicomponent plasma with account for the particle collisions [10-11]:

$$\frac{\partial f_\alpha}{\partial t} + \vartheta \frac{\partial f_\alpha}{\partial \vec{r}} - \frac{q_\alpha}{m_\alpha}(\vec{E} + \frac{1}{C}[\vec{\vartheta}, \vec{B}])\frac{\partial f_\alpha}{\partial \vec{\vartheta}} = \left.\frac{\partial f_\alpha}{\partial t}\right|_{CT}, \quad \alpha = e,h,c \tag{1}$$

$$\begin{aligned} rot\,\vec{H} &= \frac{4\pi\vec{j}}{C} + \frac{1}{C}\frac{\partial \vec{D}}{\partial t}, \\ rot\,\vec{E} &= -\frac{1}{C}\frac{\partial \vec{B}}{\partial t}, \\ div\,\vec{B} &= 0, \\ div\,\vec{D} &= 4\pi\rho, \end{aligned} \tag{2}$$

Symbols used: f_α – function of plasma distribution component (e – electron, h – buffer gas ion, c – carbon ion); $\vec{E}, \vec{H}$ – electric and magnetic field intensities; $\vec{D}, \vec{B}$ – electric and magnetic induction; q_α, m_α – particle charge and mass; ρ – charge density; $\vec{j}$ – current density; C – light speed; $\vec{\vartheta}$ – particle velocity field; $\vec{r}$ – particle coordinates.

With the admission that all interactions in plasma are between electrons, buffer gas ions and carbon particles within the whole interelectrode space, we can write the pair-wise collision integrals as follows [12]:

$$\left.\frac{\partial f_\alpha}{\partial t}\right|_{CT} = \sum_{\beta=e,c,h} \iint_V (f'_\alpha f'_\beta - f_\alpha f_\beta) |\vec{\vartheta} - \vec{\vartheta}'| d\sigma d\vec{\vartheta}', \quad \alpha = e,c,h, \tag{3}$$

where $f_\alpha f_\beta$ is the function of particles α and β distribution prior to collision; $f'_\alpha f'_\beta$ is the function of particles α and β distribution after collision; $\vec{\vartheta}, \vec{\vartheta}'$ are the velocity vectors prior and after particles collision (m/s); $d\sigma$ is the differential effective cross-section.

The mathematical heat transfer model with moving boundaries was used for forming the initial boundary conditions and accounting the change of the boundary conditions in the course of the synthesis. The Maxwell distribution was used as the function describing the velocity-wise distribution of the particles in plasma.

The presence of the collision integral (3) in the Boltzmann equations (1) significantly complicates the solution of the offered model that can be carried out with a numerical method only. For this purpose, the system of input Boltzmann-Maxwell equations (1-2) is split basing on the splitting method modification [13] into three auxiliary problems which are solved in sequence. The scheme of decomposition is shown in Fig. 1.

The first problem, basing on the heat transfer model with account for the changes in the operating area configuration, forms the energetic and spatial particle parameters at the initial moments of time in the cells of the computational grid [14]. The second problem is a system of dimensionless Vlasov-Poisson equations, the solution of which is modelled by the particle motion through the boundaries of the cells and gives the initial condition for the third problem. The solution of the third problem that takes into account the particles collisions allows the determination of the coefficients in the collisions integral and obtaining the particle distribution functions for the considered moments of time. The found particle distribution functions allow determining the areas that meet the spatial and energy conditions of the probable formation of carbon cluster groups with various bonds, the growth of which forms linear carbon chains that are closed into monocyclic rings (pentagons and hexagons) and form convex carbon structures (fullerenes, nanotubes) both in plasma and on the cathode.

IOP Conf. Series: Materials Science and Engineering **498** (2019) 012008 doi:10.1088/1757-899X/498/1/012008

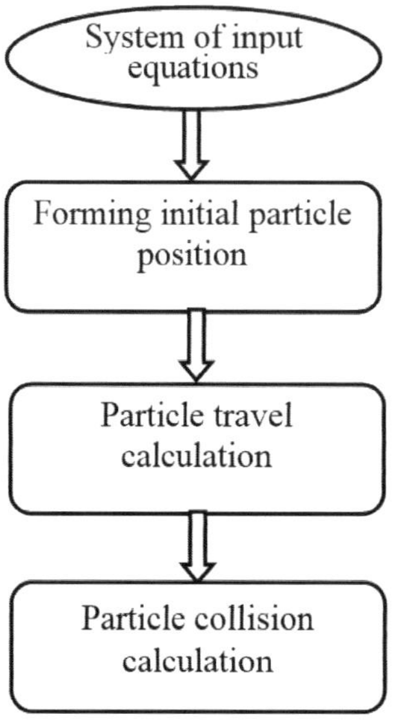

Fig. 1. Diagram of the decomposition of the input system of equations

The presence of a large amount of various particles that at the same time participate in the numerical calculation requires significant computational resources and time. Therefore, to obtain physically justified modelling results we used the large-particle method (particle-in-cell method, or PIC) [15]. PIC allows decreasing the scope of calculation and, thus, the requirements to the computational resources without compromising the calculation accuracy due to grouping one-type particles into larger ones (macroparticles), with the same ratio of the charge to the mass, as for the input ones, and the averaged velocity and direction.

The parallelization technology GPGPU (General-purpose graphics processing units) [16] was used to decrease the time and monetary expenditures for the problem numerical solution. This parallel programming technology allows reaching a rather high parallelization level on one computer and avoiding significant time losses for data transmission between computation nodes and synchronization of the calculation results. The GPGPU technology allows using general purpose graphical processors on one PC for calculations, which significantly improves the efficiency of large data processing without using a supercomputer or computational clusters. All calculations were performed on the hardware-software complex CUDA (Compute Unified Device Architecture) with the use of the nVidia CUDA technology, which enables graphical processor programming [17].

One of the special features of GPGPU is that not all similar data flows started at one moment finish calculation simultaneously. Therefore, an algorithm allowing solving the problem of parallel calculation segments synchronization was developed. For each moment of time, after the calculation of all variables on the graphical accelerator, the process control was transmitted to the central processor where the results were processed, and after that the calculation proceeded with the next algorithm iteration.

4. Results and Discussions

A number of experiments on obtaining various CNS with electric arc method in the inert gas medium (He) were performed on the laboratory facility with AICS (Automated Informational Control System) [18] to prove the adequacy of the developed complex model. The use of AICS allows the stabilization of required synthesis parameters in the real-time mode. The investigations were conducted for two regimes of synthesis without using the catalyzer: obtaining fullerenes with the maximum output $C60 \div C70$ and multiwall carbon nanotubes (MCNT). Plasma discharged was automatically maintained within the space of 1 mm between high-purity graphite electrodes (ash content < 0.05 %), with a diameter of 12 mm. The buffer gas pressure in the work chamber was set at the level of 400 Tor, the

IOP Conf. Series: Materials Science and Engineering **498** (2019) 012008 doi:10.1088/1757-899X/498/1/012008

voltage on the electrodes was 25 V. At a constant current of 120÷170 A, deposit sediment containing MCNT is formed on the cathode, and at currents of 350÷400 A, when the cathode deposit is utterly not formed, ash with a high content of fullerenes is sedimented on the walls of the reaction chamber. The output and the quality of the final product strongly depend on the set technological parameters and their stability.

The model adequacy was assessed over the results of the experimental study. The change in the cathode deposit mass depending on the process time (Fig. 2) and the change in the anode sublimation profile (Fig. 3) were assessed. The results suggest a rather good agreement of the experimental and calculated data.

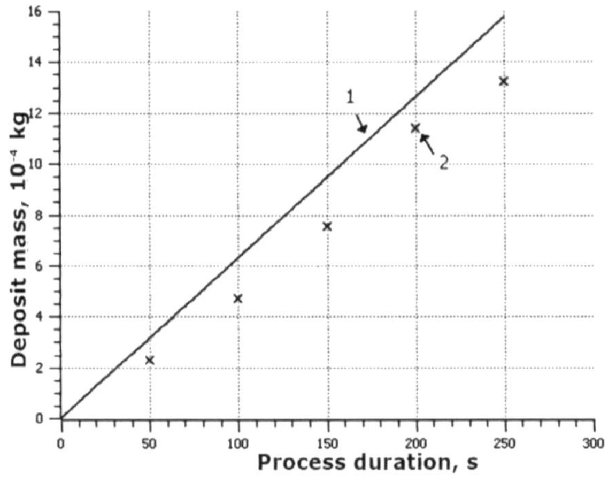

Fig. 2. Cathode deposit mass change in the mode of the MCNT synthesis vs. synthesis time: 1 – calculation with the model 2 – experimental data. Current – 150 A.

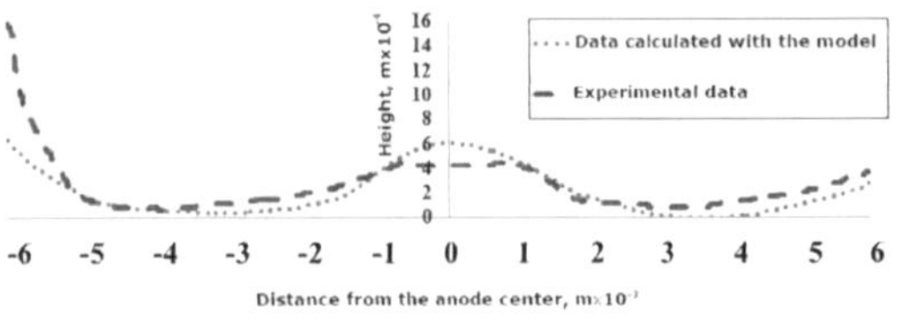

Fig. 3. Experimental and calculated data of the anode profile change in the regime of the fullerene synthesis within the synthesis period 100 s. Current – 350 A.

The model allows the study of the formation and the growth rate of the cathode deposit containing MCNT depending on the current intensity, buffer gas type and pressure in the synthesis chamber (Fig. 4, 5). The growth rate of the cathode deposit in the He environment significantly differs from that in Ar. The use of Ar as a buffer medium increases the growth rate of the deposit at similar synthesis parameters as compared with He. The analysis of numerical calculations proves that the synthesis process at high currents goes without leaving deposit sedimentations.

The absence of a buffer environment in the work area at a discharge less than 70 Torr does not result in the MCNT formation, but leads to the cathode breakdown at high currents and sedimentation of ash on the chamber walls with the admixture of modified graphite.

IOP Conf. Series: Materials Science and Engineering **498** (2019) 012008 doi:10.1088/1757-899X/498/1/012008

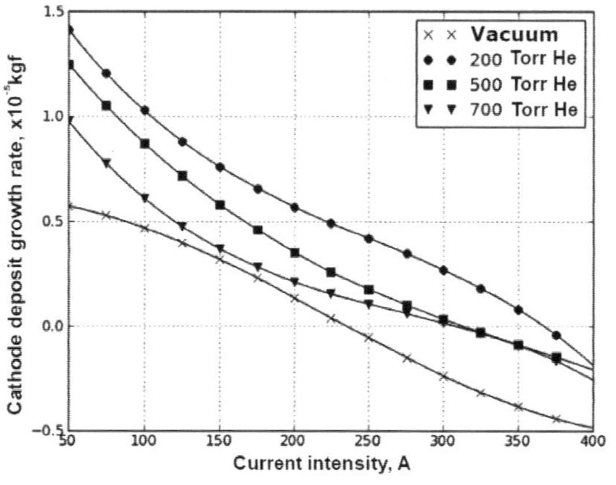

Fig. 4. Cathode deposit growth rate in the helium vs. various pressures.

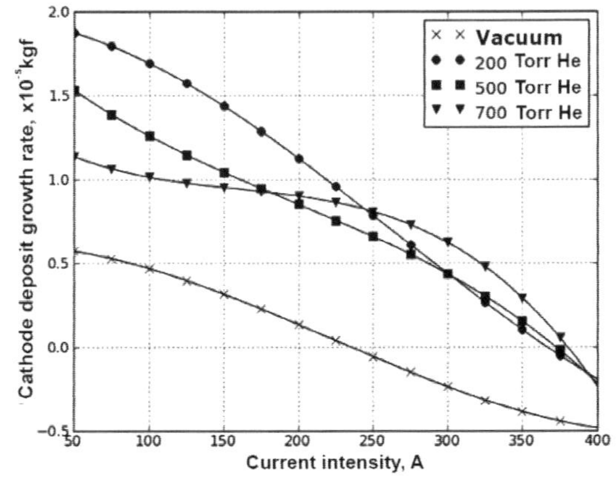

Fig. 5. Cathode deposit growth rate in the argon vs. various pressures.

The model allows the study of the cluster groups formation in plasma. The cluster groups with connections C=C–C (C3) are the basis for building pentagons and hexagons that form the spatial structures of fullerenes and nanotubes. The amount of their formations in the arc discharge plasma determines the output of the final synthesized product.

Fig. 6 shows the calculated dependences of the qualitative characteristics of stable carbon C3 cluster formation in plasma along the inter-electrode space for the MCNT synthesis regime (nanotubes regime) and fullerenes (fullerenes regime). The time interval of the numerical calculation of the total amount of the interactions shown in Fig. 6 is Δt = 360 ns.

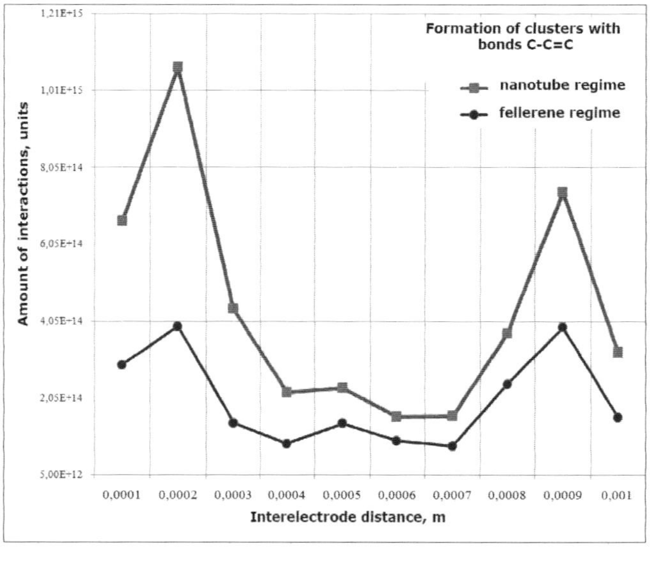

Fig. 6. The dependences of the qualitative characteristics of C3 clusters formation along the inter-electrode space for the nanotube regime (I = 150 A) and the fullerene regime (I = 350 A).

The analysis suggests that the C3 cluster formation in the considered synthesis regimes goes on in different ways. This can be explained by different parameters of the electromagnetic fields that accelerate particles, plasma temperature and different initial particle velocities. The amounts of C3 formations on the whole interval are different, but their greatest number is found in the vicinity of the electrodes in the inter-electrode space. The amount of collisions is determined by the concentrations, velocities and sizes of the particles. At the initial stage, the greatest amount of particles with the relatively low velocity is present, and this gives a large amount of collisions, some of which result in the formation of bonds and enlargement of particles. When particles move from anode to cathode, their amount decreases due to the formation of groups, and the particle concentration decreases, which results in the decrease in the total amount of collisions and bond-forming collisions. Then larger particles are accelerated by the electromagnetic field, which results in the growth of the total amount of collisions in the near-cathode area and, hence, the amount of forming bonds.

5. Conclusions

This paper presents a mathematical model based on the use of the quantum-kinetic approach based on the Boltzmann kinetic equation with account for the pair-wise particles collisions for designing the MCNT using the thermal sublimation method. The use in the model of the functions of various plasma components distribution allows the calculation of the carbon particle behaviour during the synthesis basing on the probabilistic approach due to the consideration of collective plasma phenomena: plasma oscillations, fluctuations of various characteristics, concentration and particle fluxes.

The presented technique and the algorithm of numerical solving in the considered multidimensional nonlinear problem with the application of the nVidia CUDA technology combined with the parallelization technology GPGPU allows obtaining a resource-saving solution on one calculation module. Numerical studies using the model often allow determining the effect of the main synthesis process parameters (current intensity, voltage, pressure and buffer gas) on the final product output, determine the probability areas with the maximum amount of the C-C=C clusters formation in plasma in various synthesis regimes. In the nanotube synthesis regime, the greatest amount of stable bonds is formed at the initial and final phases, which depends on the initial particle distribution and particle acceleration with electromagnetic field and necessity of their sedimentation on the cathode to form nanotubes. In the fullerenes regime, the formation of bonds goes on more even in the whole range of the space, which allows obtaining fullerenes and blowing them out of the inter-electrode space without sedimenting on the cathode.

The obtained results suggest that the developed mathematical model allows adequate description of the processes that take place in plasma during the synthesis of carbon nanostructures, taking into account temperature fields, moving electrode boundaries, buffer medium and electromagnetic field parameters, energy and spatial conditions required for the forming of stable covalent carbon bonds. The obtained results also agree well with the experimental results of other researchers. The developed mathematical model and its elements can be used at the designing of the carbon nanostructure synthesis facilities using the graphing evaporation method.

References

[1] Zhao S, Hong R, Luo Z, Lu H and Biao Yan 2011 *Journal of Nanomaterials* **2011** 346206 6 DOI:10.1155/2011/346206
[2] Mukhin I S, Fadeev I V, Zhukov M V and Golubok A O 2015 *Ultramicroscopy* **148**
[3] Sun D L et al. 2015 *The Chemical Engineering Journal* **283:9** ·
[4] Hutchison J L, Kiselev N A, Krinichnaya E P, Krestinin A V, Loutfy R O, Morawsky A P and Zakharov D N 2001 *Carbon* **39(5)** pp 761-770
[5] Ranabhat K, Pylinina A I, Skripkin K S, Sofronova E A, Revina A A, Kasatkin V E, Patrikeev L N and Lapshinsky V A 2016 *IOP Conf. Ser.: Mater. Sci. Eng.* **151**
[6] Hornyak G L 2009 *Fundamentals of Nanotechnology* (Boca Raton, FL: Taylor & Francis Group)
[7] Drexler K E 2006 *J. Comp. Theor. Nanosci.* **3(1)** pp 1-10
[8] Kamilov I K, Murtazaev A K and Aliev Kh K 1999 *Uspekhi fizicheskikh nauk* **169(7)** 773-795
[9] Shredinger E 1976 *Izbrannye trudy po kvantovoi mekhanike* (*Selected Papers on Quantum Mechanics* Moscow: Nauka) p 422
[10] Runstedtler A 2006 *Chemical Engineering Science* **61** pp 5021-5029

[11] Krishna R and Baur R 2004 *Chemical Engineering Journal* DOI:10.1016/s1385-8947(03)00149-9
[12] Abramov G V, Gavrilov A N 2012 *Vestnik Voronezhskogo gosudarstvennogo universiteta inzhenernykh tekhnologii Seriya: Informatsionnye tekhnologii, modelirovanie i upravlenie* **2(52)** (Russian) pp 71-75
[13] Geiser J 2014 *J. Appl. Math.* **2014**, Special Issue 812137 12 .
[14] Abramov G V, Gavrilov A N and Tolstova I S 2014 *Vestnik Voronezhskoi gosudarstvennoi universiteta inzhenernykh tekhnologii. Seriya: Informatsionnye tekhnologii, modelirovanie i upravlenie* (Russian) **3(61)** 67-71
[15] Bogomolov S V and Kuznetsov K V 1998 *Matem. Mod.* **10**:7 pp 93–100
[16] Abramov G, Gavrilov A and Tolstova I 2014 *Br. J. Sci., Educ. Culture* **2(6)** pp 380-387
[17] Farber R 2011 *CUDA Application Design and Development* (Morgan Kaufmann) p 336
[18] Abramov G V and Gavrilov A N 2016 *Avtomatizatsiya. Sovremennye tekhnologii* (Russian) **3** pp 10-14

Precision Machining of Silicon Substrates

V F Makarov[1], K R Muratov[1*], T R Ablyaz[1], E A Gashev[1], D M Lagunov [1] and N V Varlamov[2]
[1] Perm National Research Polytechnic University, 29 Komsomolsky prospekt, Perm, Perm krai, 614990, Russian Federation
[2] National Research Nuclear University MEPhI (Moscow Engineering Physics Institute), Kashirskoe shosse 31, 115409 Moscow, Russian Federation

[*]Corresponding author's e-mail address: karimur_80@mail.ru

Abstract. Brittle nonmetallic structural materials, such as quartz, glass, silicon, ceramics, ferrites and glass ceramics, are widely used in mechanical engineering, instrument making and radio electronics, as well as in optical, clock and jewelry industries. These materials have high hardness, strength, wear resistance and brittleness; therefore, their machining is a complicated problem. For production of piezoelectric resonators and filters, semiconductor devices, solid-state lasers, optoelectronic devices and other parts of electronics, plates made from monocrystals of quartz, lithium tantalate and niobate, silicon, germanium, sapphire, etc. are used. Semiconductor plates, which were obtained after cutting of a boule, have a number of defects, such as presence of a mechanically affected layer, nonflateness of sides, twisting and large variation of thickness. Therefore, after cutting it is necessary to carry out grinding and polishing. Grinding (abrasive finishing) is machining of semiconductor plates by means of hard finishing grinding discs (lapping tools) using abrasive flour grains. Finishing discs (lapping tools) are usually made from cast iron, glass, steel, copper or tin. The grain size of flour grains for grinding of semiconductor plates is selected from M14 to M5. Finishing machining of silicon substrates is complicated by a number of factors, such as provision of stable sizes, and, which is more important, the required microrelief of a machined surface, and unacceptability of machining marks and micro fissure on a surface. In Perm National Research Polytechnic University, the method of abrasive finishing and polishing of surfaces of silicon substrates by means of "Rast-350" flat-finishing machine was developed. The feature of the machines is in sliding movement of a tool (lap) along a nonrecurring trajectory, which has a form of a grid of complicated configuration. At that, the speeds of movement of all points of the working face of the lapping tool are the same. The studied and proposed technological recommendations for operations of finishing and polishing of silicon substrates allowed determining the main technological regimes (time, specific pressure, and density of the grid of tool marks), as well as the required abrasive material of corresponding grain size, which allowed providing the required microrelief of substrates (*Rz* 25 nm, *Ra* 5 nm) and eliminating machining marks and micro fissures on a polished surface.

1. Introduction

The modern microelectronics has quite high requirements to the quality of an effective area of semiconductor materials, in particular, silicon substrates. First of all, it is necessary to provide roughness of a machined surface on the nanometer scale. In this regard, the requirements for technology and equipment for finishing of this kind of products are increasing [1-2]. By technological parameters, grinding (finishing) is divided into preliminary and final, by structural parameters – into single-sided and double-sided, by the type of used material – into grinding with unbonded and bonded abrasive [3].

Preliminary and final grinding (finishing) consists of two processes, which differ in aims, regimes and used materials. The purpose of preliminary grinding is the high-speed smoothing of a plate and the removal of the dimensional allowance. The regimes of this process have higher frequency of rotation of a grinding disc (lapping tool) and increased specific pressure of abrasive powder on the surface of a semiconductor plate. Moreover, for preliminary grinding abrasive powders with coarser grains (M28, M14, M10) are selected. The final grinding (fine finishing) has "softer" regimes of polishing and they use abrasive powders with finer grains (M7, M5). The purpose of final grinding is further improvements of the geometry and quality of surface of machined surfaces.

Grinding by means of unbonded abrasive powder is carried out using machine tools with single-sided and double-sided processing of semiconductor plates using various suspensions and pastes. During processing, grains are unbonded, i.e. not connected with each other. Abrasive suspension creates a thin layer between the instrument (lapping tool) and a processed semiconductor plate, in which abrasive grains are rolling freely. That's why it is called unbonded abrasive material. Grinding by means of bonded abrasive material is significantly different from grinding with unbonded abrasive material, both by the physical principle of removal of a semiconductor material during polishing and by the kinematics of machining. Grinding with bonded abrasive material is carried out using machine tools with rigid axes, the feature of which is that the positions of rotation axes of a lapping tool and machined plates are constant. The main feature of this process is the structure of a grinding disc (lapping tool), which is a metal disc with applied surface diamond layer. Diamond grains are strongly attached to the surface of a grinding disc by means of special connecting binder [4].

For the improvement of the quality of machining of the surface of semiconductor materials and the decrease of the depth of mechanically affected layer, polishing is carried out. Polishing is different from grinding in the technological regime and type of an abrasive material, as well as the material of a polishing pad. The common point of these processes is the type of machining – single-sided or double-sided using unbounded abrasive material. The polishing process is carried out, generally, with soft polishing pads, which are rigid discs covered by a soft material: flannel, suède, batiste, felt, velour, cloth, silk, etc. Flour grains made from synthetic diamond, aluminum oxide, chromium oxide, silicon dioxide and zirconium dioxide (with grain size not higher than 3 μm) are used as abrasive materials. The selection of polishing cloth depends on its compatibility with the used abrasive material. Polishing cloth must be able to hold particles of abrasive material by its nap during machining of semiconductor plates. Polishing cloth must have a higher wear resistance and be inexpensive. For preliminary polishing, cloth without nap together with larger diamond grains is usually used. Cloth with nap is used for final machining using diamond grains with a size of less than 1 μm. There are several requirements for polishing pads: they must not have folds, cuts, thickened zones, ruptured threads, bulges and recesses, foreign particles and other surface defects. The wear resistance of soft polishing pads is low [5,6].

Further improvement and advancement of polishing methods for semiconductors are related to new ways of increasing the productivity, creation of polishing materials, which can provide both high quality of surface machining and good geometry of plates.

2. Problem statement

The machining of substrates before electropolishing is carried out by means of the method of abrasive finishing. "Rastr" flat-finishing machine tools (Fig. 1) are used in small-scale and large-scale production, in which it is required to frequently change small batches of parts and have wide product mix. The feature of the machines is the sliding movement of a tool (lap) along a nonrecurring trajectory, which has a form of a grid of complicated configuration. At that, the speeds of movement of all points of the working face of the lapping tool are the same [7].

IOP Conf. Series: Materials Science and Engineering **498** (2019) 012009 doi:10.1088/1757-899X/498/1/012009

a) b)

Figure 1. Flat-finishing machine tools: a) Rastr-220; b) Rastr-350.

The complicity and uniqueness of relative movement trajectories of a tool (Fig. 2) and machined surfaces of parts together with capability to adjust its parameters make it possible to effectively control machining and create a surface with the designated properties of macro- and microrelief [8].

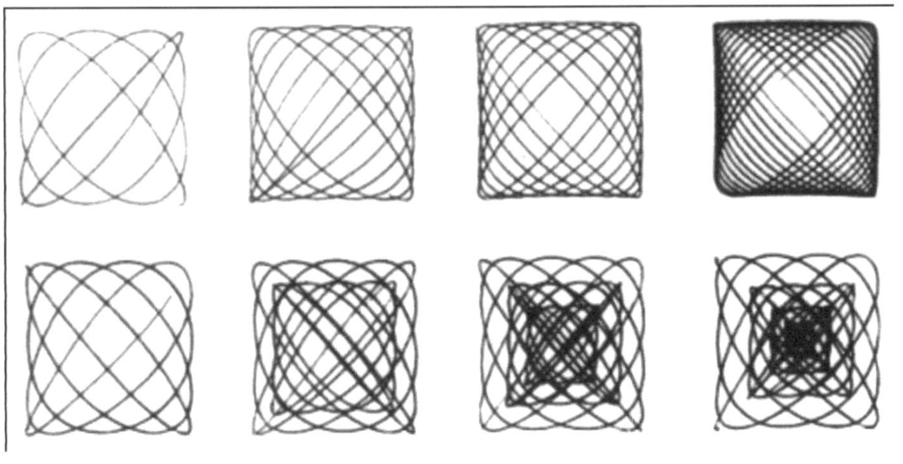

Figure 2. A trajectory of working movement of a tool (a single abrasive grain) for equipment with raster kinematics.

The capability of the selection of the shape and density of trajectory of working movement of a tool (Fig.2) allows forming the microrelief of the machined surface with optimal statistical parameters for certain properties of products. At that, the value of roughness using the parameter Rz can be achieved on the nanometer scale 35-10, Ra = 6-2 nm.

3. Discussion

In Perm National Research Polytechnic University, the trial test of the processes of finishing and polishing of silicon substrates was conducted, the results of which are presented below.

Single-sided finishing of silicon substrates was carried out using Rastr 350 machine tool using two operations. Plates with the following dimensions were used: 30 × 45 mm, 40 × 75 mm, thickness 0.7 mm. Specimens were attached by means of glue. The parameters of initial roughness of specimens are presented in Fig. 3.

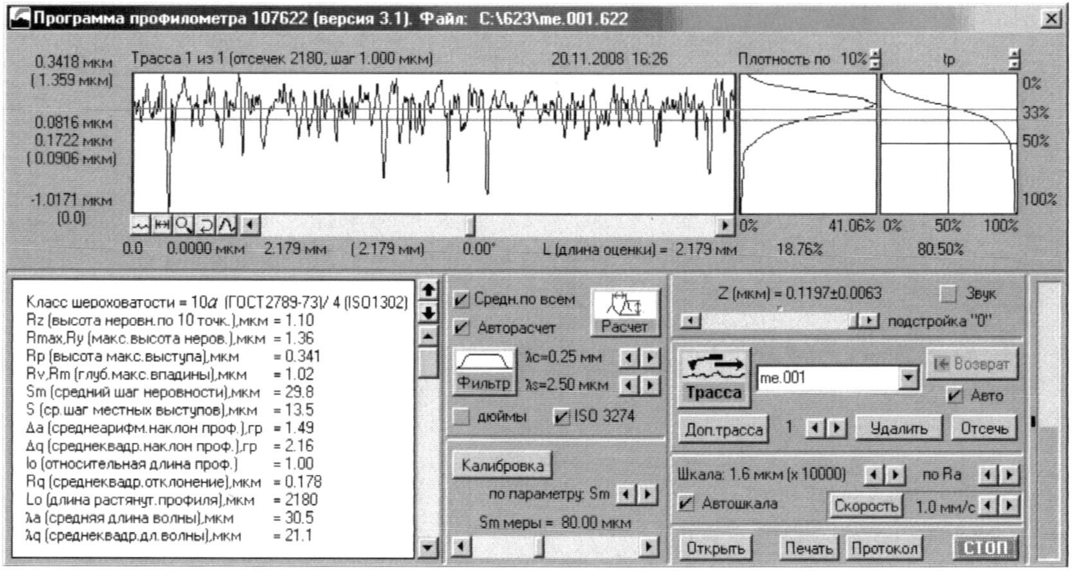

Figure 3. Initial roughness of silicon substrates.

At the first stage, finishing (grinding) was carried out using a glass lapping tool and synthetic diamond powder ASM 3/2 with addition of special liquid coolant. The specific pressure on a machined part was 10-20 kPa. The time of machining was 1.5-2 min. According to the developed recommendations, finishing was carried out using a wide grid, which allowed increasing productivity. At that, isotropic roughness by the parameter *Rz* 0.25-0.32 (Ra 0.05-0.06) was provided.

At the second stage (polishing), a composite cellulose-based lapping tool was used. Polishing was carried out using ASM 1/0 diamond paste with addition of small amount of special liquid coolant. The machining regime (pressure and time) didn't change. During polishing, a denser grid of marks of raster trajectory corresponding to misalignment of frequencies of initial oscillations of instrument of 0.5-1% was used. The results of the measurement of parameters of roughness after polishing are presented in Fig. 4. The roughness by *Rz* didn't exceed 25 nm and by Ra didn't exceed 5 nm. The overview of finished and polished substrates is presented in Fig. 5.

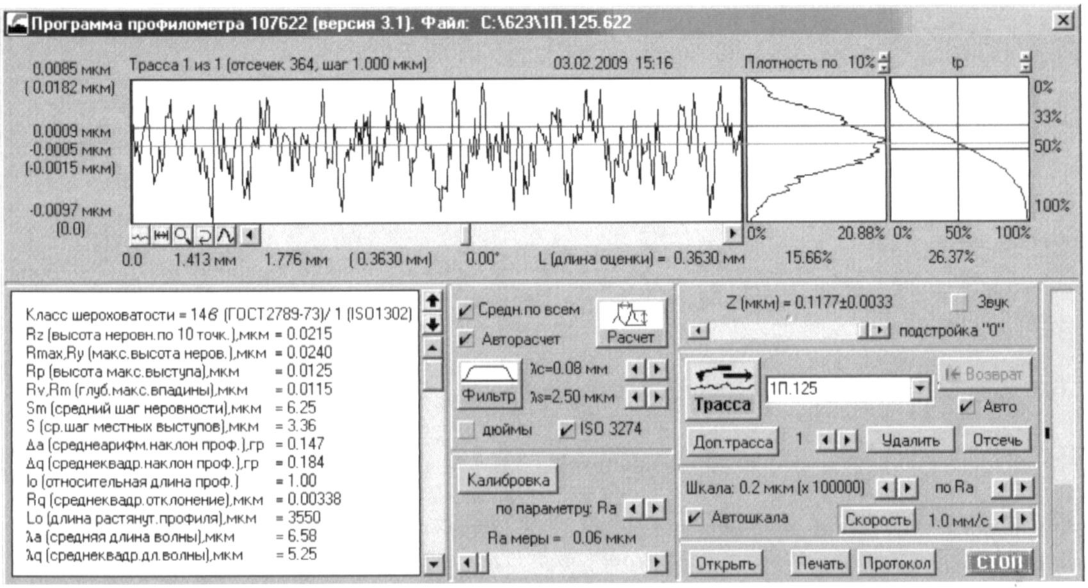

Figure 4. Roughness of surfaces of substrates after machining by means of "Rastr-350" machine tool.

Figure 5. Polished surfaces of silicon substrates.

4. Conclusion

Thus, during the operations of finishing and polishing of silicon substrates, the main technological regimes were established (time, specific pressure, density of grid of machining marks), as well as the required abrasive material with the required grain size. The proposed operation of finishing machining allowed providing the required microrelief of substrates of *Rz* 25 nm and *Ra* 5 nm and eliminating machining marks and micro fissures on the polished surface.

Acknowledgments
This study was carried out with financial support of the Ministry of Education and Science of Russian Federation under the state contract No.11.976.2017/BCh.

References

[1] Heydemann V D, Everson W J, Gamble R D, Snyder D and Skowronski M 2003 Chemi-Mechanical Polishing of On-Axis Semi-Insulating SiC Substrates *Mater. Sci. Forum* pp 457-460, 805-808
[2] Kato T, Wada K, Hozomi E, Taniguchi H, Miura T, Nishizawa S I and Arai K 2006 High Throughput SiC Wafer Polishing with Good Surface Morphology *Mater. Sci. Forum* pp 556-557, 753-756
[3] Skryabin V A, Ribakov Yu V and Igonina T I 2005 *Mashinostroitel* **7** 19-21
[4] Tam H Y, Cheng H B and Wang Y W 2007 *J. Mater. Process. Technol.* **192** pp 276-280
[5] Kang J, Hadfield M 2005 *Original Res. Article* **258** pp 2-12
[6] Kim J D, Choi M S 1995, *J. of Mater. Process. Technol.* **52** pp. 368–385
[7] Khanov A M, Muratov K R and Gashev E A 2016 Investigation of the Abrasive Lapping Process of oxide Ceramics *Oriental Journal of Chemistry* **32(1)** 391-398
[8] Hanov A M, Muratov R A, Muratov K R and Gashev E A 2010 *Stanki. Instrument* **2** 34-35

Research of influence of technological admissions on characteristics of a solid-state wave gyroscope

E Ph Pevtsov·, T A Demenkova and K A Grishunin

Moscow Technological University MIREA (Moscow State Institute of Radio-engineering Electronics and Automation), Prospekt Vernadskogo 78, 119454 Moscow, Russian Federation

·Corresponding author's e-mail address: pevtsov@mirea.ru

Abstract. The route of design and technique of simulation of a wave solid-state gyroscope with the ring resonator are developed. Simulation of dependences of gyroscope characteristics on design data and their optimization for the job of the necessary mode of movement, in particular, of resonance frequency of excitation of the second mode are executed. The technique of compensation of the mass defects resulting from admissions on technological operations of production of a wave solid-state gyroscope (technological defects) and leading to the shift of initial provision of nodal points is presented.

1. Introduction

The wave solid-state gyroscopes (WSSG) executed in design as microelectromechanical systems (MEMS) [1] possess rather high characteristics on the accuracy and working range. In combination with the low cost, considerable stability to vibrations and shock influences it does them perspective sensors for measurement of angular speed of an object in many scopes.

The effect of origin of standing waves of elastic deformations in the ring plane is the cornerstone of operation of a wave solid-state gyroscope. In the movement mode on resonance frequency in a ring there are oscillations of the second mode so that the nodal points located in two mutually perpendicular directions remain fixed. In case of rotation of the resonator with angular speed Ω, the ring on tangents are affected by Coriolis forces, the deviation of nodal points from initial situation results. Information on these movements is registered by capacitor sensors and is transferred for further processing. Theoretical bases of wave solid-state gyroscopes are stated in works [2-5].

2. Formulation of the problem

The understanding of the influence of the defects connected with the imperfection of their production is important for practical applications. Classification of defects is given in work [3] and the mathematical apparatus for calculations of their influence on characteristics of a gyroscope is offered. This work considers the problem from positions of simulation of similar effects by tools of a modern CAD.

Research problems:

-development of a route of design and techniques of modeling of WSSG with the ring resonator;

-research of the free and forced oscillations under the influence of the given forces and determination of resonance frequencies and optimization of design data of WSSG for the job of the necessary mode of movement, in particular, the resonance frequency of excitation of the second mode;

-development of techniques of compensation of the mass defects resulting from admissions on technological operations of production of WSSG (technological defects) and leading to the shift of initial provision of nodal points.

The design of WSSG is schematically presented in Figure 1. The resonator represents a thin ring which is connected to the massive basis by means of eight elastic elements.

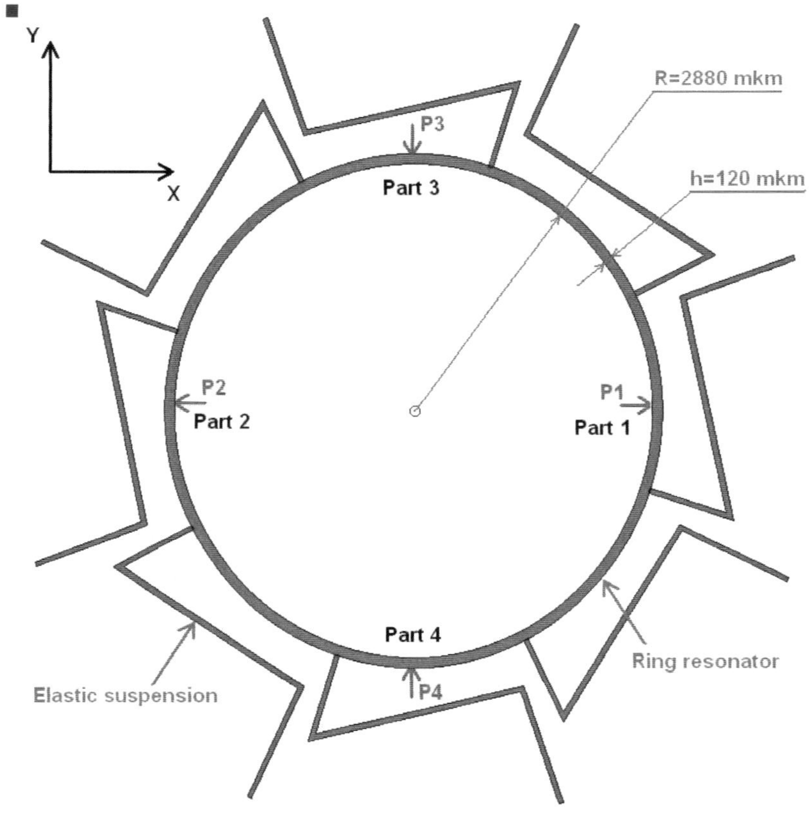

Figure 1. Design of wave solid-state gyroscope and parts of application of forces.

3. Route of design and setting simulation conditions

The work was performed on basis of tools of a package of the computer-aided design system (CAD) CoventorWare® [6], which is specially intended for design of MEMS. Reliability of the received results of simulation was controlled by their comparison to the analytical decisions received for special cases of boundary conditions and the simplified models.

The main design procedure of the design route with CAD CoventorWare is the development of the drawing of device design and a set of drawings of masks for its production on the basis of standard operations of MEMS technologies. For obtaining the relevant design decisions the plotter of the two-dimensional model (the editor of masks) Layout Editor®, the editor of properties of the constructional materials Materials Properties Data Base ® and the library of standard technological processes Process Editor ® are used.

Main operations of formation of the volume design:

- setting substrate from silicon;
- sedimentation of a layer from silicon of the set thickness on a substrate;

IOP Conf. Series: Materials Science and Engineering **498** (2019) 012010 doi:10.1088/1757-899X/498/1/012010

- etching of this layer for formation of external and internal perimeters of a ring, using masks with positive and negative photoresist;
- removal of a silicon substrate.

The result of design procedures of the description of a design is the synthesis of three-dimensional model of a gyroscope executed by tools of the Designer ® module (see Figure 2).

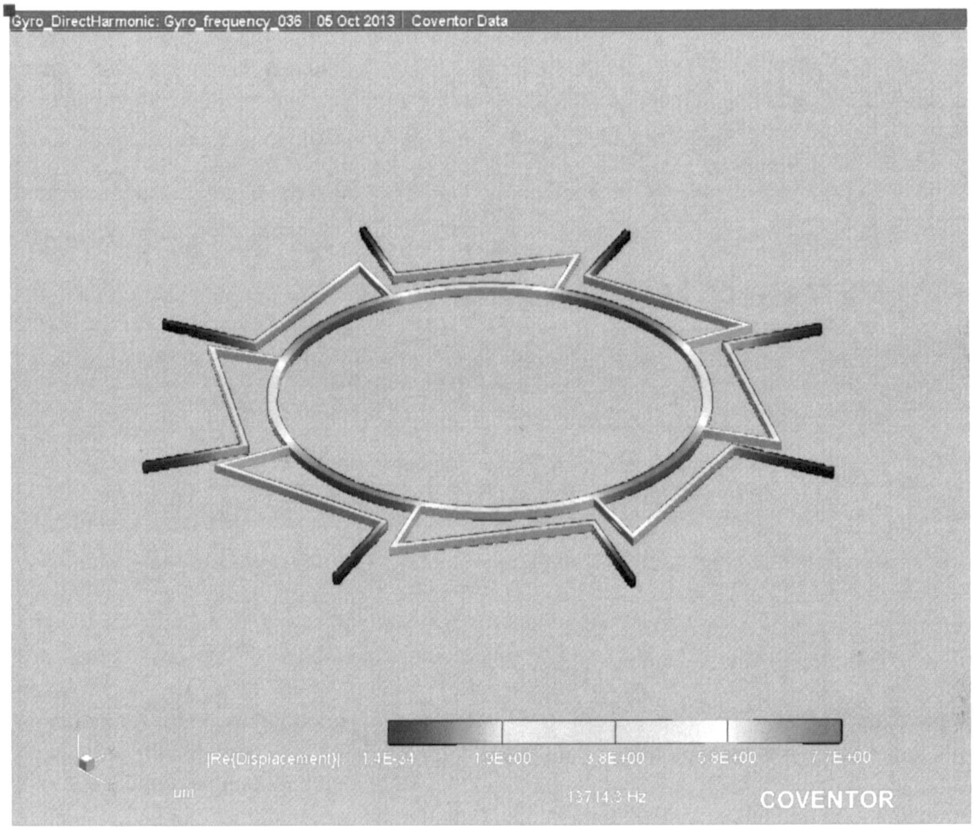

Figure 2. WSSG 3D model in the mode of oscillations excitation.

The important step preceding calculations is setting the grid for application of a method of final elements and a step of splitting. When performing this work it is experimentally shown that in the task for calculations the type of the grid "Extruded bricks" is optimum, and the step of splitting which was chosen by standard criterion of double reduction till the moment when results of calculations in the subsequent iteration differed from previous no more, than for 5%.

For simulation of the mode of movement of WSSG and excitation of oscillations of standing waves on a sensitive element of gyroscope influence of four harmonic forces P1 … P4 (see Figure 1) deforming a ring in mutually perpendicular directions was set. Value of amplitude of forces P1...P4 made $3{,}53 \cdot 10^{5}$ N, and the corresponding pressure upon sectors, exciting oscillations - $1{,}23 \cdot 10^{4}$ MPa.

4. The received results and their discussion

First, the excitement of the mode of the movement WSSG at a variation of parameters of design was simulated. The emergence of a standing wave in the ring resonator clearly demonstrates Figure 3 (counting of an angle is conducted from the positive direction of the axis OY counterclockwise, in the initial rest position the axis of symmetry of a gyroscope is turned concerning OY axis on 4,7°).

In the same figure provisions of the nodal points symmetrized on the middle of four sectors of a ring are designated.

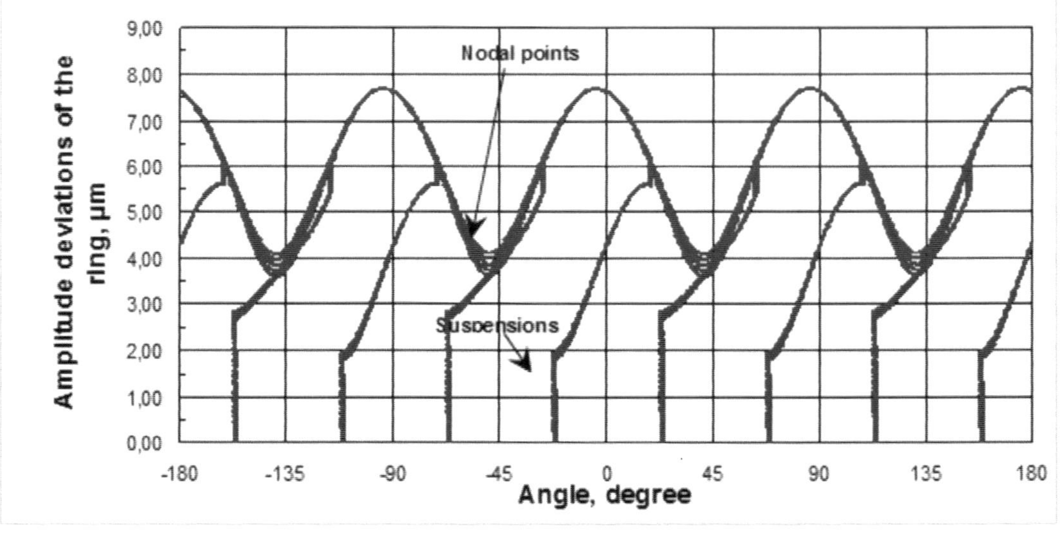

Figure 3. Deviations of points of the ring and suspensions of the WSSG resonator from initial situation in the movement mode on resonance frequency 13,6 kHz.

Examples of amplitude-frequency characteristics of gyroscope are given in Figure 4.

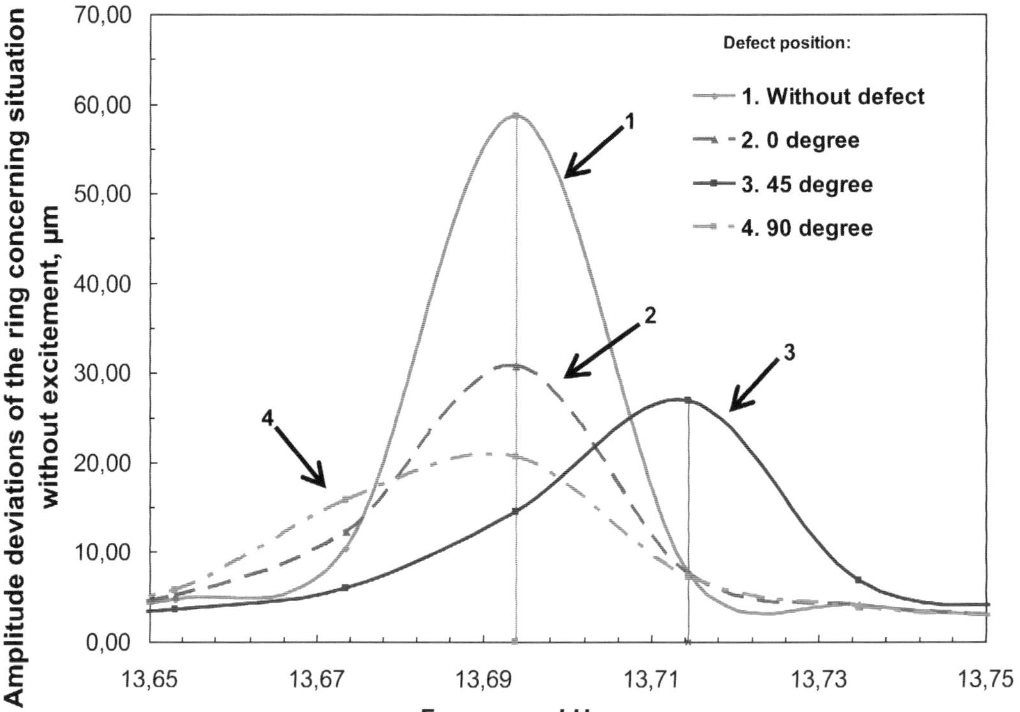

Figure 4. Amplitude-frequency characteristic of WSSG resonator ring and influence of technological defects: 1 – without defect, 2 – technological defect is located at the angle 0°, 3 – 45°, 4 – 90°.

During simulating the technique of determination of natural frequencies of the resonator depending on design data is developed. In particular value of resonance frequency for this construction free of defects of weight made 13694 Hz that with an accuracy of ten hertz matches results of the experimental studies received on a prototype of WSSG made in FTIAN [7].

For simulation of the influence of tolerances of topology masks and parameters of construction, and also the modes of technological processes of manufacture in case of simulation of construction in a ring of the resonator the defects of weight representing hollow dredging with sizes of 432x92x61 mkm³ and mass 6,1 nkg, placed in different sectors of a ring were artificially created.

Figures 4 and 5 illustrate the influence of defect of mass. Dependence of resonance frequency of the second mode of oscillations on the mass of defect is practically linear in the range of masses from 0 to 30 ng. It is shown that existence of the defects of mass located at an angle 0 and 90 degrees to OY axis doesn't influence resonance frequency of oscillations, but offsets the provision of nodal points while the defects of mass located at the angle 45 degrees in addition lead to shift of resonance frequency that can be explained by changes in elastic properties of construction.

Existence of mass defects is the negative factor reducing characteristics of WSSG because of errors of initial position of an arrangement of nodal points in the mode of the movement. For simulating of compensation of the arising distortions it is offered to enter the additional compensating mass defects. It can be done, for example, by means of laser adjustment by selective evaporation of a part of ring segment of the resonator. The purpose of such technological operation is return of points of 1-4 (fig. 2) to situation, the maximally close to initial. Results of simulation of introduction of the compensating mass defects are illustrated in Figure 5.

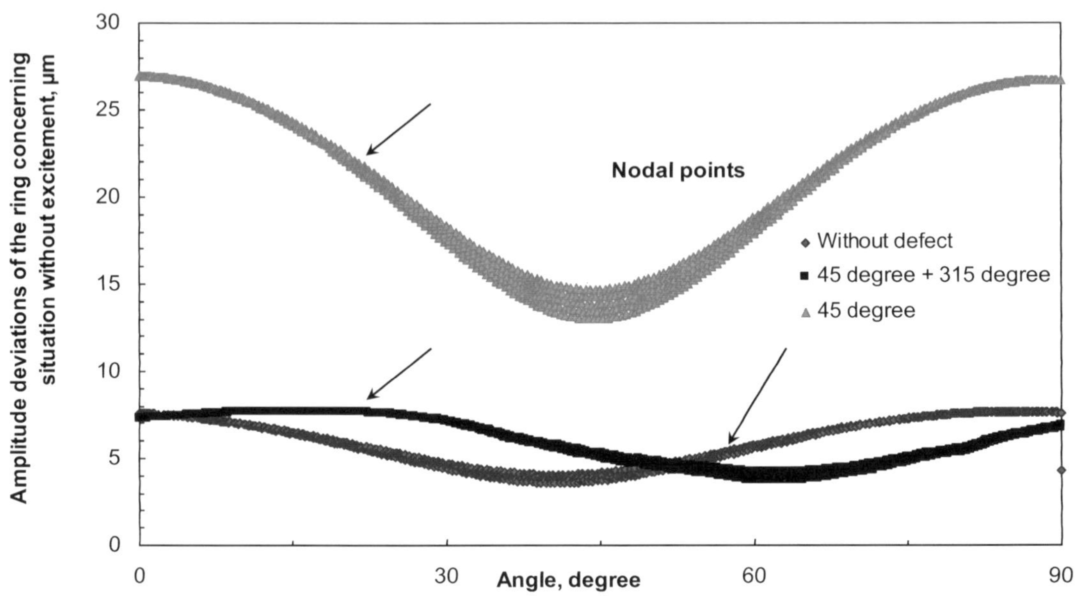

Figure 5. Influence of the compensating defects on amplitude deviations of a ring concerning situation without excitement. Arrangement of defects: 1 - without defect, 2 - 45°, 3 – the main: 45°, compensating: 315°.

From Figure 5 it is visible that distortions of a trajectory, arising in case of an arrangement of technological rectangular defect of mass in 6 ng with the center located at an angle 45° to a vertical axis can be compensated with an accuracy of 98% by introduction of the compensating defect of the same mass with the center located at the angle 315°. The developed technique of compensation can be applied to recovery of the zero provision of nodal points at other parameters of technological defects.

5. Conclusions

As a result of the simulation the decisions having essential value for increase in technical and utilization properties of a solid-state wave gyroscope with the ring resonator were received.

1. A route of design of WSSG by the CAD tools which finite project solution is the mask set for manufacture with use of standard operations of MEMS technology.

2. Technique of determination of resonance frequency of a wave gyroscope. It is shown that in case of the given parameters of construction (Rint.=2880 mkm, Rext.=3000 mkm, P=123 Pa) the value of resonance frequency received by results of simulation makes 13694 Hz that will be coordinated with the data of experimental studies of WSSG prototypes.

3. The technique of simulation of influence of mass defects on characteristics of sensitivity of WSSG and a method of optimum compensating of mass defects on the basis of deleting the compensating volumes of the fluctuating frame are offered. Results of simulation of position of the compensating defects and their masses showed that zero drift of a gyroscope can be compensated to 98,5%.

Acknowledgments

This work was supported by Ministry of Education and Science of Russian Federation, (project 8.5098.2017).

References

[1] Vasiliev A A, Sokolov A V, Pisliakov A V, Oblov K Yu, Samotaev N N, Kim V P, Tkachev S V, Gubin S P, Potapov G N, Kokhtina Yu V and Nisan A V 2016 *IOP Conf. Ser.: Mater. Sci. Eng.* **151,** 012024
[2] Gankevych Z, Kucher S and Bondar P 2010 *Gyroscope XXIV Symposium Vibrations in Physical Systems*. (Poznan – Bedlewo)
[3] Yilmaz E and Bindel D 2013 *Conference: Sensors, IEEE.*
[4] Dzhashitov V E and Pankratov V M 2010 *J. Mach. Manuf. Reliab.* **39** p 248
[5] Matveev V L, Lipatnikov V I and Alehin A V 1998 *Designing of the wave solid-state gyroscope* (M.: MSTU) p 168
[6] COVENTOR 2017. Available at: http://www.coventor.com
[7] Morozov O V, Postnikov A V, Amirov I I and Kal'nov V A 2012 *Nano- and the microsystem equipment* (Russia) **7** pp 15-19

IOP Conf. Series: Materials Science and Engineering **498** (2019) 012011 doi:10.1088/1757-899X/498/1/012011

Bit error rate evaluation in high speed communication channels

A A Garmash, D A Domozhakov, N Y Rannev

National Research Nuclear University MEPhI (Moscow Engineering Physics Institute), Kashirskoe shosse 31, 115409 Moscow, Russian Federation

Corresponding author's e-mail address: DADomozhakov@mephi.ru

Abstract. This work offers calculate technique probability errors in transmitter – receiver path during the design phase with apply theory of probability elements, which is based on results of transient process modeling. As estimation, designed method was verified by comparison his results with computer experiment results. Implementation and using variants of represented algorithm are offered.

1. Introduction

Serial transceivers are essential part of transfer data electronic systems. They provide stability and fast information exchange between devices by external communication lines.

Actual transceivers with data rate 1 Gbps and more should provide number of errors less 10^-12. With current data rates (1,25/2,5/3,125 Gbps and more) impact of phase distortion increases on transmitted data integrity.

Bit Error rate (BER) is important and obvious criterion of transceiver path work and is defined as number of errors divided by total number transmitted bits. Currently measuring devices manufacturers offer technique of measurement BER for a couple transmitter – receiver, but actual problem is estimation BER during the design phase.

Eye diagram is image of superposition all bit periods of measured signal. Estimation of correctly data transmit is provided by apply mask to eye diagram. Intersection of mask with eye diagram means error. But this method provides only rough estimate.

There are algorithms estimation BER which is based on some basic principles. Comparison of main methods is represented in Table 1. Methods, which is based on results SPICE modeling, require long time modeling. Example, confirmation of BER level less 10^-12 require modeling transmit more 10^12 bit. Statistical methods provide high precision of estimation and high calculate speed. Algorithm types is possible, which can have universality with modeling different transmitter – receiver path types.

Table 1. Comparison common types of evaluation methods BER.

methodology	simplicity	speed	*universality*
SPICE – modeling	+	-	+
worst-case modeling	+/-	+/-	+
using IBIS models	+	+/-	-
statistical method	+/-	+	+

2. The task of receiving data

The task of any transceiver is to trusty exchange information between two paired devices. It is necessary to control all critical nodes that are subject to changing conditions of the circuit operation and the influence of jitter. One of the main elements of the receiver, requiring attention, is the first trigger of the shift register in the receiver, which latches the input data stream and converts it into an internal data signal Q_CORE (Figure. 1).

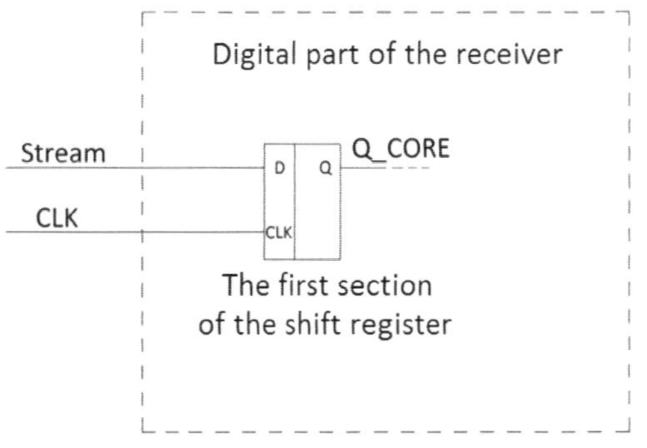

Figure. 1. The first trigger of the shift register of the digital part of the receiver. The Stream and CLK signals are a data signal and a clock signal, respectively. The phase of the CLK signal is restored from the Stream signal using the Clock and Data Recovery (CDR) block. The Stream data signal is received from the interface part of the receiver, the CLK reference signal is produced by a phase and frequency adjustment unit called the CDR. To successfully pass the data signal with the subsequent deserialization of the stream, in general, it is necessary to observe the required phase and amplitude relations between these signals.

From the central limit theorem of probability theory it is known that the sum of a sufficiently large number of weakly dependent on each other random variables having the same scope, subject to the normal distribution. Based on this situation it is assumed that the noise, which is the source of bit errors has a normal distribution in amplitude. Therefore, for the treatment of the consequences of his influence can be to use the tools of probability theory.

The parasitic elements of the topology and the housing deteriorate the parameters of the transmitted signal [4]. Therefore, when the worst bit sequences are transmitted, for example "0101" or "1110", the signal amplitude is distorted and the deterministic jitter increases [5].

Two types of causes of bit errors can be identified. First, at low values of the signal-to-noise ratio, the transmitted signal can cross the receiver switching level (Figure 2), and, consequently, the probability of an error in receiving increases.

Secondly, the necessary condition for obtaining correct data receiver is that the permitted switching timing between the data signal DATA and the clock signal CLK input to flip-flop in the digital portion of the receiver (Figure. 3).

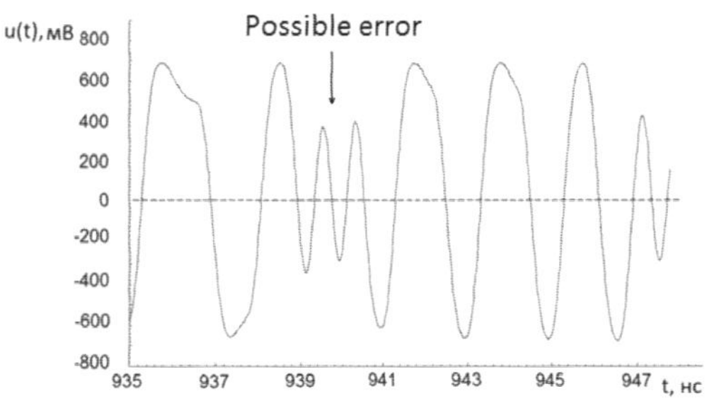

Figure 2. Amplitude distortion effect.

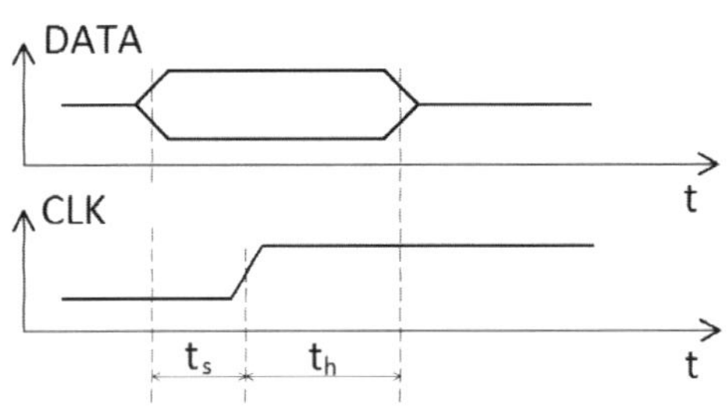

Figure 3. Setup and Hold times.

The data must arrive at the input no later than the minimum setup time (t_{s_min}) to the clock edge. Also, the data must remain at the input after the clock signal drops no less than the minimum hold time

(t_{h_min}). If one of the conditions is not met, the recording of the correct information in the first flip-flop of the shift register becomes impossible.

3. Description of the BER analysis method

During transmission, a bit error can occur regardless of two reasons: amplitude or phase distortion of the input signals. In this case, the error probabilities of these two species will be different. The total BER will be equal to the algebraic sum of the probabilities of bit errors in amplitude and phase as two independent events. The proposed algorithm for estimating BER is a separate analysis of the magnitude of the probability of errors due to phase and amplitude distortions. The analysis of the bit stream of data is carried out in regions A and B, respectively (Figure. 4).

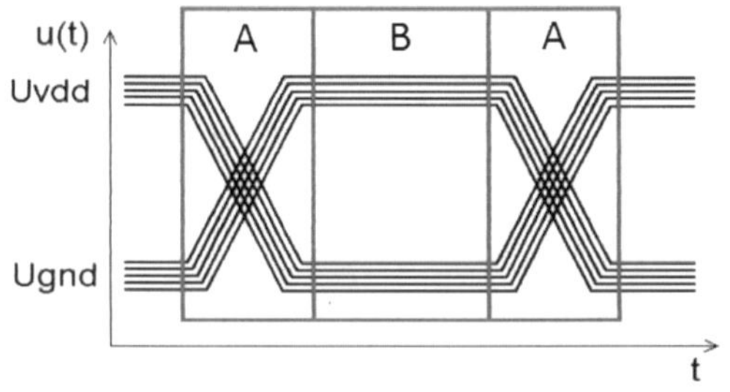

Figure 4. Areas of eye diagram analysis. As was shown above, there are various reasons for the occurrence of bit errors in the transmission of digital data. Therefore, in the process of analysis, it is expedient to divide the eye diagram into regions of influence of the described factors.

On the basis of the obtained values, separately for the zero level and the unit level, the RMS deviation of the random variable σ_A (RMS) is calculated from the signal amplitude:

$$\sigma_A = \sqrt[2]{\frac{1}{n}\sum_{i=1}^{n}(A_i - m_A)^2}. \quad (1)$$

The mathematical expectation m_A is known: this is either the logical unit level or the logical zero level. The random variable A_i is the instantaneous amplitude of the signal, n is the number of instantaneous amplitude measurements. The number of samples is predetermined. Thus, the probability density function is determined for each of two cases:

$$P_{DATA}^{A}(A) = \frac{1}{\sigma_A\sqrt{2\pi}} \exp\left(-\frac{(A - m_A)^2}{2\sigma_A{}^2}\right), \quad (2)$$

where A is the instantaneous amplitude of the signal.

Next, the probability density is integrated over the range corresponding to the case of error generation at the intersection of the receiver response level for each case: (2a) for the "one" level and (2b) for the "zero" level. The probabilities obtained are multiplied by the number of bits in the pattern and added (3).

$$BER_A^0 = \int_{Atr}^{+\infty} P_{DATA}^{A}(A)dA, \quad (2a)$$

$$BER_A^1 = \int_{-\infty}^{Atr} P_{DATA}^{A}(A)dA, \quad (2b)$$

$$BER_A = BER^0 + BER^1, \quad (3)$$

Where BER_A^0 – the probability of an error in the transfer of zero bit, BER_A^1 – the probability of an error in the transmission of a one bit, BER_A – the total error probability, Atr – the threshold level of the receiver trigger switching.

BER analysis by phase. In Figure. 5 illustrates the distribution of data phases P_DATA and phases of the clock signal P_CLK at the input of the receiver. In this case it is necessary to satisfy the condition under which the data and clock edges are arranged relative to one another no closer than t_{s_min} - minimum setup time of the data signal. Analysis the relative position of the fronts of these signals to meet the minimum hold time t_{h_min} is similar.

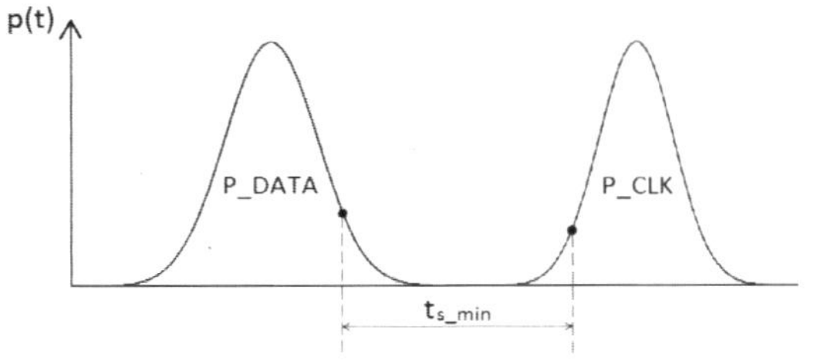

Figure 5. Probability distribution of the phases of the data signal and the clock signal. The probability distribution of signal fronts is described by means of a normal distribution. The occurrence of an error can be considered as the simultaneous occurrence of two random events, for which the condition of time relations between the arrival of the clock signal and the data signal is not satisfied.

The RMS signals DATA and CLK, respectively, have the form:

$$\sigma_{\mathrm{DATA}} = \sqrt[2]{\frac{1}{n}\sum_{\mathrm{i=1}}^{\mathrm{n}}(A_{\mathrm{i}} - m_{\mathrm{T}})^2}, \quad (4)$$

$$\sigma_{\mathrm{CLK}} = \sqrt[2]{\frac{1}{n}\sum_{\mathrm{i=1}}^{\mathrm{n}}(A_{\mathrm{i}} - m_{\mathrm{T}})^2}, \quad (5)$$

где m_T – mathematical expectation, the moment of arrival of the signal front corresponding to an ideal grid of clock signals.

The distribution of the probability density of the arrival times of the DATA and CLK signal fronts, respectively, is:

$$P_{\mathrm{DATA}}^{T}(t) = \frac{1}{\sigma_{\mathrm{DATA}}\sqrt{2\pi}} \exp\left(-\frac{(t - m_{\mathrm{t}})^2}{2\sigma_{\mathrm{DATA}}{}^2}\right), \quad (6)$$

$$P_{\mathrm{CLK}}^{T}(t) = \frac{1}{\sigma_{\mathrm{CLK}}\sqrt{2\pi}} \exp\left(-\frac{(t - m_{\mathrm{t}})^2}{2\sigma_{\mathrm{CLK}}{}^2}\right), \quad (7)$$

where t is the arrival time of the signal front.

RMS deviation of the phase of the data signal and the clock signal are reduced to the total RMS by the formula:

$$\sigma = \sqrt[2]{\sigma_{\mathrm{DATA}}{}^2 + \sigma_{\mathrm{CLK}}{}^2}, \quad (8)$$

It is assumed that the phase of the data is unchanged, and the total RMS has a phase of the clock signal. In this case, BER is the probability of the edge crossing of the CLK signal of the minimum thresholds of the setup and hold times t_{s_min} and t_{h_min}. The resulting BER can be calculated from the formula:

$$BER_{\mathrm{T}} = 1 - \int_{\mathrm{t_{s\_min}}}^{T - \mathrm{t_{h\_min}}} P_{\mathrm{CLK}}^{\mathrm{T}}(t)\mathrm{dt}, \quad (9)$$

where T is the duration of one bit of data.

In integrated circuits, bitstream transmission produces deterministic jitter, partly due to intersymbol interference. In this case there is the background of the length of several bits results in a variation of the duration of a single bit interval (Figure. 6).

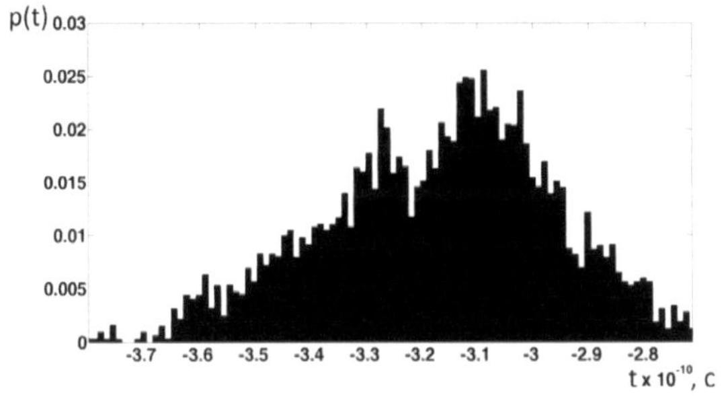

Figure 6. Distribution of the probabilities of the phases of the data signal and the clock signal taking into account the deterministic jitter. As can be seen in the figure, the distribution has two maxima, which should be taken into account during the analysis. This distribution depends on the pattern of the transmitted bit code.

In this case, the sophisticated model takes into account the difference in the lengths of bit intervals, depending on the sequence of transmitted data. The algorithm analyzes the transmitted code, making bit patterns that take into account the history of several data bits. For example, if the length of the pattern is 5, the total number of templates will be 32. In this case, the length of the last bit depends on the sequence of the preceding four bits. Taking the data stream as infinite, the algorithm calculates the probability of the appearance of one or another template, giving them a weighting factor P_{j}:

$$P_{\mathrm{j}} = \frac{D}{N} * M_{\mathrm{j}}, \tag{10}$$

where D is the length of the pattern in bits, N is the total number of transmitted bits, M_{j} – The number of repeats of the jth template in the sending of the transmitted bits.

In this case, the probability density function for each of the cases will have the form:

$$P_{\mathrm{DATAj}}^{\mathrm{T}}(t) = \frac{1}{\sigma_{\mathrm{DATA}}\sqrt{2\pi}} \exp\left(-\frac{(t-m_{\mathrm{tj}})^2}{2\sigma_{\mathrm{DATA}}{}^2}\right), \tag{11}$$

where m_{tj} – the mathematical expectation for each of the distribution variants.

Also, as it was described earlier, the RMS phases of the data signal and the clock signal are reduced to the total RMS according to the formula (8).

The presence of two or more distribution peaks since the arrival of data bits means that in the formula (4) will be used to its limits of integration according to each of the data signal distribution maxima. In this case, the RMS for each of the distributions is assumed to be the same. Then integration is performed according to the formula (11) for each of the patterns:

$$\mathrm{BER}_{\mathrm{Tj}} = 1 - \int_{\mathrm{t_{s\_minj}}}^{\mathrm{T} - \mathrm{t_{h\_minj}}} \mathrm{P}_{\mathrm{CLKj}}^{\mathrm{T}}(\mathrm{t})\mathrm{dt}, \tag{12}$$

where $t_{\mathrm{s\_minj}}$ и $t_{\mathrm{h\_minj}}$ – minimum setup and hold times corrected according to the offset of the data signal distribution center for each of the cases.

Thus, the BER in the signal phase will be the algebraic sum of the BERs for each of the patterns, taking into account their weighting factors:

$$BER_{\mathrm{T}} = \sum_{\mathrm{j=1}}^{\mathrm{n}} BER_{\mathrm{Tj}}, \tag{13}$$

where n – the number of distribution maxima.

The resulting path BER is a BER amount, calculated by the signal amplitude and phase of the signal:

$$BER = BER_{\mathrm{T}} + BER_{\mathrm{A}} \tag{14}$$

4. Check and use of the methodology.

The technique was tested on the basis of SPICE modeling of the path connecting the transmitter and the receiver operating on the SpaceWire interface at bit frequencies up to 3.125 Gbit / s.

A model of an ideal input trigger of a receiver with a fixed switching level and minimum setup and hold times was created. At its input signal is fed to the data encoded in 8B / 10B and clock signal with a known value of RMS phase and amplitude noise. The length of the bit interval was 320 ps, which corresponds to a bit frequency of 3.125 Gbit / s. After that, the number of erroneously received bits was counted. The obtained data were compared with the results of the prediction of the error rate according to the developed method. A series of computer experiments with a gradual increase in the complexity of the model and the value of jitter. The results are recorded in Table. 2. Thus, A is a model with zero setup and hold times of the receiver's input trigger and with a clock signal drop located in the middle of the bit interval. Model B takes into account the times t_s and t_h, model B - additionally takes into account the different positions of the clock edge within the bit interval.

It was found that if the signal arriving at the receiver satisfies the electrical specifications of the standard in amplitude, then the amplitude noise does not make a significant contribution to the total BER.

Table 2. The results of the verification of the proposed BER evaluation methodology.

m, ps	σ, ps	The calculation formula	Predicted BER*10^(-3)	Modeled BER*10^(-3)
160	55	$BER_T = 1 - \int_{t_{s\_min}}^{T - t_{h\_min}} P_{CLK}^{T}(t)dt$	23,5	24,8
160	55		49,9	55,4
160	65		94	102,6
160	70		160,9	161,5

The sample size was 5000 bits, the setup time was -22 ps and -12 ps, the hold time was -31 ps and -56 ps for writing the unit and zero, respectively.

As can be seen from the table, as part of the verification error by BER estimation method described above is an average of 10%, which is sufficient to assess the order of magnitude of BER at the stage of the transceiver design. Also, with an increase in the sample size, the error decreases.

5. Conclusion.

The described technique is implemented as a program in the MATLAB environment. Before working in the environment files are imported simulation results. This technique can also be used with the VERILOG model, which receives data and clock signals, and generates an analog voltage equal to BER. This method allows us to estimate the BER value with high accuracy, without resorting to long-term SPICE modeling, regardless of the embodiment of the transmitter-receiver path.

References

[1] Buchs K, Zabinski P and Coker J 2008 *VLSI Test Symposium* **26** p 51
[2] Miller M 1994 *Hewlett-Packard Journal* 29
[3] Garmash A, Dubinskij A, Domozhakov D and Kobylyatskiy A 2013 *Questions of radio electronics* **2** p 79
[4] Mitić D 2012 *Serbian Journal of Electrical Engineering* 361
[5] Blankman A 2012 Teledyne *Lecroy Everywhereyoulook* 6

Formation and optical properties of hybrid organic-inorganic $MAPbI_3$ perovskite films

T Zelenyak[1*], V Kinev[1], P Rezepov[1], O Korolik[2], A Mazanik[2], M Tivanov[2], N Doroshkevich[3], A Lavysh[4], V Gevorgyan[6], A Tameev[5], A Vannikov[5], V Turchenko[3,6] and P Gladyshev[1]

[1] State University "Dubna", Dubna, Moscow region, Russia
[2] Belarusian State University, Minsk, Belarus
[3] Joint Institute for Nuclear Research, Dubna, Moscow region, Russia
[4] Yanka Kupala State University of Grodno, Grodno, Belarus
[5] Institute of Physical Chemistry and Electrochemistry. A.N Frumkin, RAS, Moscow, Russia
[6] Russian-Armenian (Slavonic) University, Erevan, Armenia

*Corresponding author's e-mail address: tatyana.zelenyak@bk.ru

Abstract. A rapid progress in the development of solar cells based on hybrid organic-inorganic perovskites $CH_3NH_3PbI_3$ ($MAPbI_3$) is observed in recent years, and power conversion efficiency as high as 22.1 % has been reported. However, a low stability is the main drawback of these materials, which impedes their practical use for solar energy conversion. This work is devoted to the synthesis of $CH_3NH_3PbI_3$ films from CH_3NH_3I and PbI_2 precursors and spectroscopic investigation of their stability under high-intensity laser illumination.

1. Introduction.

The first successful attempt to prepare semiconductor organic-inorganic perovskite $MAPbI_3$ (MA = CH_3NH_3) for solar energy conversion has been reported in 2009 [1]. This class of semiconductors has attracted a lot of attention due to high light absorption, band gap energy close to the optimal value, sufficiently large lifetime of non-equilibrium charge carriers (up to 1 μs), possibility of solution based synthesis [2]. Solar cells based on ABX_3 perovskites have reached power conversion efficiency of 22.1 % for a short time [3].

Low-temperature wet methods (spin-coating, drop casting, dip coating, doctor blading, screen-printing, spray methods) are used most commonly to synthesize perovskite materials [4]. Solution based methods of synthesis of $MAPbI_3$ thin films can be divided into one- and two-stage. In the first case, the films are formed by mixing of methylammonium iodide (MAI) and lead iodide PbI_2 solutions. The two-stage approach includes deposition of PbI_2 layer at the first stage, which is followed by its transformation to $MAPbI_3$ by MAI intercalation. Morphology and quality of the films prepared using the one-stage method are determined mainly by the film shrinkage during the perovskite crystallization and removing of the solvent. The main problem of the two-stage method is expansion of PbI_2 film during intercalation of MAI [5].

A low stability of hybrid organic-inorganic perovskite films is known as the main impediment for their practical use in photovoltaics [4]. This makes extremely important revealing degradation mechanisms and looking for the ways to reduce it. In this work, we analyze transformation of optical properties of $CH_3NH_3PbI_3$ films under influence of optical illumination with different power density.

2. Experimental

2.1 Film preparation

$CH_3NH_3PbI_3$ films were prepared using PbI_2 and CH_3NH_3I precursors according to the previously described procedures [6-11]. All reagents were purchased from Sigma Aldrich.

One-stage method. 0.395 g of CH_3NH_3I and 1.157 g of PbI_2 (99%) were dissolved in 2 ml of N,N – dimethylformamide (DMFA, C_3H_7NO). The obtained solution was stirred with a magnetic stirrer for 12 h at 60 °C and then spin-coated on the cleaned FTO glass substrate (25x25 mm^2) under normal conditions. Rotating speed was 500 RPM for the first 10 s, 2000 RPM for the second 10 s and 5000 RPM for the last 20 s. To avoid crystallization of lead iodide, the working solution was maintained at the temperature from 60 to 80 °C before deposition. The obtained films were annealed for 30 min at 100 °C on a hot plate, which resulted in formation of $CH_3NH_3PbI_3$.

Two-stage method. To obtain PbI_2 solution, 1 g of PbI_2 was dissolved in 2 ml of DMFA (purity better than 99%) and stirred at 120 °C for 1 h. The prepared solution was kept at 80 °C until its deposition. 40 μl of solution was spin-coated on FTO glass substrate (500 RPM for 10 s, 2000 RPM for 10 s, and 4000 RPM for 20 s). The obtained PbI_2 film was annealed for 15 min at 70 °C on a hot plate.

CH_3NH_3I solution was prepared immediately before its use by a careful stirring of CH_3NH_3I in anhydrous isopropyl alcohol (purity better than 99%, 10 mg/ml) for 15 min at room temperature.

The prepared PbI_2 film was dipped into CH_3NH_3I solution for 60 s, washed with anhydrous 2-propanol to remove unreacted CH_3NH_3I and finally annealed for 30 min at 80 °C on a hot plate.

Since prepared $CH_3NH_3PbI_3$ films demonstrate well-known fast degradation during ambient storage (Figure 1a), they were covered by a thin (0.15 mm) glass coverslip with a thorough sealing of the end faces with a silicone hermetic immediately after synthesis.

2.2 Characterization

X-ray diffraction analysis was performed with an Empyrean diffractometer (PANalitical, Cu Kα emission). UV-Vis-NIR transmission spectra were measured in the 400 – 1100 nm range with an UNICO 2804 (UNICO, USA) single-beam spectrometer.

Raman and photoluminescence (PL) spectra were measured at room temperature using a Nanofinder HE confocal spectrometer (LOTIS TII, Belarus-Japan) with 532 nm DPSS laser as an excitation source. Back-scattered light without analysis of its polarization was dispersed with a spectral resolution of 2.5 cm^{-1} (0.1 nm) and detected with a cooled CCD-matrix. 50x Olympus lens (numerical aperture of 0.8) was used to focus laser beam on the surface of $CH_3NH_3PbI_3$ film. Spectral calibration was done using a built-in gas-discharge lamp providing accuracy better than 2.5 cm^{-1} (0.1 nm).

PL decay kinetics were measured in the time-correlated single photon counting mode. PDL 800-B laser with LDH-P-C-470 laser head (PicoQuant, Germany) was used providing excitation at wavelength of 467 nm with pulse duration and frequency of 73 ps and 10 MHz, respectively. Low power density in these experiments (10 W/m^2) ensured their non-destructive character. PL decay detection system included PMA-182 photodetector with TimeHarp200 device (PicoQuant, Germany) for time-correlated single photon counting measurements. The obtained kinetics were analyzed taking into account that the measured PL intensity $I(t)$ is the convolution of the instrument response function (IRF, $g(t)$) and "true" PL decay function $F(t)$:

$$I(t) = g(t) \otimes F(t) = \int_0^t g(t')F(t-t')dt' \qquad (1)$$

3. Results and discussions

Our experiments have demonstrated that two-stage method is preferable to prepare $CH_3NH_3PbI_3$ films, because it provides better control of their properties and better uniformity. Therefore, the films prepared using two-stage method were chosen for investigation of optical properties.

XRD analysis at room temperature demonstrates the tetragonal crystal structure of $CH_3NH_3PbI_3$ films (Figure 1). Differential scanning calorimetry reveals a transition from tetragonal to cubic phase at 53–57 °C.

Typical absorption spectra of PbI_2 (reference) and $CH_3NH_3PbI_3$ films are presented in Figure 1b. PbI_2 films demonstrate a pronounced absorption edge at 515 nm, which corresponds to the band gap energy of 2.4 eV [13]. Transformation of PbI_2 to $CH_3NH_3PbI_3$ results in a shift of absorption edge to appr. 780 nm in agreement with [14,15]. Absorption spectra of $CH_3NH_3PbI_3$ films point to a significant light scattering related to formation of large $CH_3NH_3PbI_3$ crystals [13]. Spectra plotted in the absorbance – energy coordinates (inset in Figure. 1b) allowed determination of band gap energy of $CH_3NH_3PbI_3$ (≈1.50 eV), which is in a good agreement with literature data [16,17].

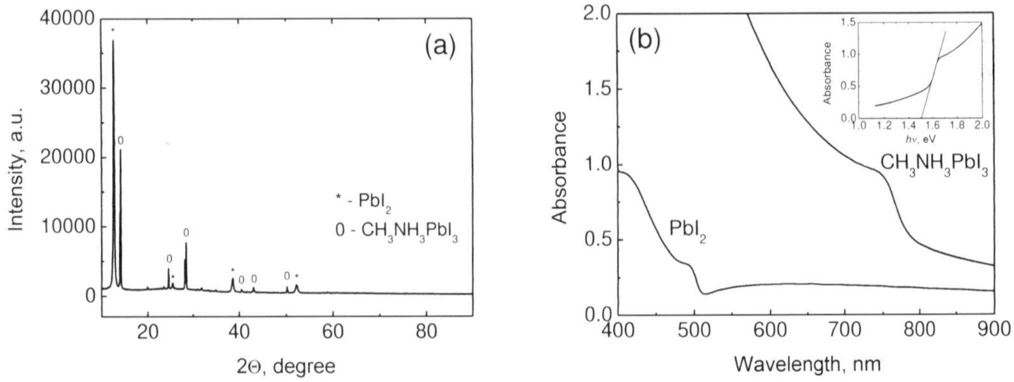

Figure 1. XRD pattern of unsealed $CH_3NH_3PbI_3$ film (a); UV-Vis-NIR absorption spectra of $CH_3NH_3PbI_3$ and PbI_2 films (b).

The prepared $CH_3NH_3PbI_3$ films reveal an intensive PL peak with maximum at appr. 0.77 μm (Figure 2a), which corresponds to interband radiative charge recombination [16,18].

Measurements of PL decay kinetics enabled one to determine the lifetime of non-equilibrium charge carriers equal to 25 ns (Figure 2b).

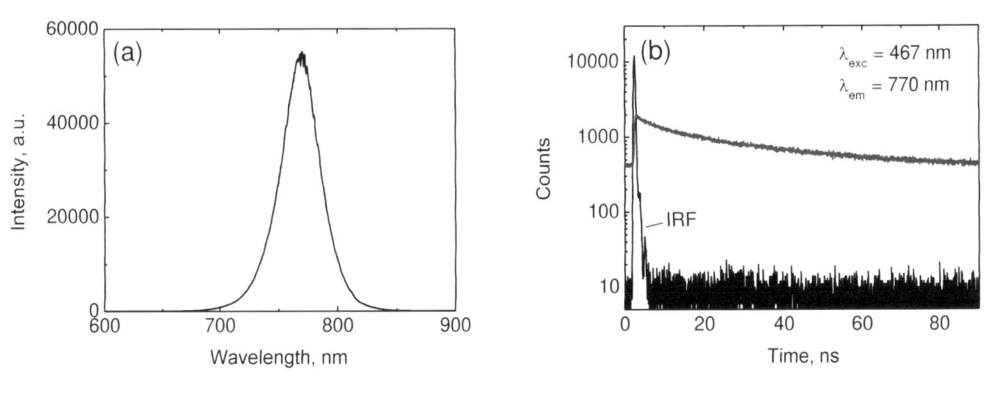

Figure 2. Room temperature PL spectrum (a) and decay kinetics (b) for $CH_3NH_3PbI_3$ film.

IOP Conf. Series: Materials Science and Engineering **498** (2019) 012012 doi:10.1088/1757-899X/498/1/012012

It is a matter of common knowledge that the intensity of interband radiative recombination decreases with increasing imperfection degree of material due to the increased probability of competing non-radiative recombination channel. Spectral position of PL maximum is determined by the electron density of states (DOS). In particular, the presence of pronounced tails of DOS (a large Urbach energy) leads to a red shift of PL maximum, which was observed for perovskites by many authors [19,20]. Hence, analysis of PL intensity and peak position enables one to track variation of perovskite film quality under influence of different impacts. Available literature data regarding Raman spectroscopy of $CH_3NH_3PbI_3$ are rather contradictory, and there are some discrepancies between the theoretical and experimental data [18]. Nevertheless, Raman spectroscopy makes it possible to detect PbI_2 as the degradation product of $CH_3NH_3PbI_3$ layers.

In our experiments, optical exposure of the films was carried out in the following sequence. It is known that illumination of the $CH_3NH_3PbI_3$ films can give rise to increase in intensity of their photoluminescence, which is usually related to change in the charge state of defects acting as centers of non-radiative recombination [21]. Therefore, the films were subjected to light soaking using a low-intensity illumination prior to study of influence of high-power light exposure on their properties. For this purpose, fifty PL spectra were measured sequentially with the focusing of the laser beam at the same point on the film surface. The signal acquisition time in each measurement was 5 s at the optical power of 0.6 μW. The experiments have shown that such exposure leads to a gradual two-fold increase in PL intensity with subsequent attainment of saturation. Then the film was subjected to a high-intensity optical exposure for 30 s, and PL spectrum was measured again. The power of each subsequent optical exposure was increased, whereas the mode of PL spectrum measurement after each exposure was the same (signal acquisition time of 5 s and optical power of 0.6 μW). Maximum optical power used was 600 μW.

Figure 3a presents PL maximum position and intensity for the $CH_3NH_3PbI_3$ film depending on the power of preceding optical exposure. As is seen, dependence of peak position is characterized by a pronounced minimum. It means that exposure of the films with a relative low power does not deteriorate, but even improves the quality of the material. At the same time, exposure with a greater power makes material more defective, which correlates with the decrease in the PL signal intensity and appearance of the bands corresponding to PbI_2 in the Raman spectra (Figure 3b).

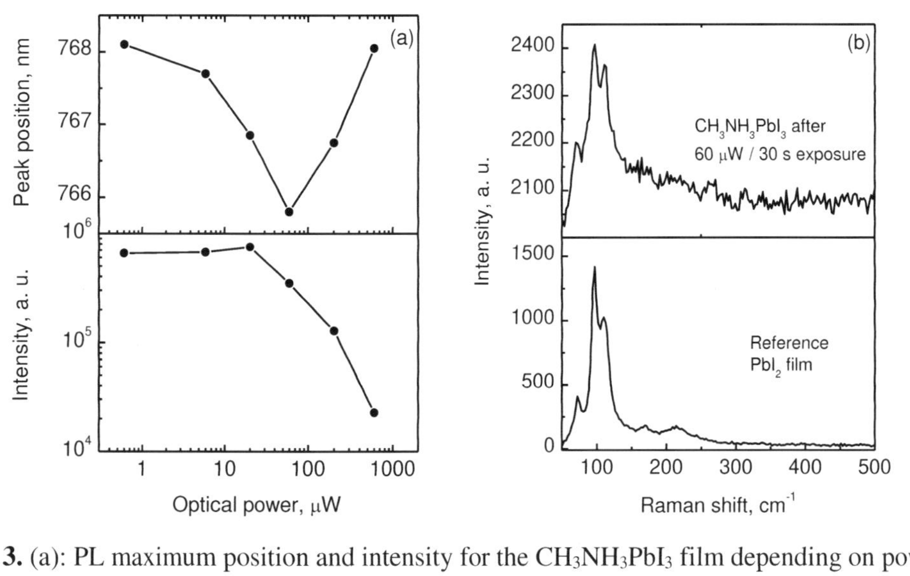

Figure 3. (a): PL maximum position and intensity for the $CH_3NH_3PbI_3$ film depending on power of preceding optical exposure; lines are given for convenience; (b): Raman spectra of $CH_3NH_3PbI_3$ film subjected to 60 μW / 30 s optical exposure and reference PbI_2 film.

There are different hypotheses regarding mechanisms of light soaking induced improvement of properties of $CH_3NH_3PbI_3$ films. In particular, it was reported that illumination can result in a reversible structural transformation of $CH_3NH_3PbI_3$ lattice [22]. However, taking into account high densities of optical power in our experiments, one can suppose that the observed effect is related rather to optical annealing of the film.

Knowing the diameter of the light spot on the sample surface during measurement of PL spectra (≈0.8 μm), one can estimate the ranges of optical power density corresponding to the observed changes in film properties. So, the quality of the $CH_3NH_3PbI_3$ film deteriorates only when power density exceeds 10^8 W/m^2, whereas the optical impact on the films with power density less than $4 \cdot 10^7$ W/m^2 improves their quality.

4. Conclusion

$CH_3NH_3PbI_3$ films were characterized with different spectroscopic techniques. It was revealed that exposure of the $CH_3NH_3PbI_3$ films with a monochromatic light ($\lambda = 532$ nm) with intensity less than $4 \cdot 10^7$ W/m^2 results in a blue shift of interband photoluminescence, which is related to decrease in the Urbach energy and points to improvement of film perfection. At the same time, film exposure at a power density more than 10^8 W/m^2 leads to decrease in intensity of PL band, its red shift and appearance in the Raman spectra of bands corresponding to PbI_2 phase. Thus, the ranges of light intensity corresponding to both degradation and improvement of the $CH_3NH_3PbI_3$ films quality are determined.

Acknowledgements

This work was supported by the Russian Science Foundation, grant no. 15-13-00170. O.K., A.M., and M.T. acknowledge financial support by the Belarusian Republican Foundation for Fundamental Research (project F16MS-015).

References

[1] Kojima A, Teshima K, Shirai Y and Miyasaka T 2009 *Am. Chem. Soc.* **131** pp 6050–6051
[2] Liu D Y and Kelly L 2014 *Nat. Photonics.* **8** pp 133–138
[3] Bae S, Park JS, Han IK, Shin TJ and Jo WH 2017 *Solar Energy Materials & Solar Cells.* **160** pp 77–84
[4] Gladyshev P P, Yushankhai V Y and Syurakshina L A 2015 *Ivan. Gos. University Press.* 676 pp 426-556
[5] Zhao Y and Zhu K 2016 *Chem. Soc. Rev.* **45** pp 655-689
[6] Heo J, Im S, Noh J, Mandal T, Lim C, Chang J, Lee Y, Kim H, Sarkar A, Nazeeruddin M, Grätzel M and Seok S 2013 *Nat Photonics.* **7** pp 487–492
[7] Xiao M, Huang F, Huang W, Dkhissi Y, Zhu Y, Etheridge J, Gray-Weale A, Bach U, Cheng Y-B and Spiccia L 2014 *Angew. Chem., Int. Ed.* **53** pp 9898-9903
[8] Della Gaspera E, Peng Y, Hou Q, Spiccia L, Bach U, Jasieniak J and Cheng Y 2015 *Energy.* **13.** pp 249-257
[9] Liang K, Mitzi D. B and Prikas M T *Chem. Mater.* **10** pp 403-411.
[10] Burschka J, Pellet N, Moon S J, Humphry-Baker R, Gao P, Nazeeruddin M K and Gratzel M 2013 *Nature* **499** pp 316-319
[11] Im J-H, Kim H-S, Park N-G 2014 *APL Materials.* **2** 081510
[12] O`Connor D V and Phillips D 1984 *Time-correlated Single Photon Counting N.Y.: Acad. Press* 283 p.
[13] Chen Q, Zhou H, Hong Z and Luo S 2014 *J. Am. Chem. Soc.* **136** pp 622-625
[14] Jeon Y-J, Lee S, Kang R, Kim J-E, Yeo J-S, Lee S-H, Kim S-S, Yun J-M and Kim D-Y 2014 *Subject Areas.* **4** 6953-1–6953-8
[15] Duyen H C et al 2014 *APL Materials.* **2** 091101
[16] Shi D, Adinolfi V, Comin R, Yuan M and Alarousu E 2015 *Science.* **347** pp 519-522

IOP Conf. Series: Materials Science and Engineering **498** (2019) 012012 doi:10.1088/1757-899X/498/1/012012

[17] Abdelhady A L, Saidaminov M I, Murali B and Adinolfi V 2016 *J. Phys. Chem. Lett.* **7** pp 295–301
[18] Park B-W, Jain S M, Zhang X, Hagfeldt A, Boschloo G and Edvinsson T 2015 *ACS Nano, ACS Nano*. **9** pp 2088–2101
[19] Yang D, Yang R, Ren X, Zhu X, Yang Z, Li C and Liu S 2016 *Adv. Mater.* **28** pp 5206–5213
[20] Shao Y, Xiao Z, Bi C, Yuan Y and Huang J 2014 *Nature Communications* **5** 5784
[21] Zhao C, Chen B, Qiao X, Luan L, Lu K and Hu B 2015 *Adv. Energy Mater.* **5** 1500279
[22] Gottesman R, Gouda L, Kalanoor B S, Haltzi E, Tirosh S, Rosh-Hodesh E, Tischler Y and Zaban A 2015 *J. Phys. Chem. Lett.* **6** pp 2332–2338.

IOP Conf. Series: Materials Science and Engineering **498** (2019) 012013 doi:10.1088/1757-899X/498/1/012013

Optimized immonochromatographic system for antigen determination based on monoclonal antibody conjugates with quantum dots

S Ibragimova[1*], S Dezhurov[2], V Vrublevskaya[3], O Morenkov[3], V Moroz[1], R Gorshkova[1],D Krilsky[2], A Vasiliev[1,4], P Gladyshev[1]

[1] State University "Dubna", Dubna, Moscow region, Russia
[2] Researcher Research Institute of Applied Acoustics, Dubna, Moscow region, Russia
[3] Institute of Cell Biophysics, Pushchino, Russia
[4] National Research Centre "Kurchatov Institute", Moscow, Russia

*Corresponding author's e-mail address: ibragimova21@yandex.ru

Abstract. Optimized immonochromatographic system for the determination of antigens of viruses based on conjugates of monoclonal antibodies with multilayer colloidal quantum dots (QDs) as fluorescent labels was developed. The main attention was paid to the choice of the structure and synthesis of stable QDs with optimal spectral properties for analysis, to the selection of the conditions of synthesis of antibody conjugates with QDs, to the production of test strips for immunochromatographic analysis, and to the development of fluorescent reader design. It is shown that the optimal region of QDs fluorescence lies in near infrared range (700 – 800 nm). The stable QDs CdTeSe/CdS/CdZnS/ZnS and their conjugates with monoclonal antibodies for detecting Aujeszky's disease virus were synthesized, which fluorescent in this spectral region. Using these QDs, limit of detection of antigen equal to 30 pM was obtained.

1. Introduction.

Immunochromatographic analysis (IChA) is one of the most productive methods of immunodiagnostics, commonly known as lateral flow immunochromatographic assays [1]. Theory and practice of immunological analysis methods are described in detail in monograph [2] and reviews [3, 4]. The basis of the immune diagnosis is the interaction of pathological biological agents (PBA), such as antigens (Ag), or proteins - markers of disease, with an antibody (Ab) to this PBA. Immunochromatographic methods allow selectively detect specific proteins in a few minutes. In all variants of IChA the basic elements of the method are the test strips, regardless of the using approaches. Despite the apparent simplicity of the design, the test strips are the result of modern high technology, including the production of monoclonal and polyclonal antibodies to specific PBA and the use of unique porous membranes.

Researchers undertake permanent efforts to improve the sensitivity and reliability of IChA. In this work, a direct (sandwich) method of IChA using conjugate Ab-QDs deposited on a conjugate pad was investigated. Antibodies specific to the target Ag were immobilized on the test line, and anti-species antibodies specific to primary antibodies were immobilized on the control line. After the contact of sample containing the PBA with the membrane, the PBA binds to conjugate Ab-QDs. Then the immune complex moves into test zone, where it binds to specific antibodies, forming a "sandwich" Ab-Ag-Ab-QDs. The excess of unbound conjugate binds then to anti-species antibodies of the control line. Thus, the registration of two lines on the test strip means positive result of the test.

The list of antibodies usable for the production of conjugates with markers of the analytical process and the list of determined proteins extends continuously. Therefore, the development of a new analytic platforms and technologies of IChA enabling the diagnostics of different PBA is important. The authors proposed a new analytical platform IDDRI [5], important feature of of this platform is the use of QDs as markers of monoclonal Ab. According to [6], the application of QDs as markers in IChA was described for the first time in 2010. The comparison of the sensitivities of immunochromatographic tests using identical immunological reagents showed that the limit of detection visible by naked eye was 10 times lower in the case of the application of QDsCdTe (at excitation of a UV-lamp) than in the case of gold nanoparticles [**Ошибка! Закладка не определена.**7]. Later in [8], the detection limits of quantum dots and colloidal gold on the membrane for immunoassay analysis were compared. It was shown that the use of quantum dots allows reducing by a factor of 24 the limit of detection of chloramphenicol compared to test systems with colloidal gold. In [9] antibiotics detection limits ranged from 8 to 200 pg / mL

It should be noted that in IChA with QDs as fluorescent markers, the fluorescence of albumin blocked nitrocellulose membranes reducing the ratio signal/background is not always fully taken into account. To improve the analytical system it was necessary to optimize spectral characteristics of the QDs (that is their architecture and methods of synthesis) and design of a fluorescent reader, including the excitation source and the detector of the fluorescent signal. This activity was the purpose of the recent work.

2. Experimental

2.1. Materials

QDs CdTeSe/CdS/CdZnS/ZnS were synthesized according to the methods described in [10]. Monoclonal antibodies ADV 34/2 to gB glycoprotein Aujeszky's disease virus were obtained according to [11]. In work were used N-hydroxysuccinimide NHS ($\geq$ 97%, Sigma-Aldrich), N,N-dimethylformamide DMFA (99.8%, Sigma-Aldrich), N-ethyl-N′-(3-dimethylaminopropyl) carbodiimide hydrochloride EDC·HCl (Sigma Aldrich), triethylamine TEA ($\geq$99.5%, Sigma-Aldrich, NaCl (AR, Chimmed), KCl (AR, Chimmed), Na_2HPO_4 (AR, Labtex), KH_2PO_4 (AR, Labtex), diethanolamine DEA (99,9%, Sigma-Aldrich), ultracentrifuge filters Amicon, tween -20 ($\leq$3.0% water, Sigma-Aldrich).

2.2. Preparation of conjugates of quantum dots with monoclonal antibodies (MAbs)

Phosphate-saline buffer was used as the working buffer to produce conjugates: pH= 7,8 (137 mM NaCl, 2,7 mM KCl, 10 mM Na_2HPO_4 и 1,76 mM KH_2PO_4). The QDs were activated by EDC and NHS. Necessary amounts of EDC and NHS compounds are prepared in the form of 1.64% and 1.21% solutions in DMF, respectively. After 15-20 minutes of QD activation, the required amount of the antibody solution is added at permanent stirring to the QDs. The mixture is incubated for 1.5 - 2 hours at room temperature in the dark. After conjugation the solution was stabilized by the addition 1/10 of the total volume of conjugate of 10% DEA (diethanolamine) and left another 20 minutes in the dark. After stabilization, the conjugate was concentrated by ultracentrifugation in Amicon tubes at g = 16.1. The final step is the purification of conjugates from unbound antibodies using gel filtration on column with a superose 6.

2.3. Preparation of immunochromatographyc strip
Immunochromatographic test strips were manufactured by the modified technology presented in the reference [15]. Monoclonal antibodies ADV 34/2 to the gB glycoprotein of the virus of Aujeszky's disease are immobilized on the test zone, and we used goat antibody instead of the rabbit antibody to mouse immunoglobulin as anti-species control antibodies.

2.4. Test of the QDs conjugates with monoclonal antibodies
The lysate PRV-k is used as antigen (hereinafter lysate). Phosphate-buffered saline with pH=7.4 is used as eluent (137 mM NaCl, 2,7 mM KCl, 10 mM Na_2HPO_4 and 1,76 mM KH_2PO_4), containing 2% tween-80. Lysate is diluted in buffer in a ratio sufficient for dye the test line. The desired amount of conjugate solution is applied to the test strip. Then 50 µl of eluent with antigen is applied. When the eluent passes all analytical zones, we added 50 µl of eluent, and after 3-5 min we added additionally 50µl of buffer. After 20 minutes on completion of the elution the test result is interpreted.

2.5. The study of fluorescence spectra of the test and control zones
We used in the study a specially made fluorescent reader based on USB Digital Microscope EL-Micro-2 (analogue BW-400X of Chinese firm Brigh well Technology LTD) with a modified optics, fluorescence was excited by 450 nm LED with the light filter KS-1, providing fluorescence registration above 700 nm.

The RGB color-photomatrix is used in the microscope. Digitizing module processes the RGB-signals and gives the brightness in each channel in a range from 0 to 255 units. These values are divided by 255 and dimensionless quantities of signals are received in conventional units (CU) in the range of 0 - 1.

2.6. Investigation of conjugate distribution along the test strip
Scan along the test strip was carried out at the fluorescence excitation at 470 nm using the device, based on USB-650 Red Tide firm Ocean Optics (350-1000 nm) spectrometer.

3. Results and discussion
Bioconjugation QDs and Ab can be made by various methods, for example by electrostatic and hydrophobic interactions or by the formation of covalent bond. It is important to obtain strong fixation of antibodies to the QDs, to prevent the dissociation of the antibody - QDs conjugate in the immunochromatographic process. In this connection, covalent immobilization of antibodies is used in the work for the preparation of conjugates of quantum dots with an antibody [12]. For this purpose, initial QDs are modified to form on surface carboxylic groups [13]. Immobilization protein to carboxyl groups is carried out under mild conditions by carbodiimide method to form a covalent bond [14] according to the method described above.

In addition to the selection of efficient monoclonal Ab to target Ag and preparation of stable conjugates with QDs for sensitive immunochromatographic system, it is extremely important to provide optimal combination of spectral properties of QDs, considering the parasitic fluorescence of albumin blocked nitrocellulose membrane, and spectral characteristics of fluorescence reader (excitation wavelength and QD fluorescence registration). In this regard, the background luminescence of albumin blocked nitrocellulose membranes was measured; it was shown that it almost doesn't fluoresce in the near infrared region above 700 nm. This leads us to a conclusion that it is expedient to use for IChA QDs with fluorescence in this spectral range. For this purpose QDs with required fluorescence wavelength were synthesized and their hydrophilic derivatives containing surface carboxyl groups necessary for conjugation of protein were obtained. Conjugates of monoclonal Ab with appropriate spectral properties for these modified QDs were prepared. Consequently chromatography strips for the determination of Aujeszky's disease virus (the virus glycoprotein gB) were developed. To further optimize the immunochromatographic system including reader and the strip, distribution QDs-markers along with immunochromatographic strip was investigated after completion of the process (Figure.1). The test strip scan (Figure.1) shows that the conjugates are not

IOP Conf. Series: Materials Science and Engineering **498** (2019) 012013 doi:10.1088/1757-899X/498/1/012013

adsorbed on the analytical membrane of test strip in the area of window of immunochromatographic cassette except test and control zones. Thus excessing conjugate is localized predominantly on the absorbent pad, as well as in the area, where the analytical membrane is glued to the sample and the absorbent pads.

Investigation of the fluorescence spectra of the test and control zones, and background fluorescence outside these zones at various fluorescence excitation wavelengths makes possible the determination of optimal optical characteristics of the fluorescent reader, which corresponds to excitation wavelength of 470 nm and registration above 700 nm (Figure.2).

Taking into account the results of the present study, model fluorescent reader was designed and manufactured. It includes corresponding LEDs, light filter and photomatrixes (Figure.3). Signals from photomatrixes are processed using special software on a personal computer. Obtained results are shown in Figure. 4, and the resulting calibration is presented in Figure.5. There is a linear relationship (Figure. 5) between the analytical fluorescent signal and concentration of antigen up to a concentration of 160 nM. This enables quantitative determination of antigens in the samples

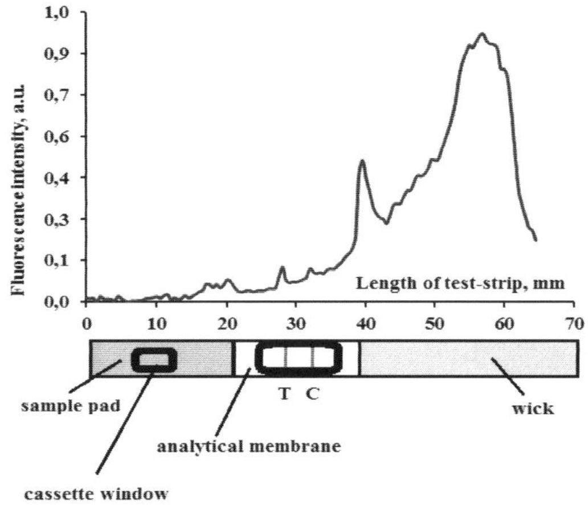

Figure 1. Investigation of conjugate distribution along the test-strip after completion of IChA (scanning at 700 nm).

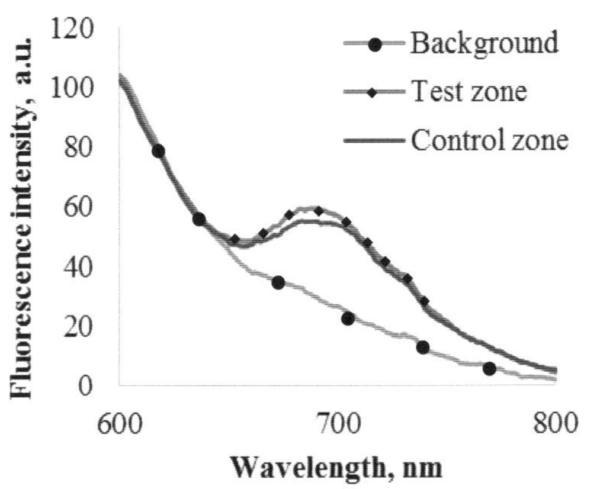

Figure 2. The fluorescence spectra of the test and control lines, and background (fluorescence excitation at 470 nm).

Figure 3.Model of fluorescence reader.

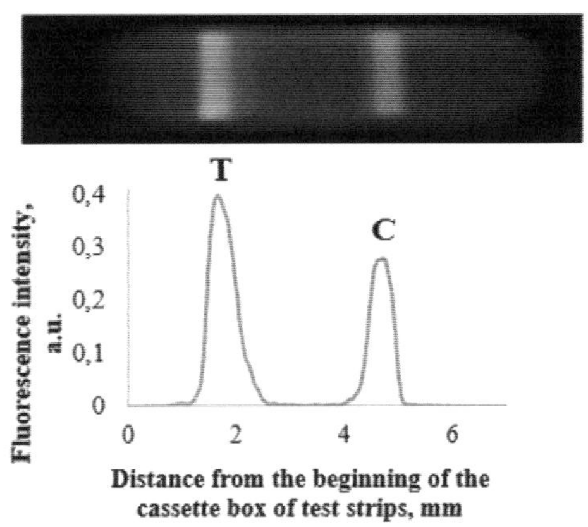

Figure 4. Fluorescent scan of test-strips: T – the test and C – the control zone.

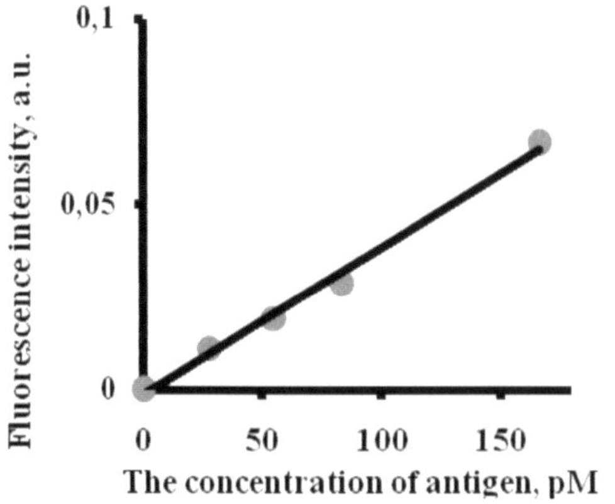

Figure 5. The fluorescence intensity of the test zone as a function of antigen concentration.

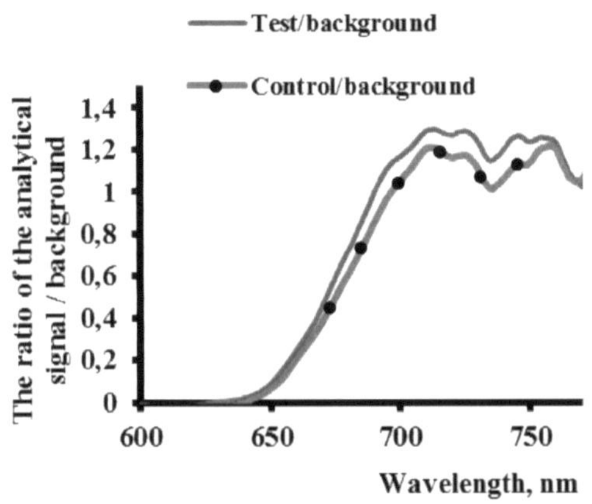

Figure 6. Analytical signal / background ratio as a function of fluorescence detection range.

In addition, the study of analytical signal / background ratio for the test and control zones was conducted (Figure.6). Figure. 6 shows that the optimal range of fluorescence registration for the QDs is the spectral region from 700 to 750 nm.

4. Conclusions

Optimized immunoassay system using QDs with fluorescence in near infrared region was developed. The optimal spectral characteristics of the fluorescence reader were chosen using the strips after IChA. The maximum ratio of the analytical signal / background was observed at wavelengths of 700-750 nm. Calibration curve for the quantitative determination of antigens was drown. The minimum concentration of antigen detection in the sample at this stage work is 30 pM.

Acknowledgments

This work was supported by the Russian Science Foundation, grant no. 15-19-00229.

References

[1] Wong R C and Tse H Y 2009 *Lateral Flow Immunoassay* DOI10.1007/978-1-59745-240-3_11 (NewYork: HumanaPress) p 223

[2] Egorov A M, Osipov A P, Dzantiev B B and Gavrilov E M 1991 *Theory and practice of immunoassay* (Moscow: High School) p 288

[3] Piven N V 2007 *Immunopathology, Allergology, Infectology* **2** 6-22

[4] Koivunen M E and Krogsrud R L 2006 *Labmedicine* **37** 490-497

[5] Gladyshev P P, Vasiliev A A, Morenkov O S, Vrublevskaya V V, Tumanov Yu V, Boldyrev A N, Dezhurov S V, Krilsky D V and Ibragimova S A 2016 *Papers of the LI International scientific and practical conference No. 1* **45** pp 22-49

[6] Goftman V V, Speranskaya E S and Goryacheva I Yu http://www.chem.msu.su/rus/books/2013/inorgworkshop/075.pdf

[7] Bai Y, Tian C, Wei X, Wang Y, Wanga D and Shi X 2012 *RSC Adv* **2** 1778–1781

[8] Taranova N A, Berlina A N, Zherdev A V and Dzantiev B B 2011 *Proceedings of the 2nd International School "Nanomaterials and nanotechnology in living systems. Safety and nanomedicine "* 102

[9] Dzantiev B B 2014 Russian Academy of Sciences Department of Chemistry and Materials Science, report of SCAC RAS 2013

[10] Dezhurov S V, Trifonov A Yu, Lovyginc M V, Rybakova A V and Krylsky D V 2016 *Nanotechnologies in Russia* **11** 337–343

[11] Zaripov M M, Morenkov O S, Siklodi B, Barna-Vetro I, Gyiingyiisi-Horvath A and Fodor I

1998 *Res. Virol.* **149** 29-41

[12] Gao X, Yang L and Petros J 2005 **16** 63–72

[13] Loginova Y F, Dezhurov S V, Zherdeva V V, Kazachkina N I, Wakstein M S and Savitsky A P 2012 *Biochemical and Biophysical Research Communications* **419** 54–59

[14] Wang J, Huang X, Zan F, Guo C-G, Cao C and Ren J. 2012 *Electrophoresis* **33** 1987-1995

[15] Vrublevskaya V V, Afanasyev V N, Grinevich A A, Skarga Yu Y, Gladyshev P P, Ibragimova S A, Krylsky D V and Morenkov O S 2017 *Journal of Virological Methods* **240** 54–62

IOP Conf. Series: Materials Science and Engineering **498** (2019) 012014 doi:10.1088/1757-899X/498/1/012014

Thermal stability of tantalum nitride based thin film resistors

S A Shostachenko*, R V Zakharchenko, R V Ryzhuk and S V Leshchev

National Research Nuclear University (MEPhI), Kashirskoe shosse 31, 115409 Moscow, Russian Federation

*Corresponding author's e-mail address: SAShostachenko@mephi.ru

Abstract. Tantalum nitride thin films were deposited on Al2O3 substrates by the dc-magnetron sputtering technique. The nitrogen content in the argon/nitrogen flow varied from 5 to 50%. Structural properties were studied using X-ray diffraction. The ratio of Ar:N2 was 4:1; the ratio of Ta:N became 1:1. Sheet resistance depends on thickness and is in the range of 20 – 80 Ω/□ due to thickness 100 – 50 nm. The TaN films deposited at a nitrogen/argon ratio of 20% show the thermal stability of the resistance in the 25–400°C temperature range. Sheet resistance degradation was ~ 5%. The TCR value was determined in the range of 25 – 300 C and was equal to -21 ppm/K.

1. Introduction

To create compatible components of microwave electronics materials with unique properties, CMOS technology is required. Tantalum nitride may serve as an example of such a material [1]. It is chemically inert, resists corrosion and has a near-zero temperature coefficient of resistivity [2].

The difference in stoichiometry is quite a common feature of tantalum nitride films because of their defective structure [3]. Thus, microstructure, electrical properties, etc. will vary depending on the conditions of film formation. Typically, tantalum nitride films are made by CVD [4,5] and PVD [6-12] methods.

In this paper, we studied the thermal stability of tantalum nitride thin films. Electrical and structural properties were observed using sheet resistivity and X-Ray diffraction. Thermal stability was tested in the temperature range 300 – 600 by measurement of sheet resistance. In addition, we measured IVT and defined TCR.

2. Experiment

Tantalum nitride films were deposited by reactive magnetron sputtering of the installation Kurt J. Lescker PVD 250. The ratio of nitrogen / argon was varied in a range from 10 to 30%. Al2O3 was used as a substrate. The operating pressure in the installation was 8.4 e^{-4} Torr. Sputtering was conducted for power stabilization mode, power is maintained equal to the value of 200 watts. The surface resistance was measured by 4-probe method on Jandel RM-3000. The phase composition of the films was determined by the X-ray diffraction method on Rikagu Ultima IV. Temperature stability was tested in the temperature range 200 - 600 C in an atmosphere of air on Modular RTP-600S.

The temperature coefficient of resistance was determined by using IVT measurements on the installation Agilent B1500 [13]. For this test, the films with TLM pattern were formed; Ti films (250 nm) were used as ohmic contact. Resistance measurements were carried out in the temperature range of 25 - 300 C. The resulting dependence was approximated by the least squares method.

3. Results and Discussion

3.1 Structural properties

The films were deposited by reactive magnetron sputtering. Al_2O_3 was used as a substrate. To a nitrogen concentration of 5% and 10% of the total flow of the carrier gas, irreproducible parameters of the film were achieved. This can be explained by the poor stoichiometry and phase instability of the film composition [11]. The ratio of nitrogen atoms to atoms of tantalum in the film was 1:4 and 1:2, which are 5% and 10% of the total nitrogen carrier gas flow, respectively. For 20% of the content of total nitrogen carrier gas flow, the ratio of nitrogen and tantalum atoms of approximately 1:1 was obtained. The X-ray diffraction measurements are shown in Figure 1.

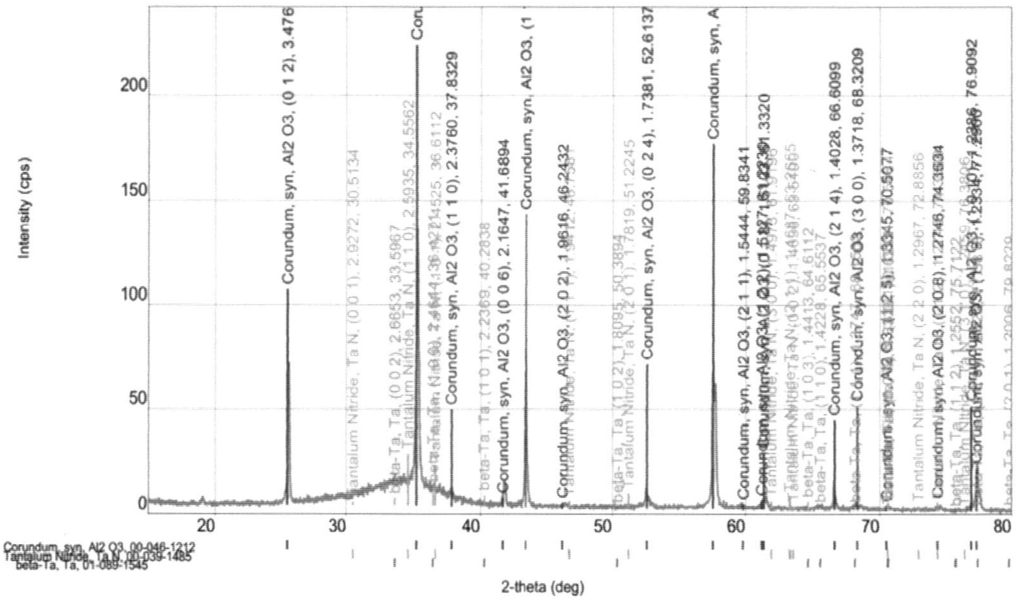

Figure 1. X-ray diffraction spectrum (2θ scan mode)

3.2 Electrical properties

For TaN films obtained from the ratio of Ta: N = 1:1, the dependence of the surface resistance of the thickness (Figure 2) was obtained.

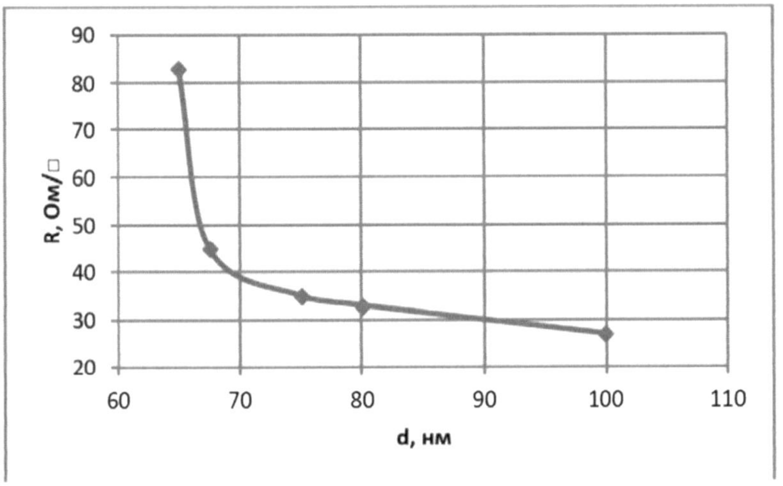

Figure 2. The experimental dependence of the surface resistance of the thickness

For TCR measurements of solid tantalum nitride film, the TLM test was formed by photolithography followed by chemical etching. The ohmic contacts to the resulting structures were formed with the Ti (200) electron beam evaporation. On the obtained test, specimens of current-voltage characteristics were measured at temperatures of 25-300 C (in steps of 50 C). The obtained data was fitted by the least squares method and was calculated with TCR. Its value was -21 ppm / K, which is the best result for a given thickness and application method according to [2].

3.3 Thermal stability

To determine the thermal stability in the annealing furnace, rapid thermal annealing was carried out in air at temperatures of 200-600 C with increments of 100 C for 2 minutes. With increasing the annealing temperature, resistance is also increased. At 400 C, annealing temperature resistance did not increase exceeding 5%, whereas increase to 500 C changed resistance by 10%, and after 600 ° C the resistance value dramatically increased by 64%. Thus, it is possible to conclude that the resistors are stable when used in normal conditions.

4. Conclusion

Thin films of tantalum nitride were deposited by reactive magnetron sputtering with different ratios of nitrogen and argon flow. By X-ray diffraction, the composition of Ta films was determined: N = 1:1. Temperature coefficient of resistance was determined, which is equal to -21 ppm / K. The resulting value corresponds to the best results for a given method of obtaining films [2]. In addition, the thermal stability of the films was tested during an annealing temperature of 500 C with an increase in resistance not higher than 10%.

Acknowledgements

This study was carried out on equipment of the MEPhI Shared-Use Equipment Center «Heterostructural microwave electronics and wide band gap semiconductor physics».

References

[1] Riekkinen T, Molarius J, Laurila T, Nurmela A, Suni I and Kivilahti J K 2002 *Microelectronic Engineering* **64** pp 289–297

[2] Cuong N D, Kim D J, Kang B D and Yoon S G *Materials Science and Engineering B* 135

pp 162–165
[3] Lin J C and Lee C 2000 *J. Electrochem. Soc.* **147** 2 713–718
[4] Hieber K 1974 *Thin Solid Films* **24** pp 157–164
[5] Tsai M H, Sun S C, Tsai C E, Chuang S H and Chiu H T 1996 *J. Appl. Phys.* **79** 6932–6938
[6] Schauer A, Roschy M 1972 Thin Solid Films **12** pp 313–317.
[7] Sun X, Kolawa E, Chen J, Reid J and Nicolet M A 1993 *Thin Solid Films* **236** pp 347–351.
[8] Oku T, Kawakami E, Uekubo M, Murakami M, Takahiro K and Yamaguchi S 1996 *Appl. Surface Sci.* **99** 4 pp 265–272
[9] Stavrev M, Fischer D, Wenzel C, Dreschen K and Mattern N 1997 *Thin Solid Films* **307** 1–2 pp 79–88
[10] Nakao S, Numata M and Ohmi T 1999 *Jpn. J. Appl. Phys.* **38** 2401–2405
[11] Nie H B,Xu S Y, You L P, Yang Z, Wang S J and Ong C K *Appl. Phys., Mat. Sci. Proc A* **73** pp 229–236
[12] Lee W H, Lin J C and Lee C 2001 *Mater. Chem. Phys.* **68** (1–3) pp 266–271
[13] Shostachenko S, Vanukhin K and Ryzhuk R 2015 *Physics Procedia* **72** pp 419-424

IOP Conf. Series: Materials Science and Engineering **498** (2019) 012015 doi:10.1088/1757-899X/498/1/012015

Evaluation of commercial PIN diodes as gamma radiation dosimeters

M Y Nikiforova and B I Podlepetsky

National Research Nuclear University MEPhI (Moscow Engineering Physics Institute), Kashirskoe shosse 31, 115409 Moscow, Russian Federation

Abstract. The influence of gamma radiation on the current-voltage characteristics of silicon PIN photodiodes S1223 from Hammamatsu was investigated for their use as gamma radiation dose rate sensitive elements. The transfer characteristic and the dependence of the current sensitivity on the absorbed dose rate in the range from 6.7 Gy(Si)/h to 124 Gy(Si)/h were studied. The range of reverse voltages corresponding to the highest current sensitivity of PIN diodes is determined. An analytical expression of the transfer characteristic for the operating mode with maximum sensitivity to gamma radiation is obtained.

1. Introduction

The gamma radiation dosimeters are used for monitoring low and high dose rate in scientific research, industry and medicine. The sensitive elements current is changed in proportion to the total ionized dose under the influence of radiation [1]. The value of the total ionized dose is determined by multiplying the dose rate by the exposure time.

Among the large number of small-sized and low-cost radiation sensors, the PIN diodes have the lowest power consumption, high sensitivity, high reproducibility and stability of characteristics, high quantum efficiency and ability to operate without external voltage [2].

The purpose of this study was to evaluate the possibility of using the commercial PIN photodiodes S1223 as the sensitive elements of gamma radiation dosimeters. In this work, several silicon PIN photodiodes were tested under gamma radiation. The photodiodes were exposed to gamma radiation from microfocus X-ray tube "БС5-Mo" and the induced reverse current as a function of the dose rate was determined.

PIN photodiodes consist of an intrinsic high-ohmic semiconductor layer realized between n^+ and p^+ heavily doped semiconductor regions [3]. The induced photocurrent of the PIN photodiodes exposed to ionizing radiation depends on dose rate, active area of the PIN photodiode, width of the depletion layer, orientation of the photodiode with respect to the radiation source, energy of the radiation source and ambient temperature. Since PIN diodes have a wider depletion layer than PN diodes, their sensitivity to irradiation is higher.

2. Devices Under Test

Silicon PIN photodiodes S1223 provide the following features: excellent linearity with respect to incident light, wide spectral response range, mechanically rugged, compact and lightweight and long life [4]. They also feature high-speed response, high sensitivity and low noise.

IOP Conf. Series: Materials Science and Engineering **498** (2019) 012015 doi:10.1088/1757-899X/498/1/012015

The main technical characteristics of S1223 PIN diodes are the photosensitive area size of 2.4×2.8 mm^2, the operating temperature range from -40 to +100 °C, dark current at reverse voltage 20 V of 0.1 nA, as well as maximal dark current of 10 nA.

3. Test Setup

During irradiation, the samples were isolated in a special lead irradiation chamber that is impermeable to optical and gamma rays, which contain the gamma radiation source. The samples were installed in a special contacting device. The connection with a measurement system was carried out using a twisted pair. The X-ray tube was controlled by the measurement system, i.e. the dose rate and the time of exposure of gamma radiation to the PIN diodes were set. A computer was connected to the measurement system. For PC connectivity of the measurement system a USB cable was used.

Software and test equipment have provided real-time monitoring of the measured current-voltage characteristics of the samples and displaying on the monitor screen graphs of current versus voltage. Software allows also displaying on the monitor of the current at predetermined voltage values and saving the data to a file.

4. Test Description

Eight test samples were irradiated at the room temperature of about 21 °C during radiation exposure, using a gamma ray source.

The dose rate was measured during the exposure to gamma radiation because the radiation induced photocurrent occurs while the PIN diode is irradiated.

The samples were exposed to six different dose rates: 6,4 Gy(Si)/h, 7,8 Gy(Si)/h, 14 Gy(Si)/h, 25 Gy(Si)/h, 50 Gy(Si)/h, and 124 Gy(Si)/h.

The current-voltage characteristics of the samples were measured in the range from 0 to -90 V.

The dark current was changed from approximately 0.17 nA at 0 V up to 0.36 nA at -90 V.

5. Results

S1223 PIN diodes have been exposed to gamma irradiation to investigate the effect of gamma radiation on their current-voltage characteristics. The highest sensitivity to gamma rays by PIN diodes in consequence of radiation exposure is determined.

The reverse current as a function of dose rate at the voltage of 0, -10, -20, -50, -70, and -80 V is shown in Fig. 1. This figure let us observe that increasing the dose rate leads to the increase of the reverse current for all the reverse voltages and zero voltage.

The current sensitivity is equal to the reverse current per dose rate. The current sensitivity as a function of the dose rate and the reverse voltage is depicted in Fig. 2. This figure shows that the current sensitivity increases with increasing reverse voltage and decreasing dose rate.

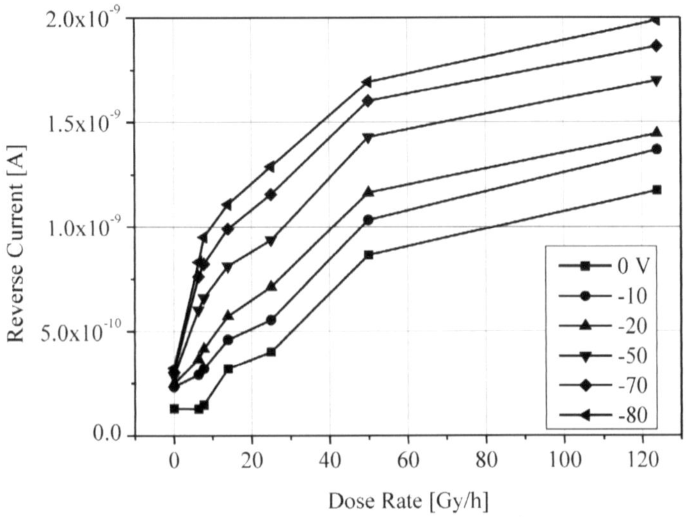

Figure 1. Reverse current versus dose rate at 0 V and the reverse voltage of -10, -20, -50, -70, and -80 V.

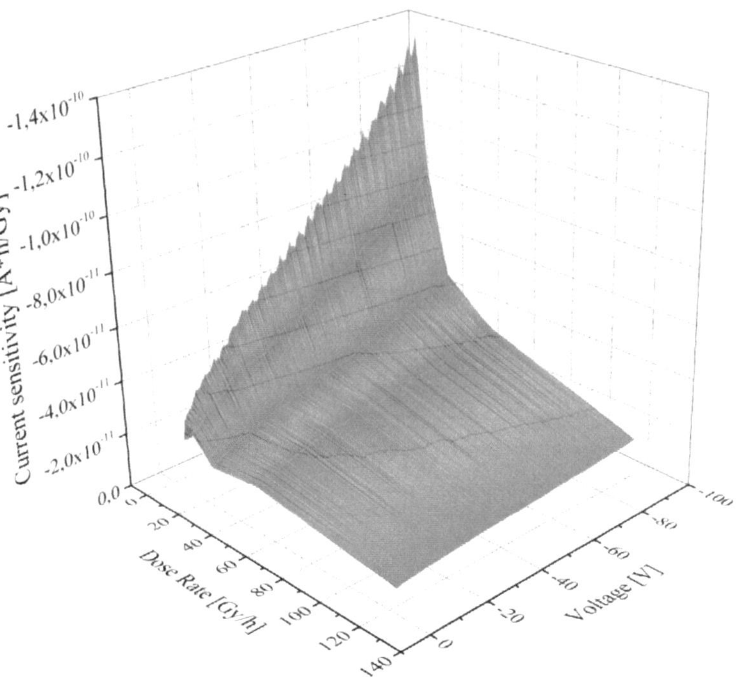

Figure 2. Current sensitivity as function of the dose rate and the reverse voltage.

Figure 3 shows the dependences of the current sensitivity on the dose rate for voltages of 0 V, -10 V, -20 V, -50 V, -70 V, and -80 V. This figure let us observe that increasing dose rate leads to the decrease of the current sensitivity for all the reverse voltages and zero voltage.

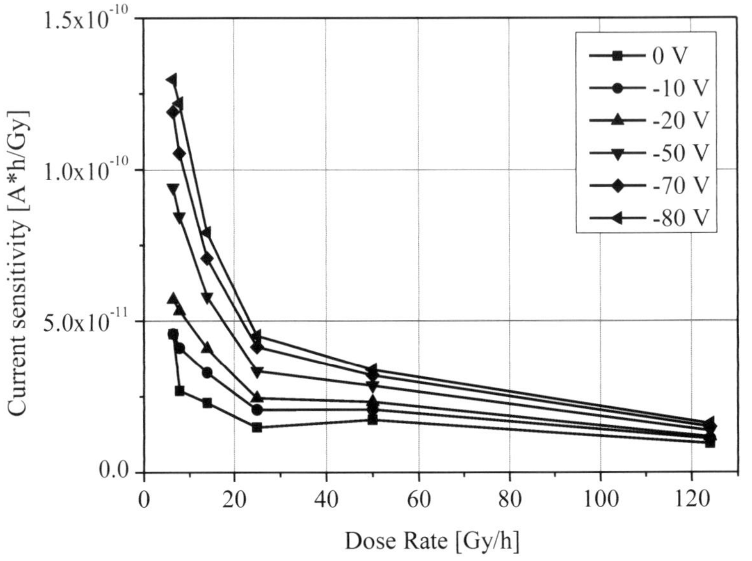

Figure 3. Current sensitivity versus dose rate at 0 V and the reverse voltage of -10, -20, -50, -70, and -80 V.

Hence, the maximum current sensitivity of PIN diodes corresponds to reverse voltages of more than -70 V.

6. Relation Between Reverse Current and Dose Rate

It is possible to derive a mathematical expression for the dependence between the reverse current and dose rate, using the obtained results.

The experimental data of reverse current as function of the dose rate H at the reverse voltage of -80 V (fig. 4) has been fitted with the various power functions. The best accuracy has a fitted equation of the PIN diodes reverse current I:

$$I = a \cdot H^b,$$

where a and b are the approximation coefficients. The values of both coefficients are equal to $b = (29.4 \pm 1.9) \cdot 10^{-2}$ and $a = (5.05 \pm 0.3) \cdot 10^{-10}$ A$\cdot$(h/Gy)^b. The reduced chi-square for this approximation is 0.14205963, coefficient of determination equals 0.98023117.

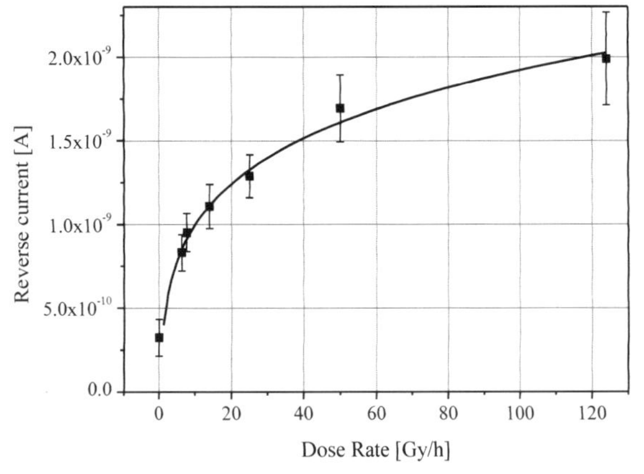

Figure 4. Reverse current as function of the dose rate at the reverse voltage of -80 V.

7. Conclusions

In this work, the silicon PIN photodiodes S1223 from Hammamatsu were irradiated with gamma rays at six different dose rates: 6,4 Gy(Si)/h, 7,8 Gy(Si)/h, 14 Gy(Si)/h, 25 Gy(Si)/h, 50 Gy(Si)/h, and 124 Gy(Si)/h.

The current-voltage characteristics of PIN diodes are measured. The transfer characteristic and the dependence of the current sensitivity on the absorbed dose rate were determined using of these current-voltage characteristics.

Acquired results have shown that the reverse current increases in direct proportion to the dose rate. A transfer characteristic and the dependence of the current sensitivity on the dose rate in the range of reverse voltages from 0 V to -90 V were obtained. It is found that current sensitivity increases with increasing reverse voltage and decreasing dose rate. The range of reverse voltages corresponding to the highest sensitivity of PIN diodes is determined. For the reverse voltage of -80 V, the dependence of the reverse current on the dose rate was approximated by a power function. Numerical values of the approximation coefficients are obtained. Thus, the analytical expression of the transfer characteristic for the operating mode with maximum sensitivity to gamma radiation is obtained.

References

[1] Knoll G F 2000 *Radiation Detection and Measurement* (New York: John Wiley & Sons)
[2] Alexander D R 2003 *IEEE Trans. Nucl. Sci.* **50** (3) 565–82
[3] Andjelkovič M S and Ristič G S 2013 *Nuclear Technology and Radiation Protection* **28** (1) 73–83
[4] The S1223 PIN photodiode: https://www.hamamatsu.com/jp/en/product/category/3100/4001/4103/S1223/index.html

Effect of the active mode NMOS-transistor irradiated on formation of surface defects

K A Petukhov, V D Popov*

National Research Nuclear University MEPhI (Moscow Engineering Physics Institute) Kashirskoe shosse 31. 115409 Moscow, Russian Federation

*Corresponding author's e-mail address: wdpopov@mail.ru

Abstract. The results of experiment on irradiation CMOS integrated circuits are presented at dose rate 0,1 rad/s with in electric and passive modes. Two stages of surface defects formation in both cases were observed.

1. Introduction

The two stages of surface defects formation in *n*MOS-transistors of CMOS integrated circuits at low dose rate described in [1, 2]. These integrated circuits were irradiated by the gamma-rays of Cs^{137} at passive mode of *n*MOS-transistors. The first stage of surface defects formation is described by an exponential dependence and have any saturation [3]. The second stage has some delay and saturated too [1]. Also, some useful information is presented in articles [6-8]. The purpose of this paper is the investigation of surface defects formation at Si-SiO_2 interface under active mode and low dose rate gamma irradiation.

2. Metod

The density of surface defects (interface traps) at the $Si\text{-}SiO_2$ interface was calculated from the slope value drain-gate characteristics *I-U* [4]. As for the experiment were used measurements *I-U* curve on sloping plot drain-gate character transconductance *n*MOS, consequently equation was used:

$$I_d = k(U_G - U_0)^2, \tag{1}$$

where I_d is the drain current; k is the effective transconductance; U is the gate voltage and U_0 is threshold voltage.

The linear dependence can be written as:

$$\sqrt{I_d} = \sqrt{k}\,(U_G - U_0\,), \tag{2}$$

Because the parameters *n*MOS are not known, we analyzed relative changes transconductance:

$$\frac{\sqrt{k(0)}}{\sqrt{k(D)}} = \frac{\sqrt{\frac{\mu_S(0)}{2} * \frac{W}{L} * C_{ox}}}{\sqrt{\frac{\mu_S(D)}{2} * \frac{W}{L} * C_{ox}}}, \tag{3}$$

IOP Conf. Series: Materials Science and Engineering **498** (2019) 012016 doi:10.1088/1757-899X/498/1/012016

where *k(0)* and *k(D)* is s transconductance before and after irradiation at dose *D*, *W* and *L* is width and length of the channel NMOS, C_{ox} is the gate capacitance per unit area.

Of the relative change transconductance, has been described relative value surface electron mobility μ_s, in channel *n*MOS-transistors:

$$\sqrt{\frac{k(0)}{k(D)}} = \sqrt{\frac{\mu_s(0)}{\mu_s(D)}}, \qquad (4)$$

where $\mu_s(0)$ and $\mu_s(D)$ are values surface carrier mobility before and after irradiation dose *D*.

Surface mobility depends on the change interface trap concentration, and is described [5]:

$$\mu_s(D) = \frac{\mu_s(0)}{1+\alpha\Delta N_{it}}, \qquad (5)$$

where ΔN_{it} is change interface trap concentration; $\alpha = (7{,}0 \pm 1{,}3)10^{-13}$ cm^2. Thereby, change interface trap in this paper, defined as equations:

$$\Delta N_{it} \cong 1{,}43 \cdot 10^{12} \left(\frac{\mu_s(0)}{\mu_s(D)} - 1\right). \qquad (6)$$

3. Experiment

In this experiment three CMOS integrated circuits *CD4069UBCN* in which characteristics of 9 MOS transistors were measured were used. This integrated circuits was irradiated by Cs^{137} gamma source at dose rate P = 0,1 rad/s. The irradiation was in active and passive modes. In case of using active mode it was given at input and supply voltage U=12 V. In passive mode all pins of the integrated circuit were grounded during irradiation.

Characteristics *n*MOS-transistors were measured before and after each step irradiation. During experiment were obtained the average values of the changes of the electrons mobility, which have been determined by statistical techniques. The function of the normalized electron mobility on the dose irradiation is presented in fig. 1.

During the experimental study, the calculated values of the density interface traps after each stage of irradiation was received. The experimental results are presented in fig. 2, where interface trap density as a function of dose is shown.

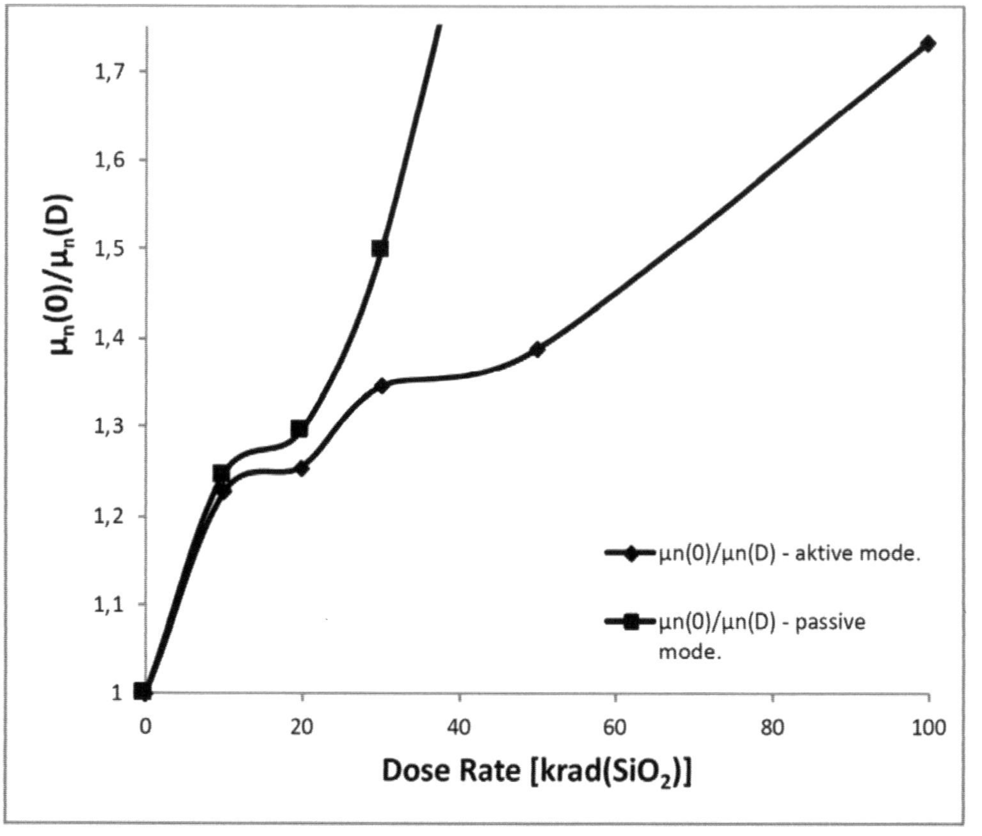

Figure 1. Dependence of normalized electron mobility versus dose irradiation.

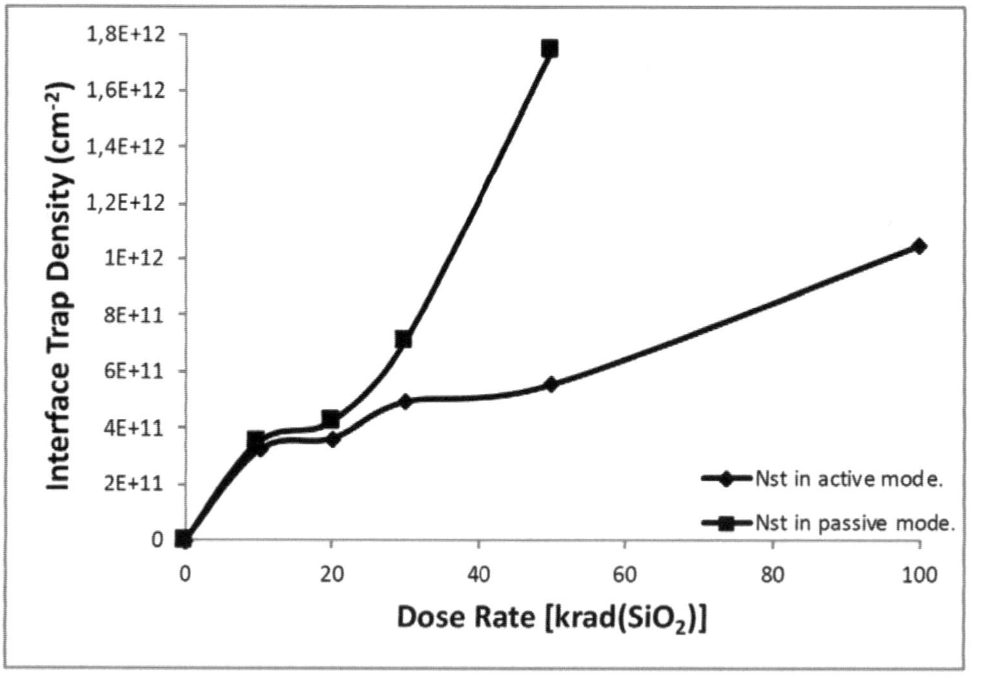

Figure 2. Interface trap density versus dose irradiation.

IOP Conf. Series: Materials Science and Engineering **498** (2019) 012016 doi:10.1088/1757-899X/498/1/012016

4. Conclusion

Two stage of interface trap formation is observed in active mode and in passive mode irradiation. The first stage of the formation of surface defects is practically independent of the electrical mode and a true effect. In the second step this process becomes dependent on the dose rate and time. This leads to an additional increase in the density of surface defects. It is observed that in active mode the rate of formation of surface defects is reduced.

Acknowledgement

I am grateful to Professor V. S. Pershenkov for useful discussion.

References

[1] Popov V D 2016 *Fizika i Tekhnika Poluprovodnikov* **50** 3 p 349

[2] Popov V D and Vin C K 2014 *Proceedings of the 13th International Conference on Physics of Dielectrics* (St/-Peterburg. Russia) **2** p 234

[3] Rashkeev S N, Cirba C R, Fleetwood D M, Schrimpf R D, Witzak S C, Mishez A and Pandelides S T 2002 *IEEE Trans.on Nucl. Sci.* **49** 6 p 2650

[4] Emel'anov V V, Zhukov Y N, Loshkarev V V and Meshurov O M 1995 *Vopr.Atom.Nauki and Tech. Ser. Fizika Rad. Vozdeistv. on Radioelectron. Apparat.* 3-4 p 44

[5] Sexton W and Schwank J R 1985 *IEEE Trans. on Nucl. Science.* **NS-32** 6 p 3975

[6] Boychenko D V, Kalashnikov O A, Karakozov A B and Nikiforov A Y 2015 *Russian Microelectronics* **44** 1 pp 1-7

[7] Zebrev G I and Drosdetsky M G 2016 *IEEE Transactions on Nuclear Science.* **63** 6 2895-2902.

[8] Pershenkov V S, Sogoyan A V and Telets V A *IOP Conference Series: Materials Science and Engineering* **151** 1

IOP Conf. Series: Materials Science and Engineering **498** (2019) 012017 doi:10.1088/1757-899X/498/1/012017

Comparison of hardware and timing penalties for eliminating SRAM failures

L A Shchigorev[1,2,*] and I I Shagurin[1,2]

[1] National Research Nuclear University MEPhI (Moscow Engineering Physics Institute), Kashirskoe shosse 31, 115409 Moscow, Russian Federation
[2] JSC "Module", 3 Eight March 4th Street, 125190 Moscow, Russian Federation

*Corresponding author e-mail address: l.shchigorev@module.ru

Abstract. This article provides a quantitative estimation of timing and hardware penalties, which are the implication of different SRAM methods for failure elimination. Different ways of building fault-tolerant blocks of SRAM are discussed using error-correcting codes and self-testing – self-repair units. As criteria for evaluating hardware costs, the additional chip area required for the placement of fault tolerance is considered, the timing penalties are determined by the growth of memory access time. Comparative analysis of obtained results is made. It allows estimating the effectiveness of the considered methods of fault tolerance improvement.

1. Introduction

The reliability of modern digital systems is largely determined by the accuracy of the information stored in their storage devices. Therefore, in highly reliable systems, different methods are used to eliminate the effects of failed elements of the memory blocks. Most often, the following methods are used:

- Using error correction codes (ECC) units.
- Using built-in self-test and self-repair units in memory blocks.

The most commonly used technique of ECC is the Hamming code [1], which provides a correction value of one stored data bit of the information word. Words, which are submitted for storing in memory, are encoded with the introduction of additional bits that provide the detection and correction of erroneous bits. While reading, the decoding of the selected memory word is made with error correction, and the correct value is provided. This way you can parry a single failure in the stored word. The additional hardware penalties, in this case, are due to the increase in the bit size of the stored coded words and the inclusion of ECC encoders and decoders. An estimation of hardware costs for the implementation of Hamming encoder-decoders is introduced here [2].

The BISR implementation requires the addition of self-test units (STU) and self-repair units (SRU). If failed elements are detected during the self-test of the memory, the BISR unit connects the redundant units instead, which are placed on the crystal in the form of additional rows or columns of the memory matrix. This method is used at the stage of production control of chips in order to increase the yield of integrated circuits. To eliminate failures during the operation process, it is necessary to initiate the self-test and self-repair sessions of SRAM blocks.

The easiest way of replacing failed memory elements is to shift the data bits by one position. Figure 1 shows an example of performing such a shift in the reading channel of the word by using N multiplexers "2 in 1", where N is the bit width of the corrected word, FE – damaged element, RE –

redundant element. Thus, recording channel has the elements selectors to "1 in 2", which upon detection of a failure element shift the corresponding bits to the adjacent element. The control signals for the multiplexers and selectors are formed by the block of SRU according to the results of previously executed self-test.

BISR method is effectively used in the case of a single failure (Single Event Upset – SEU), when the corrected word (or words) has to replace only one bit. A frequently used variant of the double failure correction, when for each word two redundant elements are supplemented, one of which serves as the correction of the low part and the other for correction of the high part. In this case, it is possible to replace two failed bits, if they are located in different halves of the word.

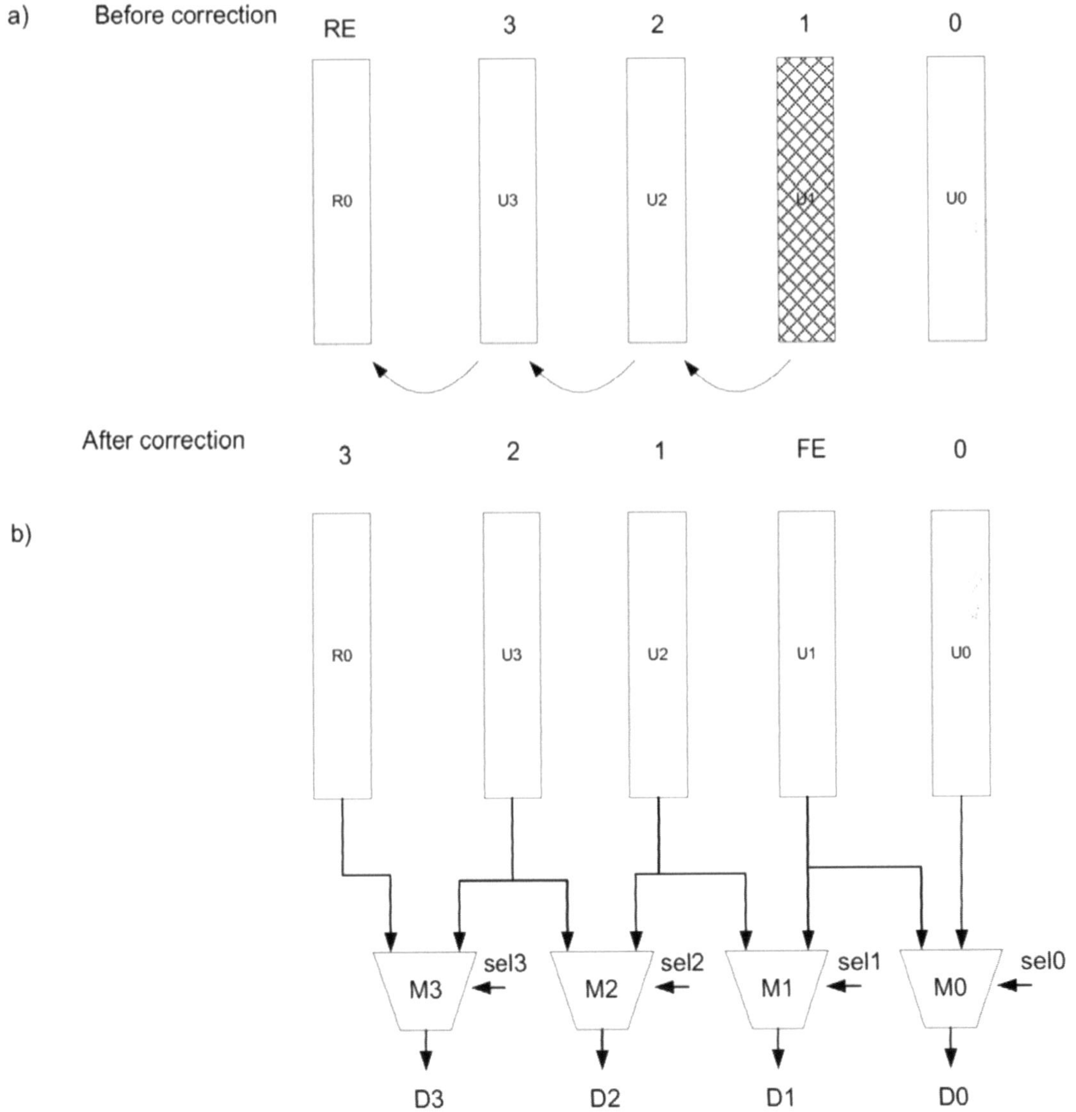

Figure 1. Replacing a failed memory element by shifting the bit numbers (a) and the scheme implementation of the shift in the read channel using multiplexers (b)

For correction using the self-repair method of multiple failures (Multiple Cell Upset – MCU), the use of multiple redundant elements for each word stored in the memory is required. The required number of redundant elements m is determined by the number of correcting bits – maximum number bits of the word to be corrected. In this case, for correction of N-bit words when the number of failures is m, N multiplexers "(m+1)-to-1" [3] are required. Figure 2 shows an example of enabling the multiplexers in the read channel memory for correction of the failure with multiplicity m=3 in the word having N=4 bits. The channel of writing includes the selectors "1-to-(m+1)", which provides turning off of the failed element while maintaining an appropriate bit in one of the redundant elements.

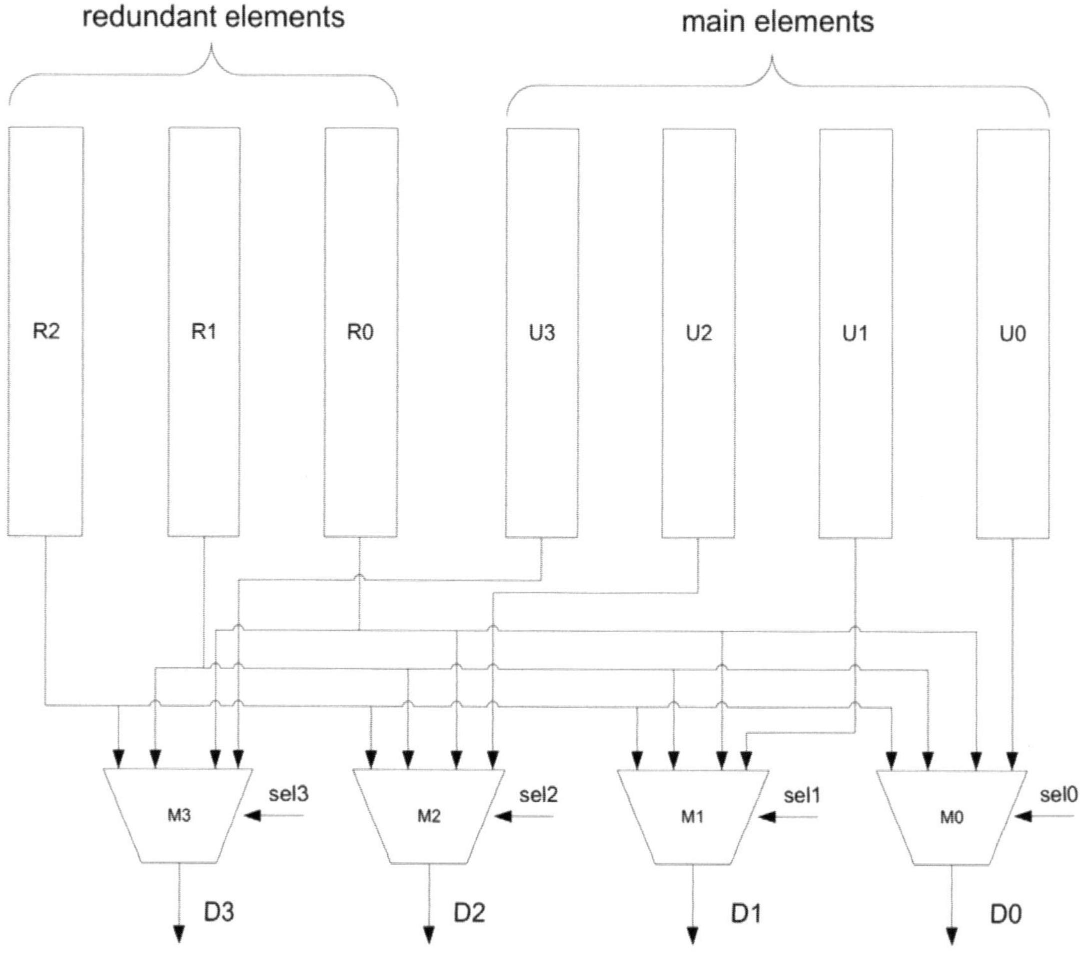

Figure 2. Implementation scheme of the multiple replacement of faulty memory elements by using multiplexers (read channel)

The additional hardware penalties during the BISR implementation caused by the introduction of the necessary number of redundant memory elements include multiplexers, which provide a replacement of the failed elements. These hardware costs lead to an increase in the area occupied on chip block memory, as well as to increased energy consumption.

In addition to hardware costs, the use of ECC methods and BISR induces an increase in data access time that is due to both to the increase of the geometrical sizes of the memory blocks and the associated parasitic capacitance, and the added combinational logic to the data communication scheme– multiplexers and encoder-decoders.

The purpose of this study was to obtain and compare quantitative estimates of time and hardware costs arising from the use of different methods of parrying of elements failures in blocks of static RAM.

2. The estimation of area and timing penalties

For the investigation, the most popular data widths were selected: 8 bits, 16 bits, 32 bits and 64 bits. The capacity of each block is 4096 (4K) words.

For eliminating single failures using ECC method the encoders-decoders of Hamming correcting single failures (ECC-SEC). Eliminating failure elements by the BISR method can be performed by replacing failed elements by a shift scheme (BISR-SHIFT) or by multiplexing bits from the redundant and the main elements in the read channel (BISR-MUXm). In the embodiment of BISR-SHIFT, each word had 2 redundant elements to parry failures in low and high parts of the word. In implemented embodiments BISR-MUXm the selected number of spare elements is m = 1, 2, and 4. Therefore, the option BISR-MUX1 provides eliminating of single failures, options BISR-MUX2, BISR-MUX4 allow to fend off the 2-fold and 4-fold failures. Table 1 shows the number of additional bits in the stored input word to implement different variants of failure elimination.

Table 1. Number of additional bits to implement different methods of failure eliminating

	Word width, bit			
Method of eliminating	8	16	32	64
ECC-SEC	4	5	6	7
BISR-SHIFT	2	2	2	2
BISR-MUX1	1	1	1	1
BISR-MUX2	2	2	2	2
BISR-MUX4	4	4	4	4

To evaluate the performance of different variants of fault-tolerant RAM blocks, their synthesis was carried out using CAD Cadence Encounter RTL Compiler during use of CMOS technology with library with the 28 nm process. For the comparative assessment of RAM blocks parameters of a similar organization, which do not have any features for parrying of failures (NC – not corrected), the above-mentioned figures were synthesized. Table 2 shows the data on the chip area required to implement the control options of blocks of SRAM with a capacity of 4K words at different bits of stored words. Figure 3 is a diagram showing the relative area increase of SRAM blocks for the cases of applying the above options to improve resiliency.

Table 2. Memory area without any features for parrying of failures

	Word width, bit			
	8	16	32	64
Area, um^2	10077	15688	26911	49356

Note that the estimations for the BISR method do not take into account the hardware cost for the implementation of the units of STU and SRU, as there are various algorithms and means of functional diagnostics of memory, the analysis of which is beyond the scope of this paper. The capacity of implementing self-test procedures can be found in [4-6], different ways of self-repair are described in [7-12].

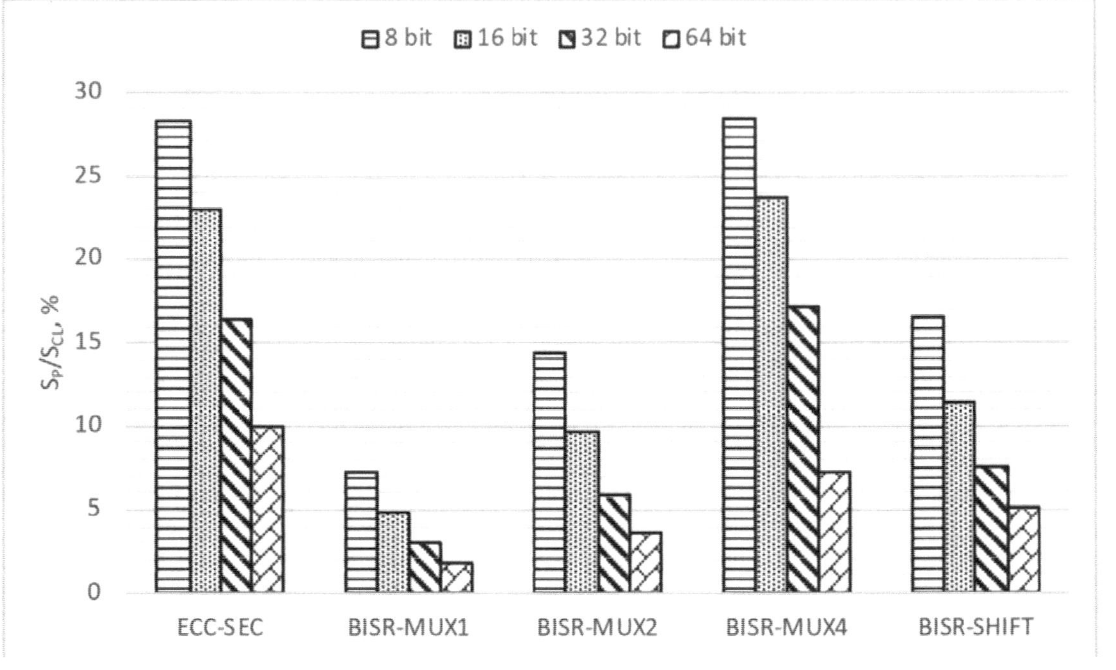

Figure 3. The relative increase of the chip area of SRAM blocks using different variants of failure eliminating

In the simulation of the synthesized variants of the SRAM blocks, the access time of the sample data with different values of the supply voltage U_{dd} and the temperature T °C is determined. The simulation was performed for the operating conditions specified in Table 3.

Table 3. Memory area without any features for failure elimination

Shorthand	U_{dd}, V	T, °C
LPLT (Low Power, Low Temperature)	0,9	-40
LPHT (Low Power, High Temperature)	0,9	125
TT (Typical Power, Typical Temperature)	1,0	25
HPLT (High Power, Low Temperature)	1,1	-40
HPHT (High Power, High Temperature)	1,1	125

Figure 4 shows the obtained estimations of the maximum access time values of sampling (data reading) with implementation of different methods for failure elimination.

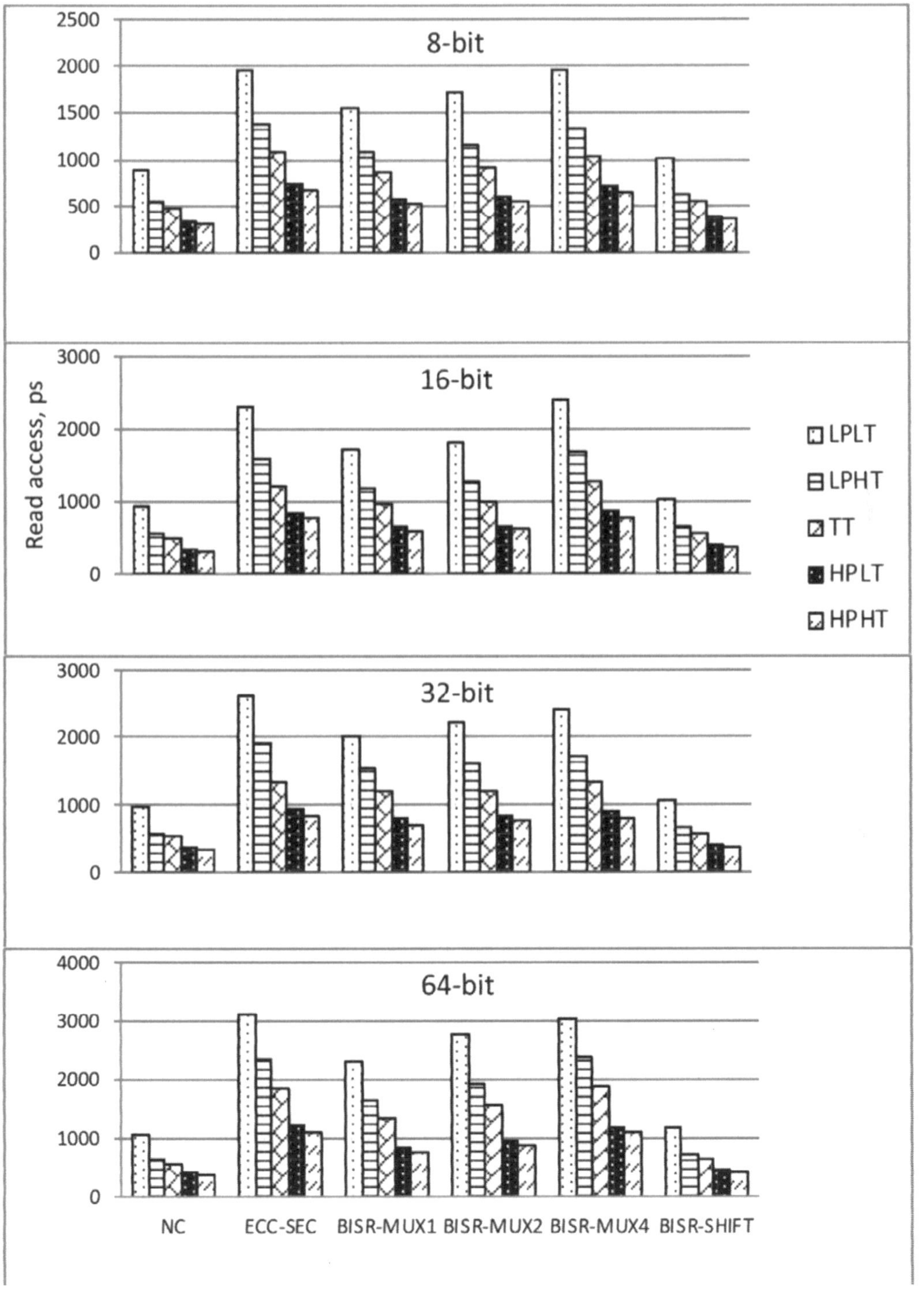

Figure 4. Access time for SRAM blocks with different word width

3. Analysis of obtained results

The analysis of the results of the synthesis and simulation of different variants of SRAM blocks implementation allows to draw the following conclusions:

- The relative cost of chip area with implementation of the methods for failure eliminating depends on the word width N and the number of additional bits. With the increase of N, the additional cost square is substantially reduced. With the value N = 32 or more bits, the increase in chip area does not exceed 17%.
- Access time depends significantly on the supply voltage and operation temperature. The maximum value of access time is observed at a lower supply voltage and temperature (worst case). The maximum and minimum access time in the range of the supply voltage U_{dd} = 0,9...1,1 V and temperature T = -40 to 125 °C can differ in 2,5...3,5 times.
- Access time increases with the word width N. When N changes from 8 to 64 bit access time increase for the cases of NC and BISR-SHIFT is about 20%, and for the other SRAM – 1.5 ...1.9 times.
- With the implementation of the discussed methods of failure eliminating access time increases by 10...20% for cases of BISR-SHIFT, 2.5...3.0 times for ECC-SEC and BISR-MUX4 2.1...2.5 times for BISR-MUX2 and 1.7...2.2 times for BISR-MUX1.
- For single error elimination, BISR-SHIFT provides the access time of 1.5 times less, yet it takes 2 times more area compared to BISR-MUX1. The ECC-SEC method has 2...3 times larger access time and 1.8...2.1 times larger area compared to the BISR-SHIFT but does not require the addition of STU and SRU blocks.
- The BISR-MUX4 and ECC-SEC methods have close values of access time and additional occupied area. BISR-MUX4 provides failure elimination with multiplicity m = 4 but requires the addition of STU and SRU blocks.

4. Conclusion

The investigated methods of SRAM failure elimination have each their own features that determine the capabilities and effectiveness of their application. When choosing a method of increasing fault tolerance, it is necessary to account for such factors as the mode of SRAM operation, the requirements for its characteristics, chip area, access time, energy consumption and the nature of the occurring faults. BISR can be used if the mode of SRAM operation allows for the implementation of testing and repairing sessions. If the SRAM operation conditions require an on-speed correction of failures in the process of continuous operation of the system, the failures can be eliminated with the application of the ECC method. However, the use of ECC leads to higher hardware and time penalties, especially when responding to multiple failures. These penalties degrade the performance of SRAM and systems in which they operate. Therefore, the BISR technique is more efficient in cases where its use is allowed in the operation mode of the system.

The quantitative estimations of the parameters for the fault tolerant SRAM blocks implementing the methods of the ECC, BISR, obtained in this paper, allow you to choose the most effective elimination method with consideration of the required characteristics and type of the failures occurred.

References

[1] Hamming R 1950 Error detecting and error correcting codes *Bell Syst. Tech. J.* **29** 147-60
[2] Elchin I, Shagurin I, Osipenko P and Vasilegin B 2006 *Izvestiya vysshikh uchebnykh zavedenii. Electronika*. **4** 65-70
[3] Nicolaidis M, Achouri N and Boutobza S 2003 *Design, Automation and Test in Europe Conf. and Exh. Proc.* **1** 590-5
[4] Vardanian V and Zorian Y *IEEE Int. Workshop on Mem. Tech. Design and Testing. Proc.* **1** 62-7
[5] Hamdioui S, Van de Goor A and Rodgers M 2002 *IEEE Int. Workshop on Mem. Tech. Design and Testing. Proc.* **1** 95-100
[6] Papavramidou P and Nicolaidis M 2016 *IEEE Trans. on Comp.* **65** 2284-98

IOP Conf. Series: Materials Science and Engineering **498** (2019) 012017 doi:10.1088/1757-899X/498/1/012017

[7] Shchigorev L and Shagurin I *IOP Conf. Series: Materials Science and Engineering* **151**
[8] Lu S, Tsai C and Hashizume M 2016 *IEEE Trans. VLSI Sys.* **24** 2726-34
[9] Lu S, Yang C, Hsiao Y and Wu C 2010 *IEEE Trans. VLSI Sys.* **18** 184-93
[10] Shchigorev L 2016 *Problems of Advanced Micro- and Nanoelectronics Sys. Dev.* **3** 178-85
[11] Tseng T, Li J and Chang D 2006 *Design, Automation and Test in Europe Conf. and Exh. Proc.* **1** 51-6
[12] Kim I, Zorian Y, Komoriya G, Pham H, Higgins F and Lewandowski J 1998 *Int. Test Conf. Proc.* **1** 1112-1119

Pipelined implementation of the IP-block for direct tracking on configurable Zynq 7020 SoC platform

R A Seliverstov and I I Shagurin

National Research Nuclear University MEPhI (Moscow Engineering Physics Institute), Kashirskoe shosse 31, 115409 Moscow, Russian Federation

Abstract. The possibilities that occur in using specialized hardware for the accelerated execution of a number of computational steps in solving the problem of spatial localization of an object using the modern algorithm Semi-Dense SLAM are considered. As a hardware platform for implementing the localization system, a programmable system is used on a Xilinx Zynq-7020 chip, which consists of two ARM Cortex-A9 processor cores and an array of programmable logic (FPGA). An estimate of the amount of hardware resources required to perform spatial localization is presented. It is shown that the application of the developed computational pipeline allows to significantly increase the localization rate. The results of the research can be used to design navigation systems for unmanned aerial vehicles and autonomous vehicles, as well as augmented reality systems.

1. Introduction

One of the primary tasks in the development of AUVs and other mobile robotic systems is to provide and maintain their robust navigation in the surrounding space. To solve this problem are used methods and algorithms for Simultaneous Localization and Mapping SLAM [1]. The existing methods for performing SLAM can be divided into two groups:
- methods based on the usage of Reference Points (Feature-Based SLAM),
- direct methods which take advantage of the full image (Direct SLAM).

The major concept behind the first group of methods is to divide the geometric evaluation process into two consecutive steps: first, the extraction of a set of features from the input image and then calculation of the position of the camera as a function of the location of those features. This distinction imposes an important limitation: information about the straight or curved edges of the surrounding space, which does not contain data corresponding to a given set of features, is discarded, although it may be useful for improving the overall accuracy of the algorithm. The main disadvantages of these methods are the high requirements for computational resources for hardware implementation of the processes of finding, identifying and evaluating the position of landmarks in three-dimensional space.

SLAM methods based on full images or direct methods (Direct SLAM) allow to avoid the main drawback of Feature-Based SLAM because in Direct SLAM, the camera position and the geometry of the surrounding scene are determined based on information about all pixels of the input image. Direct methods also allow the construction of complete and accurate maps of the surrounding space [2]; however, this requires the use of high-performance graphics accelerators (GPU) for real-time operation.

In this paper, the problem of spatial localization of an object is solved using one of the newest modification of classical direct methods – the Semi-Dense SLAM, which is based on the use of incomplete images. This approach allows to track the movement of an autonomous vehicle with high precision using the most informative parts of the image (a subset of pixels for which the intensity gradient is higher than a certain threshold value and for which it is convenient to track the position of the robot in space), and at the same time makes it possible to consistently build accurate and large-scale maps of the surrounding area [3].

2. The Semi-Dense SLAM algorithm for spatial localization

Figure 1 shows a block diagram of the Semi-Dense SLAM algorithm.

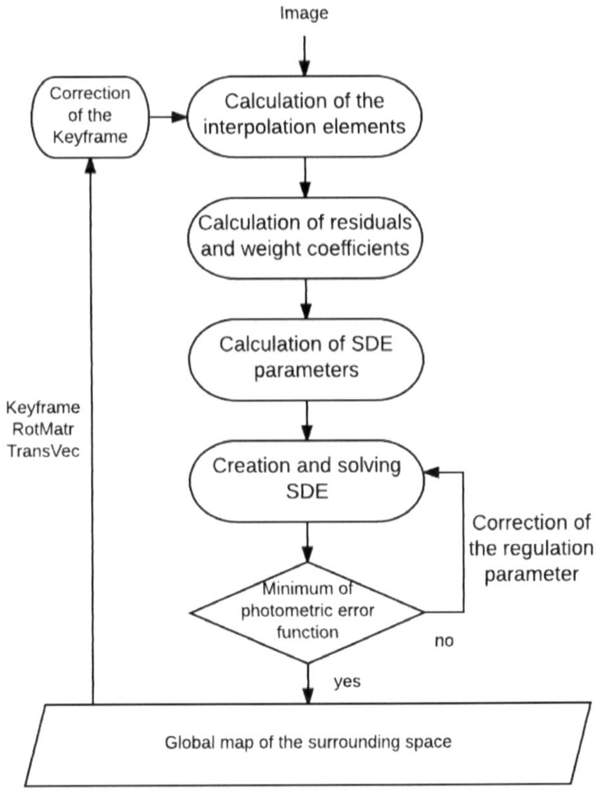

Figure 1. Block diagram of the Semi-Dense SLAM algorithm.

The algorithm is based on the use of a Keyframe, a special data structure that is stored in external memory and contains an image (with pixels determined by the x and y coordinates), as well as the image depth $\boldsymbol{D}_i$ and the depth variance σi [1]. A new Keyframe is formed after the camera is moved, when the distance between it and the current frame coming from the video camera exceeds a certain threshold value. At the stage of map construction, the previous Keyframe is decommissioned and added to the Keyframe graph. The global map of the surrounding space is represented in the form of a graph, the vertexes of which are KFs, whilst the edges are the parameters describing the rotation operations (the RotMat matrix) and the transition (TransVec vector) that the camera needs to take from the previous position to the next one, calculated by the algorithm (from one vertex of one graph to the other).

To describe the current position of the camera, an element of the Lie algebra is used - vector ξ_{ij}, which contains the RotMat matrix and the TransVec vector, which determine the location of the current frame in relation to the previous frame. The value of this vector corresponds to the position of the minimum of the normalized photometric error function (1):

$$E_p(\xi_{ij}) = \sum \left\| \frac{r_p^2(\boldsymbol{p},\xi_{ij})}{\sigma_{r_p}^2(\boldsymbol{p},\xi_{ij})} \right\| \quad (1)$$

where r_p is the residual, which is calculated as the difference in pixel intensities of the current frame $\boldsymbol{I_i}$ ***(p)*** and the corrected Keyframe $I_j(p, D_i(\boldsymbol{p}), \xi_{ij})$

$$r_p(\boldsymbol{p}, \xi_{ij}) = I_i(\boldsymbol{p}) - I_j(p, D_i(\boldsymbol{p}), \xi_{ij}) \quad (2)$$

σi – depth variance calculated with inverse depth $V_i = 1/Di$

$$\sigma_{r_p}^2(\boldsymbol{p}, \boldsymbol{\xi_{ij}}) = 2\sigma_I^2 + \left(\frac{\partial r_p(\boldsymbol{p}, \xi_{ij})}{\partial D_i(\boldsymbol{p})}\right)^2 V_i(\boldsymbol{p}) \quad (3)$$

$D_i(p)$ – depth value for the described subset of pixels.

All operations are performed only for the selected subset of pixels $\boldsymbol{p}$ with an intensity value above a certain threshold.

At first, the Keyframe correction step is executed by changing the pixel projection in accordance with the current position of the camera. At the correction stage, the image contained in the KF is also calibrated taking into account the technical characteristics of the camera.

The minimum of the error function is calculated using an iterative optimization process using the Levenberg-Marquardt method [4]. At each iteration, the value of the position vector $\boldsymbol{\xi_{ij}}$ is determined, at which the local minimum of the error function [3] is reached. The increment of the vector $\delta\xi^{(n)}$ for which $E_p(\boldsymbol{\xi}_{ij})$ reaches a local minimum is calculated for the n-th iteration by a system of differential equations:

$$\delta\boldsymbol{\xi}^{(n)} = -(\boldsymbol{H} + \lambda\boldsymbol{I})^{-1}\boldsymbol{J}^T r(\boldsymbol{\xi}^{(n)}) = -(\boldsymbol{J}^T\boldsymbol{J} + \lambda\boldsymbol{I})^{-1}\boldsymbol{J}^T r(\boldsymbol{\xi}^{(n)}) \quad (4)$$

where $\boldsymbol{J}$ is the matrix of the first derivatives of the error function (Jacobian), $\boldsymbol{H}$ is the matrix of the second derivatives of the error function (Hessian), $\boldsymbol{I}$ is the identity matrix, $\lambda \geq 0$ is the regulation parameter that improves the convergence of the algorithm by increasing the monotonicity of the function $E_p(\boldsymbol{\xi}_{ij})$. Elements of the Jacobian are calculated in accordance with expression:

$$\boldsymbol{J} = \left.\frac{\partial r(\epsilon \circ \xi^{(n)})}{\partial \epsilon}\right|_{\epsilon=0} \quad (5)$$

where ϵ is the linearization parameter. To calculate Hessian, we use the approximation $\boldsymbol{H} = \boldsymbol{J}^T\boldsymbol{J}$. The necessary data for the formation of the Jacobian are determined at the stage of calculating the interpolation elements, where the pixel intensity increments dIx, dIy are calculated for the x, y coordinates and interpolated intensity value $\boldsymbol{C}$.

In the presence of substantial fluctuations of intensity, large values of the error function arise, which slow down the process of finding the minimum. To reduce the effect of such differences, normalization is obtained with the use of the matrix of such weighting coefficients $\boldsymbol{W}$ that limit the intensity values [5].

Taking into account the weighting coefficients, the SDE for finding the minimum of the error function is:

$$\delta\xi^{(n)} = -(\boldsymbol{J}^T\boldsymbol{W}\boldsymbol{J} + \lambda\boldsymbol{I})^{-1}\boldsymbol{J}^T\boldsymbol{W} r(\xi^{(n)}) = -\boldsymbol{A}/(\boldsymbol{B} + \lambda\boldsymbol{I}) \quad (6)$$

where the parameters $\boldsymbol{A}, \boldsymbol{B}$ on the right-hand side of equations are determined by the expressions

$$\boldsymbol{A} = \boldsymbol{J}^T\boldsymbol{W} r(\xi^{(n)}),\ \boldsymbol{B} = \boldsymbol{J}^T\boldsymbol{W}\boldsymbol{J} \quad (7).$$

To form the SDE, the Jacobian and the parameters ***A, B*** are calculated, which are determined in the previous step by the values of interpolation elements, residuals and weighting coefficients.

The system of equations is solved with different values of the regulation parameter λ. The number of iterations depends on the size of the image. In the event of a global minimum of the photometric error function, the new position of the camera $\boldsymbol{\xi_{ij}}$ is fixed and added to the global map of the surrounding space. Otherwise, the value of λ is corrected and the solution of the system of equations is recalculated. For optimal choice of the values of λ, the technique described in [6] is used.

The iterative process of finding $\delta\xi^{(n)}$ is finished in the following cases:

1) After performing the maximum number of iterations N (in this work, for the resolution of 320x240 pixels, the value N = 50 is assumed);

2) If the increment $\delta\xi^{(n)}$ obtained at the last iteration turns out to be less than the predefined threshold value;

3) If the change in the error function obtained at the last iteration is less than a certain value (for example, 0.1%).

Fulfilment of these conditions ensures good convergence of the algorithm and reduces the number of iterations required to calculate the next camera position.

Calculations performed at various stages of the algorithm can be made using a universal processor (computer) or using specialized hardware blocks. In [7], an implementation of the Semi-Dense SLAM algorithm is described, which utilizes a processor with specialized hardware blocks for calculating the Jacobian and residuals. In the presented article, the possibility and efficiency of using the hardware-implemented pipeline for calculating all necessary parameters for the formation of a linear system of equations are considered, the solution of which allows to calculate the position of the object in the surrounding space.

3. The architecture of the spatial localization system

For the hardware implementation of the spatial localization algorithm, it is necessary to use a system with the following architecture:

- high-performance processor for solving the system of differential equations (SDE) with correction of the regulation parameter, general control of the system units, performance of necessary service functions;
- an array of programmable logic for the implementation of specialized blocks that perform the calculation of the parameters for the system of equations;
- sufficient memory to store a global map of the environment and necessary service information;
- a set of interface devices that provides interconnection between the system units and external devices.

As a hardware platform for the implementation of the proposed architecture, a programmable System on Chip Zynq-7020 FPGA SoC from Xilinx [8] is used, which meets the specified requirements to the greatest extent. Zynq-7020 contains two high-performance 32-bit Cortex-A9 processor cores with a maximum operating frequency of up to 766 MHz and an FPGA-type programmable logic array containing 85,000 logic cells (LUTs) and 106400 triggers. The SoC additionally contains interface blocks that allow the implementation of the high-speed AXI bus, 220 DSP blocks for performing multiply-accumulated operations, 4.9 Mb memory blocks and some other specialized units. The used SoC is implemented as part of the Digilent's ZedBoard development board, which also contains a dynamic RAM of 1 MB. This RAM is used as the external memory for the implemented spatial localization system.

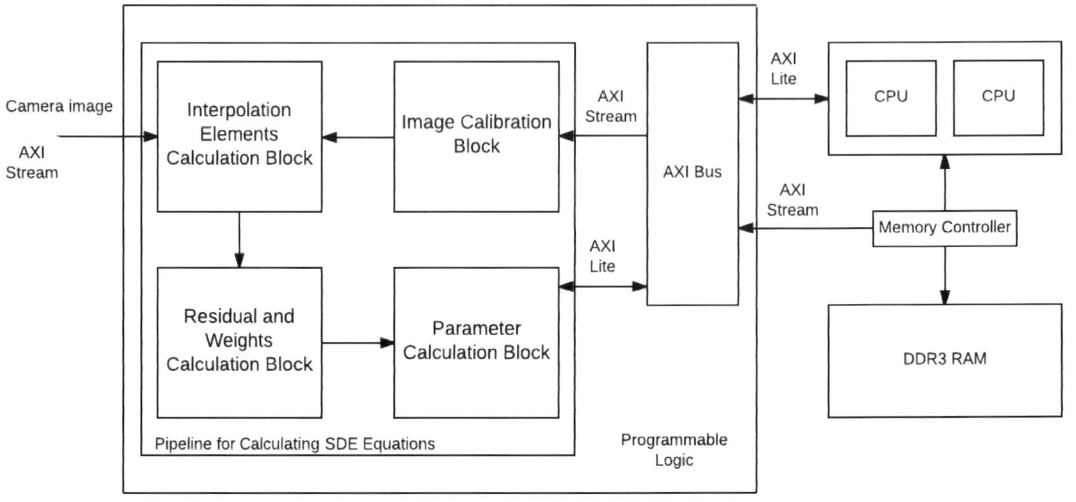

Figure 2. Architecture of the spatial localization system.

Figure 2 shows the architecture of the spatial localization system. Two Cortex-A9 processor cores perform the functions of a central processing unit (CPU), providing the overall control over the operation of the localization system and the solution of the SDE. A computational pipeline is implemented on programmable logic to calculate the parameters of the SDE. Separate stages of the pipeline perform correction of the selected Keyframe, calculation of interpolation elements, determination of residual and weighting coefficients, and calculation of the values of parameters necessary for solving the SDE. Transmission of information inside the system is provided by two types of AXI interface blocks: AXI Stream for streaming of video data and AXI Lite for transmission of control signals from the processor. The control of the external DDR3 DRAM, where the global environment map as well as the service information used during the system operation are stored, is provided by one of the CPUs with the help of a memory controller that is part of the SoC.

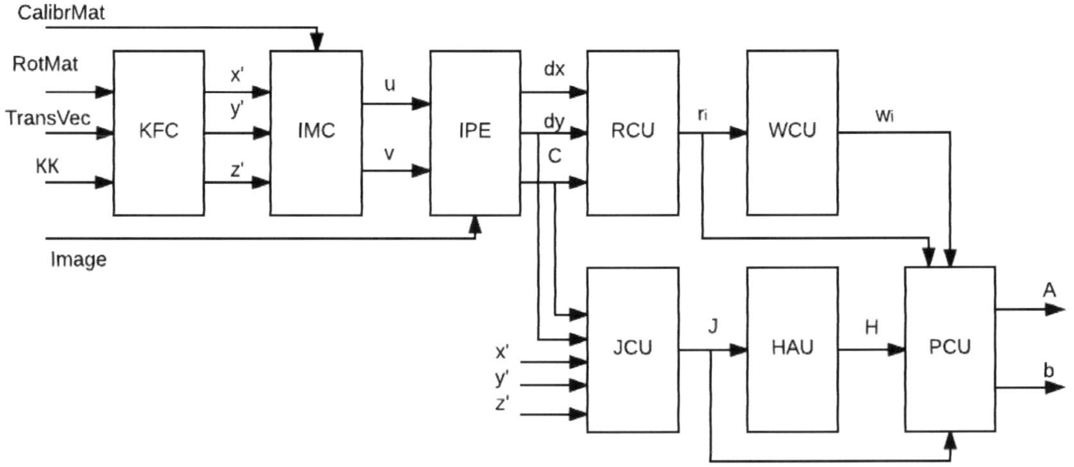

Figure 3. Microarchitecture of the computational pipeline.

Figure 3 shows the microarchitecture of the computational pipeline, which contains 8 operating units. The current Keyframe selected from DDR3, the RotMat matrix and the TransVec vector (the initial position vector ξ_{ij}), which is set by the CPU, enter the input of the Keyframe Correction Block (KFC).

IOP Conf. Series: Materials Science and Engineering **498** (2019) 012018 doi:10.1088/1757-899X/498/1/012018

In this block, the projection of the Keyframe is changed by multiplying the pixels of the Keyframe by the RotMat matrix and adding the TransVec vector. The output of the KFC generates the corrected values of the coordinates x', y' and depth z', which are fed to the image calibration block IMC. In accordance with the pinhole camera model, this block divides the coordinates x' and y' by the value of the depth z' and the resulting values are multiplied by the calibration matrix CalibrMatr. This matrix takes into account the technical characteristics of the used video camera and is stored as the service information in the system memory. A set of corrected pixel coordinates of the Keyframe (u, v), corresponding to the camera position at the current time, are sent to the output of the IMC.

Using the data calculated on the previous stage of the pipeline, the block for calculating the interpolation elements IPE [7] determines the intensity increments dIx, dIy with respect to the coordinates x and y and the interpolated intensity value C, which are then fed to the block for calculating the residual coefficients RCU. This block calculates the value of the residual coefficients $\boldsymbol{r_p}$, which is determined by the expression (2). After that, the sums of intensities and the sum of intensity squares ($\sum c$ и $\sum c^2$) of the current frame as well as the sum of the intensities and the sum of squares of the intensity of the Keyframe ($\sum c_2$ и $\sum c_2^2$) are calculated for finding auxiliary parameters of Affine transformations.

The calculation unit for the weighting coefficients WCU generates a matrix of weights $\boldsymbol{W}$, the elements of which are determined depending on the value of the corresponding residual coefficient:

$w_i = 5.0/r_i$, if $r_i < 5.0, w_i = 1$ otherwise.

The corrected values of coordinates x', y' and depth z', obtained at the output of the IMC, as well as the calculated pixel increments $dIx,\ dIy$ calculated by the IPE block are fed to the block of calculation of Jacobian JCU. This block calculates the elements of the Jacobian in accordance with the expressions:

$$J(1) = \frac{dx}{z}, J(2) = \frac{dy}{z}), \ J(3) = -\frac{x}{z^2}dx - \frac{y}{z^2}dy,\ J(4) = -\frac{xy}{z^2}dx - \frac{z^2+y^2}{z^2}dy,\ J(5) = \frac{z^2+x^2}{z^2}dx + \frac{xy}{z^2}dy,\ J(6) = -\frac{y}{z}dx + \frac{x}{z}dy \quad (7)$$

The Hessian approximation block HAU computes the approximate value of the Hessian by transposing the Jacobian and multiplying its direct and transpose matrixes. The calculated values of the Jacobian and Hessian, as well as previously derived residual and weighting coefficients, are used by the PCU parameter calculation block to determine the parameters $A,\ B$, which enter one of the CPUs where the SDE is generated and solved. The results of the SDE solution, which determine the new position of the object, are added to the global map of the surrounding space, stored in the system memory.

4. The results of modelling the spatial localization system

To evaluate the hardware resources required for the implementation of the spatial localization system using the Semi-Dense SLAM method, the system was synthesized on the Zynq-7020 SoC platform. The programmable logic provides 85000 logical cells, 4.9 MB of available BRAM memory, and 220 DSP cores. Synthesis and verification of the system model were carried out by means of EDA software Vivado HLS [9].

The model was verified by comparing the results of the work of the synthesized model and the reference model written in C++. As the benchmark test actions, the real images from the video camera and the corresponding Keyframes formed during the test run of the software algorithm on the ROS operating system were used. Test actions were applied to the inputs of the synthesized system as well as to the reference model, which performed calculations in the C Co-simulation mode (в оригинале, только Co-simulation). The results of the calculations were written to the corresponding data sets and compared to an accuracy of 10^{-4}. The testing carried out for several frame sequences confirmed the correctness of the work of the synthesized model of the spatial localization system.

Table 1 shows the number of BRAM memory cells, DSP blocks, FF triggers, and LUT logical cells used in the synthesis of the computational pipeline and its main blocks IPE, (RCU +WCU), (JCU + HAU + PCU). The distribution of resources for the implementation of synthesizable constructions that were used in the high-level description of blocks in the C++ language (mathematical operations - Expression, buffers with sequential access - FIFO, sub-blocks Instance, obtained as a result of synthesis at a given hierarchy level, memory cells, - MUX, triggers - Register) is presented.

Table 1. Hardware resources for the pipeline implementation.

	Spatial Localization Block				JCU + HAU + PCU				IPE				RCU +WCU			
	BRAM	DSP	FF	LUT	BRAM	DSP	FF	LUT	BRAM	DSP	FF	LUT	BRAM	DSP	FF	LUT
Expression				2748				2205				313				230
FIFO		25	126	1046							126	1046				
Instance	4	44	13958	20613	2	14	5573	5514		26	2287	5224		30	5498	9875
Memory	128				2								128			
MUX								2271				605				2696
Register			10424	834			6069	320			2470	40				474
Total	132	69	24508	30813	4	14	11642	10310	0	26	5483	7228	128	30	1855	13275
Utilization	47%	31%	23%	57%	1%	6%	11%	19%	0%	11%	5%	13%	46%	14%	7383	25%

The spatial localization system receives new input data from the video camera on each clock cycle. The clock speed of the system $\boldsymbol{F_t}$ is 100 MHz. According to estimates made using the Vivado HLS EDA, the localization delay determined by the signal transit time along the most critical path of the computational pipeline $\boldsymbol{T_p}$ is 10,660 cycles (about 1 ms for a given $\boldsymbol{F_t}$ value).

5. Conclusion

The use of high-speed AXI Stream interfaces and hardware pipeline for the formation of SDE makes it possible to significantly improve the speed of performing the spatial localization of objects. According to the estimates, the localization delay associated with the formation of the CDS is reduced approximately 3-fold, compared to the solution described in [7]. The proposed architecture of the computational pipeline for determining the parameters of the CDS can be effectively used in the navigation systems of unmanned aerial vehicles, autonomous vehicles, as well as in the augmented reality systems.

References

[1] Engel J, Schops T and Cremers D 2014 LSD-SLAM: Large-scale direct monocular SLAM in *European Conference on Computer Vision (ECCV)*
[2] Newcombe R, Lovegrove S and Davison A 2011 *IEEE International Conference on. IEEE* pp 2320–2327
[3] Schops T, Engel J and Cremers D *Mixed and Augmented Reality (ISMAR), International Symposium on. IEEE* pp. 145–150
[4] More J 1978 *Numerical analysis* pp 105–116.
[5] Kerl C, Sturm J and Cremers D 2013 *Proc. of IROS* pp 2100-2106
[6] Marquardt D 1963 *Indst. Appl. Mat.* **11** 2 pp 431-441
[7] Boikos K and Bouganis C-S 2016 *International Conference on Field-Programmable Logic and Applications (FPL)*
[8] Xilinx 2012 Vivado Design Suite Tutorial *UG871 (v2012.2)*
[9] Xilinx 2015 ZC702 Evaluation Board for the Zynq-7000 XC7Z020 All Programmable SoC User Guide *UG850 (v1.5)*

IOP Conf. Series: Materials Science and Engineering **498** (2019) 012019 doi:10.1088/1757-899X/498/1/012019

Temperature influence on process of Ti/Al/Ni/Au contact formation to heterostructure AlGaN/GaN.

S A Shostachenko, Y A Porokhonko, R V Zakharchenko, S V Leshchev, M M Maslov and K P Katin

National Research Nuclear University MEPHI, Moscow, Russia

Corresponding author's e-mail address: SAShostachenko@mephi.ru

Abstract. This paper is dedicated to the experimental investigation of Ohmic contacts to the n+-doped region of the AlGaN/GaN transistor heterostructure based on Ti/Al/Ni/Au metallization. The Al-Ti-N system has been assessed on the basis of available thermodynamic descriptions for binary subsystems. The effect of annealing temperature on the specific resistance of Ohmic contact was studied.

1. Introduction

GaN-based transistor devices are widely used in defense and civil industries. This includes the use of the main component of the microwave monolithic integrated circuits (RF MIC) millimeter and submillimeter bands for communication systems, radar, radio astronomy, radiometry, etc. [1]. These materials have been applied moreover in the optoelectronic industry as the basis for LED crystals along with the use of wide bandgap GaN-based semiconductors as the basis of the transistors.

Multilayer metallization is mostly used in contact microelectronics technologies because of their better parameters (conductivity, adhesion to the surface, morphology). The multilayer contacts of Ti/Al that form compounds with a low work function are often used to form an ohmic contact to the n-type semiconductor [2-4]. For standard metallization is used gallium nitride-based Ti/Al/Ni/Au.

The hypothesis of the mechanisms of Ti/Al-based ohmic contact formation to the n-GaN is based on the appearance of the Ti_xN compound, which is a conductor. It begins to form during the deposition of titanium on the n-GaN surface and in the bulk of the titanium layer metallization during thermal annealing. Ti_xN formation temperature is in the range 200÷1000°C; because of this, titanium particles sprayed by various methods can form a thin layer of Ti_xN on the surface.

2. Al-Ti-N compounds

The analysis of the features of solid-state interaction processes in Al/Ti/GaN contacts during annealing on their parameters [5] showed that the contact is transformed into a structure consisting of phases that are the products of solid-phase interaction of titanium and aluminum. The presence of nitrides, intermetallics and ternary phases along with free titanium and aluminum makes it difficult to analyze the stability of Al/Ti/GaN contacts, requiring the detection of patterns of solid-phase reactions between contacting materials and changing the phase composition of these structures in the temperature range of contact formation and their operation.

In the ternary Al-Ti-N system, the following compounds can be formed: nitrides (AlN, TiN), aluminides (AlTi, Al_3Ti, $AlTi_3$, Al_3Ti_2, Al_2Ti). To analyze the stability of the two phases in the ternary system, it is necessary to take into account all possible chemical reactions. The phases do not interact with each other, i.e. they are thermodynamically stable, if in the considered temperature range for all the reactions ΔG> 0 and a connecting line can be made between these phases in ternary phase diagram.

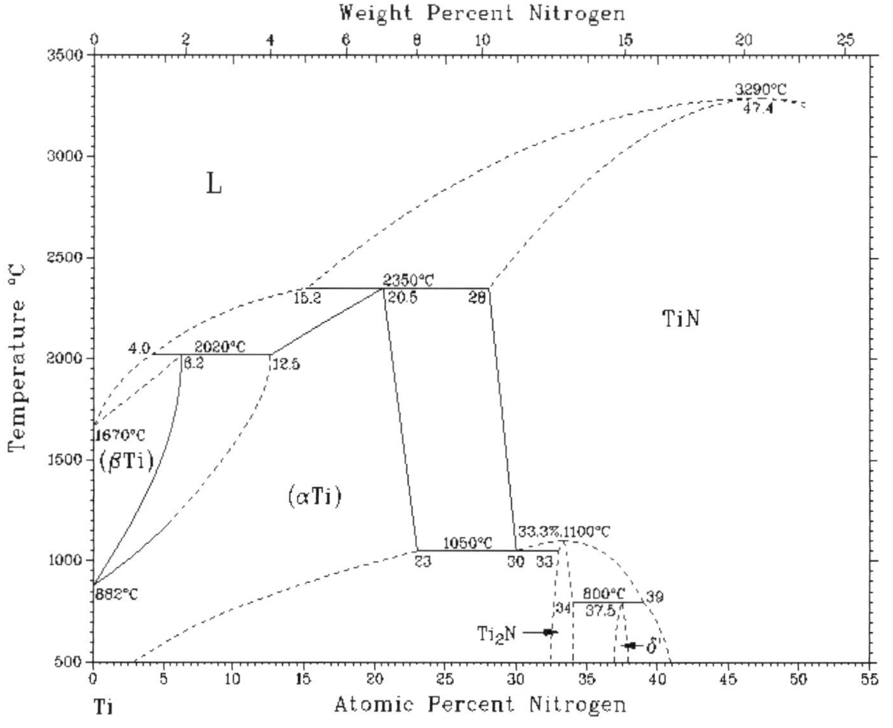

Figure 1. Titanium-Nitrogen phase diagram

According to the phase diagram (Figure 1) and [6, 7], at temperatures below 882°C, titanium has an α-Ti modification with a hexagonal close-packed lattice, at temperatures above 882°C – β-Ti with a cubic body-centered lattice. Because of this, it is necessary to consider separately the temperature ranges below and above 882°C for some compounds.

To analyze the stability of two phases in a ternary system, it is necessary to record all possible chemical interactions of these phases and calculate the temperature dependence of the change in the free Gibbs energy for each reaction. The phases do not interact with each other, i.e. are thermodynamically stable, if in the considered temperature range for all the reactions ΔG> 0, while on the ternary phase diagram, it is possible to draw a connecting line between these phases.

For example, in order to analyze the stability of titanium, aluminum, and nitrogen, it is necessary to write down the reaction equations for their interaction (Table 1).

Table 1. Gibbs free energy for reactions at different temperatures *T*.

Reaction	ΔG, KJ/(g·atom)				
	800°C	850°C	900°C	950°C	1000°C
1. Ti+N=TiN	-237	-231	-226	-222	-217
2. Al+Ti=AlTi	404	307	-64	-63	-62
3. 3Al+Ti=Al_3Ti	254	302	-128	-126	-124
4. 3Ti+Al=$AlTi_3$	-810	-889	-98	-98	-98
5. 2Ti+3Al=Al_3Ti_2	1649	1861	-117	-117	-116
6. Al+N=AlN	-206	-200	-194	-188	-183
7. Ti+2Al=Al_2Ti	121	141	-10	-0.8	-1.7

Thus, we can say that the most favorable temperature conditions for the formation of the better ohmic contacts are those up to 882°C, since free titanium goes to form a large number of different aluminides above this temperature. This reduces the possibility of the formation of titanium nitrides, which are necessary to obtain nitrogen vacancies, due to which the contact resistance decreases.

3. Experiment

We used AlGaN/GaN heterostructure. Ti/Al/Ni/Au (35/135/50/100 nm) multilayer metallization was deposited on it by resistive evaporation method with the use of the Kurt J. Lesker PVD 75 film deposition system. Annealing was performed using the rapid thermal annealing system Modular RTP-600S at a constant temperature for 30 s. Ohmic contact measurements were carried out on test structure for the transmission line method (TLM) [8]. To improve the accuracy of contact resistance measurements, we also used test structure [9]. Specific contact resistance was measured by the use of the Agilent B 1500 A semiconductor devices analyzer.

4. Results and Discussion

To investigate the effect of the annealing temperature on the properties of the contact, 8 samples were prepared. They were annealed at various temperatures in the range of 800-1000°C. The obtained values for the contact resistance are listed in Table 2. To explain the obtained results, the characteristics of Schottky barrier, which is formed in the contact, were determined. We chose the method of capacitance-voltage characteristics measurement to determine the height of the barrier φ_B. The total voltage U applied to the contact is expressed through capacity according to formula [10]

$$\varphi_B - U - \frac{kT}{q} = \frac{q\varepsilon\varepsilon_0 N_D S^2}{2C^2} \tag{1}$$

where S is area of the contact, q is elementary charge, k is Boltzmann constant, T is temperature, N_D is charge-carrier concentration, $\varepsilon = 8.9$ is permittivity of Gallium nitride, ε_0 is vacuum permittivity. Capacitance-voltage characteristics were measured for all samples in the voltage range of 0-10 V. In the first section of the capacitance-voltage characteristic ($\partial^2C/\partial U^2 > 0$), capacity of the first contact is charging. In the second section ($\partial^2C/\partial U^2 < 0$), the second capacitor is charging and capacity reaches the maximum. In this case, in the first section, the first Ohmic contact plays the role of a barrier and the second is Ohmic. Therefore, to measure the Shottky barrier in the rectifying contact, we should use the first section of the capacitance-voltage characteristic. Values of φ_B were determined for all samples by comparing the measured capacitance-voltage characteristic with the theoretical curve (3). Obtained data are listed in Table 2.

Table 2. Resistance ρ_k, transition resistance ρ_T, Schottky barriers height φ_B for contact samples, annealed at different temperatures *T*.

Sample number	*T*, °C	ρ_T, Ω·mm	φ_B, eV
1	-	-	1.21
2	800	0.35	1.04
3	850	0.32	0.918
4	880	0.07	0.903
5	885	0.28	0.95
6	900	0.67	1.05
7	950	0.77	1.08
8	1000	1.88	1.47

5. Conclusions

The formation of Ohmic contacts to the n+-doped region of AlGaN/GaN heterostructure based on Ti/Al/Ni/Au metallization through annealing was studied. The optimum temperature and annealing time are 880°C and 30 seconds, respectively. The specific resistance of the contact is 0.07 Ω·mm. The obtained resistivity values are lower than those obtained by other authors [11].

Acknowledgements

This study was carried out on equipment of the MEPhI Shared-Use Equipment Center «Heterostructural microwave electronics and wide band gap semiconductor physics» and financially supported by a Grant of the President of the Russian Federation, Grant No. MK-7410.2016.2.

References

[1] Luther P and Mohney S E 1997 *Appl. Phys. Lett.* **70** 57
[2] Kong X 2012 *J. Phys. D: Appl. Phys.* **45** 1
[3] Durbha, Pearton S J, Abernaty C R, Lee J W and Holloway P H 1996 *J. Vac. Sci. Tech B* **14** 2582
[4] Morkoc H, Strike S, Gao G B, Lin M E, Sverdlov B and Burns M 1994 *J. Appl. Phys* **76** 1363
[5] Kireyev V A 1975 *Methods of Practical Calculations in Thermodynamics of Chemical Reactions (in Russian)* (Moscow: Chemistry) p 536
[6] Samsonov G V and Vinitskiy I M 1976 *Refractory compounds (in Russian)* (Moscow: Metallurgy) p 560
[7] Hultgren R R 1963 *Selected values of thermodynamic properties of metals and alloys* (New York: Wiley–Interscience) p 963
[8] Reeves G and Harrison H B 1982 *IEEE Electr. Device Lett.* **3** 111
[9] Shostachenko S 2015 *Physics Procedia* **72** 419
[10] Sze S M and Kwok Ng K 2007 *Physics of Semiconductor Devices* (NJ, Hoboken: Wiley-Interscience) ed. 3 p 832
[11] Blank T and Gol'dberg Yu 2007 *Semiconductors* **47** 1263

The detection of nitro compounds by using MIS-sensor

A.V. Litvinov[1], N.N. Samotaev[1], M.O. Etrekova[1*] and A.A. Mikhailov[2]

[1] National Research Nuclear University MEPhI (MEPhI), Kashirskoe shosse 31, Moscow, 115409, Russian Federation
[2] Scientific production company "INKRAM", Lyublinskaya Street 151, office 222, Moscow, 109341, Russian Federation

Corresponding author's e-mail address: Etrekova@mail.ru

Abstract. It was shown the possibility of using MIS-sensors as detectors sensitive elements of the nitro-containing organic trace. The detection is carried out due to the nitro groups, formed during the reaction of pyrolysis of the nitro compounds molecules.

1. Introduction

As shown by years of international experience in the design of gas analysis systems [1, 2], the combination in a single device of high sensitivity, of simplicity of equipment, of low requirements for operator skill, of portability and of price acceptable for mass production and implementation in the security systems is a very difficult task. Therefore, the search for optimal solutions remains relevant.

A number of studies we have conducted [3, 4] show the possibility of using MIS-sensors as detectors sensitive elements of the vapor nitro compounds (for example, explosives) with the use of gas detection technology study, called thermo-redox. Its essence lays in the thermal decomposition (reaction of pyrolysis) of molecules nitro-containing explosives with subsequent detection of nitro groups (NO_2). Competitive advantages of this technology are the absence of radioactive source and carrier gas, as well as compact and easy maintenance. The disadvantage is the inability to distinguish the type of explosives and the difficulty of detection of substances with low vapor pressure such as RDX or PETN.

2. Experimental

The basis of the MIS-sensor is the structure (Fig. 1a), which is a MIS-capacitor consisting of the silicon plate (3), a layer of insulator SiO_2 (2) and palladium metal electrode of the shutter (1). Such a capacitor has the capacitance C. The sensor's temperature is maintained in the range of 100-150 °C with the help of miniature film heater (6) and stabilized due to electronic unit by using a thermistor (9). Standard dimension of the sensor in the housing is 1 cm^3 (Fig. 2).

For the manufacture of MIS-structures we have used laser technology for the deposition of thin films of metals and insulator, the essences of which consists in evaporation of the substance of the target with help focused laser beam with subsequent condensation of evaporated substances on a substrate. Used solid-state yttrium aluminum garnet laser with a radiation wavelength $\lambda = 1.06$ µm. The laser worked in the mode of modulated q-switched with pulse duration of 10 ns, pulse energy 0.1 Joule and a pulse frequency of 50 Hz.

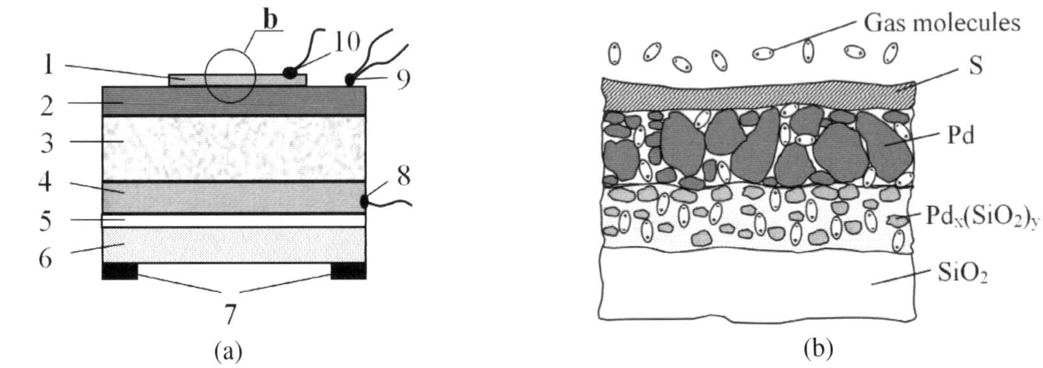

Figure 1. a) Scheme of MIS-sensor: 1 — film Pd; 2 — layer of SiO_2; 3 — silicon plate Si; 4 metal electrode; 5 — insulating plate; 6 — film heater; 7 — electrical connections to the heater; 8, 10 — the electric contacts of the MIS-sensor; 9 — thermistor.
b) Scheme of the nanostructure of the surface part of the MIS-sensor: S — surface's film (2 nm); Pd — thin film of palladium (20 nm); $Pd_x(SiO_2)_y$ — a transition layer with variable stoichiometric composition (10 nm); SiO_2 - layer of insulator

In Fig. 3 shows a typical C-U-characteristic of the MIS-sensor. Principle of operation the sensor consists in the following. During the deposition of molecules of the investigated gas on the surface of the paddle (Pd) C-U-characteristics shifted along the axis U (the dashed line). If we hold on the capacitor DC voltage, Udr, there is a capacitance change ΔC, which is converted by electronics into an analog (or digital) signal.

Figure 2. MIS-sensor's Photos

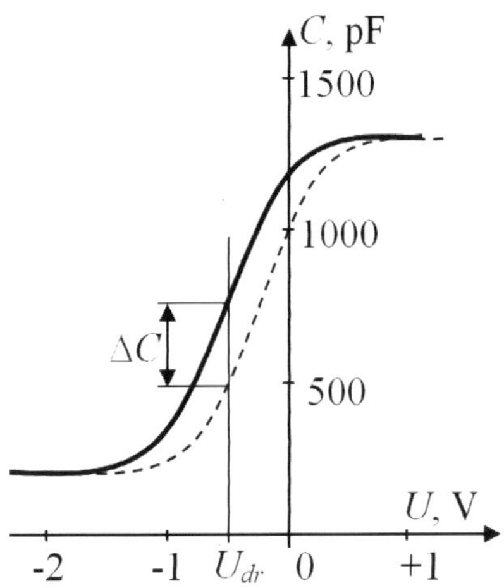

Figure 3. *C-U*-characteristics of the MIS-sensor. The dotted curve shows an example of the offset *C-U*-characteristics when exposed to the gas of interest. U_{dr} — voltage of "bias"

According to the works [5, 6, 7], the model of sensitivity of MIS-sensor is as follows. Due to the porous structure of the palladium film (Fig. 1b) of the gas molecules freely penetrates to the boundary of metal-insulator, where they create a layer of charge, which generating the changing electric field in the semiconductor. This leads to a shift of the C-U-characteristics of the sensor.

The experimental setup was as follows: the pump (outlet side of the gas path) ensured flow of atmospheric air through a reactor made of quartz glass, where the gas mixture is heated up to

200…500 °C required for pyrolysis of explosives molecules, and after cooling, was fed into the chamber with the MIS-sensor. Output signal (capacity MIS-sensor) is processed by electronics and displayed on a PC.

The MIS-sensor was calibrated using a graduated source micro flow NO_2 on a concentration of 0.25 ppm (0.46 mg/m^3) in air at the operating sensor's temperature of 150 °C. The results are shown in Fig. 4.

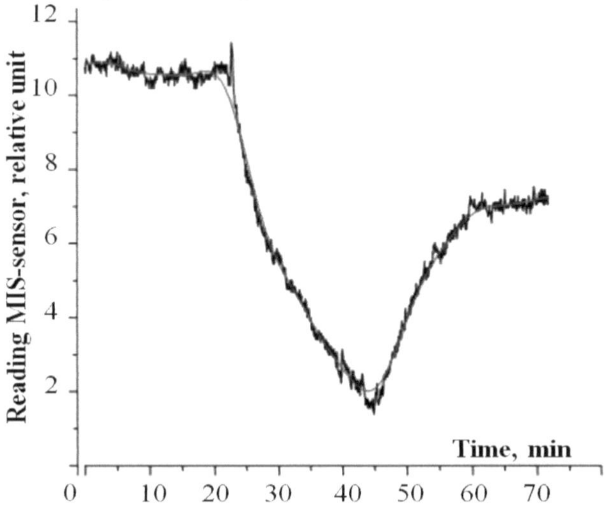

Figure 4. The results of the calibration of the MIS-sensor on NO_2 (0.46 mg/m^3 in the air)

In the first series of experiments on the experimental facility was tasked to determine the response of the sensor for heating of gas sample with help the reactor, without explosive vapors. It was also to check depend whether the sensor response on to known concentrations of NO_2 from heating temperature of the reactor. It was established that the stabilization of the readings of sensor after we enable heating of the reactor takes 30...50 min. In addition, it was shown that heating (in the temperature range 250...450 °C) gas mixture containing a known concentration of NO_2 has no effect (within the error of measurement of 15 %) on the magnitude response of the MIS-sensor mounted in the calibration. Thus, it was shown the ability to start main research using samples of explosives.

In this work, we studied the reaction of MIS-sensors on the trinitrotoluene (TNT). Samples of TNT were prepared in two versions:

Test No. 1 was a glass cylindrical container with a volume of 0.1 l, on the inner surface of which is uniformly distributed a solution of TNT in acetone of high purity, followed by removing acetone vapors;

Test No. 2 — test paper strips (2 x 5 cm) which was impregnated with a solution of TNT in acetone of high purity and carefully dried.

During the experiments with Tests No. 1 the reliable response (0.7 ± 0.3 Rel. units) managed to capture at the temperature of the gas mixture containing vapors of TNT, at least 420°C. To increase the effective area of the heating of the reactor, it was decided to optimize its design. For this the inlet nozzle of the reactor was filled with chemically pure heat-resistant glass fiber (~50% Al_2O_3, ~50% SiO_2). By the microscopic fibers in the composition of the glass molecules of TNT were holding in the gas circuit of the reactor a longer time, which increased the reaction efficiency of thermal destruction. The experiments conducted with the use of advanced reactor showed an increase the response of the MIS-sensor ceteris paribus to 1.5 ± 0.3 Rel. units. It should be noted that the containers of glass were heated to 30...55 °C in all experiments with the Tests No. 1, for increasing the vapor pressure of explosives. Indeed, according to [8], the vapors pressure of TNT at room temperature does not exceed 5...7 ppb (60 μg/m^3, or 60 pg of 1 cm^3). However, with increase in temperature for every 5°C, saturated vapor pressure of TNT has doubled [8].

During the experiments with Tests No. 2 (at the temperature of the reactor 400 °C and with application glass fiber - concentrator) was established the dependence of the response of the MIS-sensor from temperature of heating of test strips with the TNT microcrystals. The results are as follows: when heated to 30 °C, the response was 2.0 Rel. units; at 50°C - 6 Rel. units; at 70°C – 10 Rel. units (temperature measurement accuracy is ±2°C, the noise level of the signal ± 0,3 Rel. units) – Fig. 5. Note that the reaction of MIS-sensor on a clean test strips as well as the strips which were moistened with acetone high purity and subsequently were dried, were recorded.

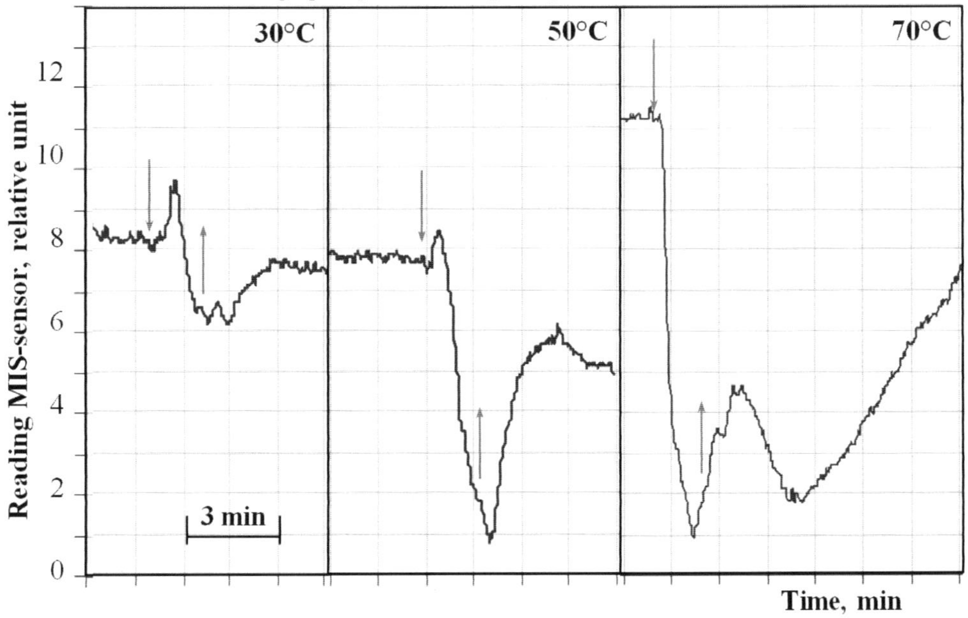

Figure 5. The dependence of the MIS-sensor's response on Tests No. 2 from the temperature of heating a sample of TNT. Arrows indicate the moments of filing and removing test strips

According to our estimates, the mass evaporated from the surface of the test strip of TNT when it is heated to 70 °C was about 1 μg.

3. Conclusions

Proved the possibility of using MIS-sensors as detector sensitive elements of the nitro-containing organic trace. Since MIS-sensors does not have sensitivity to the molecules of organic substances [9] that is why in the proposed method of the detection the sign of the presence of nitro compounds is the gas NO_2 which is emitted during the reaction of pyrolysis of the molecules of explosives.

4. References

[1] David J and Lewis A M 2008EUR 23023 EN
[2] Applications. National Institute of Justice 1999
[3] Nikolaev I N, Litvinov A V and Emelin E V 2007 Sensors and systems 5 p 66 (In Russian)
[4] Nikolaev I N and Emelin E V 2004 The measuring technique 11 p 54 (In Russian)
[5] Emelin E V and Nikolaev I N The measuring technique 2006 **5** p 68 (In Russian)
[6] Emelin E V, Nikolaev I N and Sokolov A V Sensors and systems 2005 **10** p 37 (In Russian)
[7] Nikolaev I N, Litvinov A V and Emelin E V Sensors and systems 2006 **7** p 66 (In Russian)
[8] Qxley J C., Smith J L, Luo W and Brady J 2009 Propellants Explosives Pyrotechnics **34**(6) pp.539 - 543
[9] Nikolaev I N and Onishenko P O 2006 Sensors and systems **3** p 34 (In Russian)

IOP Conf. Series: Materials Science and Engineering **498** (2019) 012021 doi:10.1088/1757-899X/498/1/012021

Enhanced performance of a dye sensitized solar cell using metallic and bi-metallic nanoparticles

K Ranabhat[1], K S Skripkin[1], K F Chernyshova[2], E A Sofronova[1], A A Revina[2], A I Pylinina[1], L N Patrikeev[1, 3] and V A Lapshinsky[1, 3]

[1]Peoples' Friendship University of Russia, Miklukho-Maklaya str. 6, Moscow, 117198, RussianFederation
[2]A.N. Frumkin Institute of Physical Chemistry and Electrochemistry of the Russian Academy of Sciences, Leninsky prospect 31, bld. 4, Moscow, 119071, Russian Federation
[3]National Research Nuclear University MEPhI, Kashirskoe shosse 31, Moscow, 115409, Russian Federation

Corresponding author e-mail address: Ranabhat.kiran@gmail.com

Abstract. Dye-sensitized solar cells (DSSCs) have emerged as a promising source to meet the energy demand as an environmentally friendly alternative. The work presented in this paper focuses on improving the efficiency of DSSCs using natural dyes and metallic and bi-metallic nanoparticles. The solar cell's performance in terms of efficiency (voltage vs. current) was tested under an indoor room light and fluorescent lamp. The experimental results demonstrate that cell with the dye from blue honeysuckle berries and Au nanoparticles as cathode produced greatest efficiency (η~9 %) under fluorescent lamp.

1. Introduction

Solar energy sources are considered one of the most promising renewable sources [1-6], since solar energy sources are clean, inexhaustible, environment friendly, plentiful and easy to utilize. Solar is considered to be profitable and safer investment for our family and business. There are various kinds of solar cells. Amongst solar photovoltaic, DSSCs are an efficient and affordable type of solar cells.

The DSSC was first reported in 1972 as a chlorophyll-sensitized zinc oxide (ZnO) electrode. Dye-sensitized solar cell (DSSC), invented by Professor M. Grätzel in 1991[7], is a most promising inexpensive route toward sunlight harvesting, owing to their simple structure, transparency, flexibility, low production cost, and wide range of application. Despite these advantages, the low efficiency of DSSCs compared to that of silicon-based cells has limited their commercial implementation [3-4].

DSSC are composed of only five components: a transparent conductive oxide (TCO) substrate, a nanostructured n-type semiconductor, a visible-light absorber dye, an electrolyte and a counter electrode (cathode) [5].

This research study focused on improving the methodology of producing dye-sensitized solar cells by the process of amalgamation with natural dyes and metallic and bimetallic nanoparticles. This research is done for maximizing cell efficiency and sustainable power production for mass manufacturing processes at the most economical cost.

2. Hypothesis

2.1. Application of metallic and bi-metallic nanoparticles in DSSC.
Observing the optical and electrochemical properties of metallic and bi-metallic nanoparticles [2], we decided to use metallic and bi-metallic nanoparticles in DSSCs manufacturing as an additive to different natural dyes and as cathode. It is assumed that use of the metallic and bi-metallic nanoparticles increase the efficiency of DSSCs.

3. Experimental Section

3.1. Dye Extraction Process
Preparation of the natural dye solution requires some skill of filtration. First around 50gm of samples (black grapes, pomegranate or blue honeysuckle berries) were grinded in a mortar until a crude mixture was observed, usually about 4-5 min. The crude mixture was then transferred slowly to a pre-assembled filtration apparatus. If the resulting mixture is very thick, i.e., no filtrate, several drops of water or acetone was used to wash the mixture and add the total liquid amount. A spatula or a glass rod can be used to swirl the mixture to give off more solution.

3.2. Preparation of nanoparticles
Metallic nanoparticles (NPs) Au, Ag and bimetallic NPs Au/Ag are synthesized by the method of "molecular assembly" using chemical (Chem) reduction reactions of Ag^+ ions and subsequently formation of metallic nanoparticles in reverse micelles in the presence of a natural pigment from the class of flavonoids - quercetin (QR) and molecular oxygen.

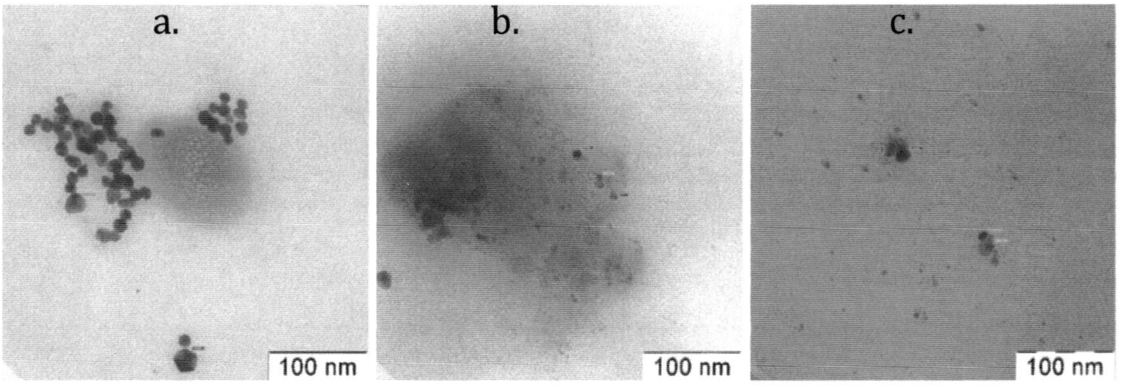

Figure 1. TEM microphotographs of metallic NPs: Au a); NPs Ag b) and bimetallic NPs Au-Ag c)

The dimensions of the nanoparticles were determined using transmission electron microscopy LEO 912AB OMEGA. The size of metallic nanoparticles (NPs) Au, Ag and bimetallic NPs Au/Ag used in this work is 1.5-20 nm, 6-12 nm and 10-15nm respectively (Figure 1).

3.3. Assembling the Dye-Sensitized Solar Cell
1. Determine the conductive side of glass by touching both of protruding leads of the multi-meter with one side of the glass as shown in Fig. 2.a. The conductive side could be identified with average resistance from 20-38 ohms.
2. Fix two sides of the plate using tape with the conductive sides facing up shown by Fig. 2.a.
3. Prepare the TiO_2 paste by adding a few drops of dilute acetic acid (0.035M) to 1 gram (about 2 table spoons) TiO_2. The resulting mixture was grinded in a mortar and pestle until a colloidal suspension with a smooth consistency (like cake icing) was observed [8].
4. Add 2-3 drops of the TiO_2 suspension onto the conductive side and spread out the TiO_2 evenly on the surface of the plate with glass rod. Leave it for 4-5 minutes in room temperature and carefully remove

the tape without perturbing the TiO_2 layer.
5. Dry the glass with TiO_2 under room temperature over 20 minutes and then heat it at 420° C for another 30 min, until the dried TiO_2 turns brown and then white again Fig.2. b-c.
6. While heating, coat the conductive side of the other piece of glass with graphite as cathode or 2-3 drops of nanoparticles and spread out evenly on the surface of plate with glass rod as cathode Fig. 2. d.
7. Cool both plates to room temperature.
8. Submerge the plate with TiO_2 face down in the dye solution and keep it for 24 hours Fig. 2.e.
9. Wash the dye layer with ethanol carefully.
10. Clamp both the plates together and apply 2 drops of standard iodine ($I^-/I3^-$) electrolyte solution onto the interface between the two plates. Allow the electrolyte to cover the surface of TiO_2 by capillary action.
11. Measure the photo-voltage and photocurrent under the indoor room light and fluorescent light using a laboratory multimeter Fig. 2. h-i.

Figure 2. Assembling of Dye-Sensitized Solar Cell

4. Results and Discussion

In this section, results from each experiment mentioned in the experimental section are presented and analyzed. All of the results were measured under standard condition (T = 298K and P= 1 atm) with the illumination of indoor light and fluorescent light.

4.1. Regular Voltage & Current Data

So the measurement of the power (voltage and current) generated by the dye-sensitized solar cells was carried out under two different conditions: Indoor light and fluorescent lamp. The main aim was to investigate the highest combination of voltage and current produced by solar cells. These solar cells were made with dye extract from different fruits samples and metallic & bimetallic nanoparticles were used as additive to dye and as a cathode. Each sample's voltage and current were averaged when tested under each light source for 3 minutes. The result obtained for solar cells with various dyes used, the various dyes with NPs additive and those obtained for the various dyes with NPs as cathode are presented in the Figure 3.

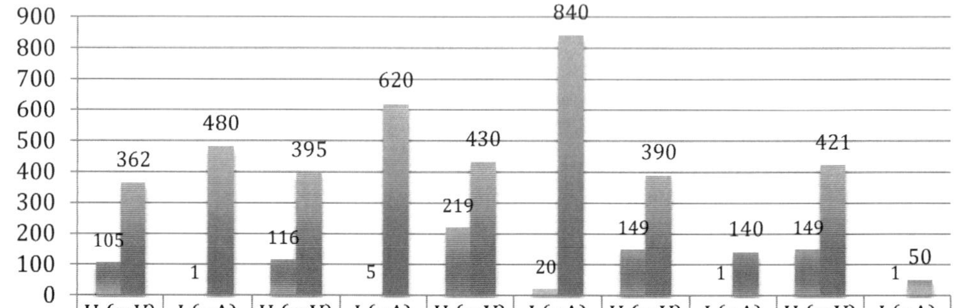

	Black Grapes		Pomogaranate		Blue honeysuckle berries		Black Grapes + NPs Ag additive		Black Grapes + NPs Au,Ag additive	
	U (mV)	I (μA)	U (mV)	I (μA)	U (mV)	I (μA)	U (mV)	I (μA)	U (mV)	I (μA)
Room light	105	1	116	5	219	20	149	1	149	1
Lamp	362	480	395	620	430	840	390	140	421	50

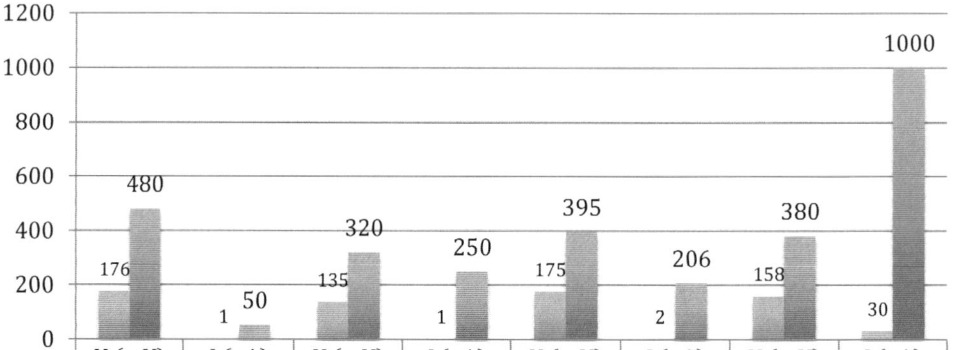

	Blue honeysuckle berries + NPs Au additive		Black Grapes + NPs Ag cathode		Pomogaranate + NPs Ag as cathode		Blue honeysuckle berries + NPs Au as cathode	
	U (mV)	I (μA)	U (mV)	I (μA)	U (mV)	I (μA)	U (mV)	I (μA)
Room light	176	1	135	1	175	2	158	30
Lamp	480	50	320	250	395	206	380	1000

Figure 3. Results of regular photo current-voltage production for DSSC

The current and voltage data of each solar cell was tested 5 times and measurements were carried out under indoor light and fluorescent lamp.

The voltage produced from the cell with blue honeysuckle berries + NPs Au additive 480 mV and the current from the cell with blue honeysuckle berries + NPs as cathode 1000 μA are maximum.

4.2. DSSC Power and Efficiency Formula and Calculations

The maximum efficiency of a solar photovoltaic cell is given by the following equation:

$$\eta_{max}(maximum\ efficiency) = \frac{P_{max}(maximum\ power\ output)}{E\ (Incident\ radiation\ flux) * A_c(Area\ of\ collector)}$$

The intensity (power) of the incoming light from the fluorescent lamp has been calculated from the outset to be 21 W/m^2. The area of the conductive glass slide coated with TiO_2 is approximately 1cm x 2cm = 2cm^2, i.e. equals to 0.0002 m^2.

The efficiency of all the solar cells were measured using the formula above and the result is tabulated in Table 1. The comparison of highest power per surface area and efficiency is shown in the Figure 4. The maximum power output and maximum efficiency was from cells with blue honeysuckle berries + NPs additive about 1.9 W and 9.05% respectively.

Table 1. Results obtained for the power output and conversion efficiency for the solar cell exposed to a fluorescent lamp.

	I (A)	U (V)	SA (m^2)	P cell (I*V)/SA	Efficiency (η)
Black Grapes	0.0005	0.3620	0.0002	0.8688	4.1371
Pomegranate	0.0006	0.3950	0.0002	1.2245	5.8310
Blue honeysuckle berries	0.0008	0.4300	0.0002	1.8060	8.6000
Black Grapes + NPs Ag additive	0.0001	0.3900	0.0002	0.2730	1.3000
Black Grapes + NPs Au, Ag additive	0.0005	0.4210	0.0002	1.0525	5.0119
Blue honeysuckle berries + NPs additive	0.0001	0.4800	0.0002	0.1200	0.5714
Black Grapes + NPs Ag (cathode)	0.0003	0.3200	0.0002	0.4000	1.9048
Pomegranate + NPs Ag (cathode)	0.0002	0.3950	0.0002	0.4069	1.9374
Blue honeysuckle berries +NPs Ag (cathode)	0.0010	0.3800	0.0002	1.9000	9.0476

Our hypothesis and experiments was that to use metallic and bi-metallic nanoparticles in DSSCs manufacturing and to know how the metallic and bi-metallic nanoparticles affect the efficiency of a dye-sensitized solar cell. The result we obtained shows that using metallic and bi-metallic nanoparticles increases the efficiency of a dye-sensitized solar cell. More studies will be carried out to investigate the best method to improve the cell efficiency and performance.

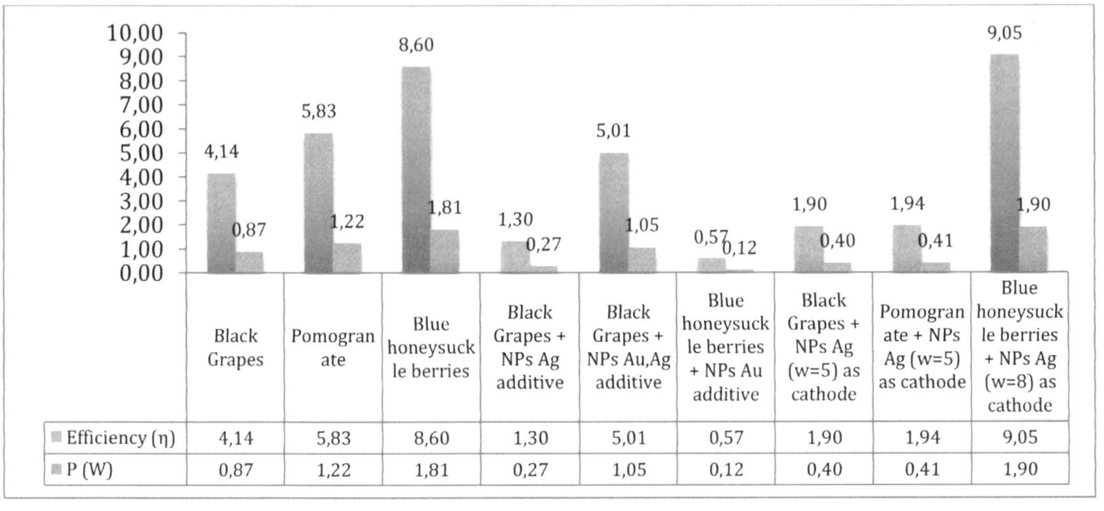

Figure 4. Results obtained for the power output and conversion efficiency for DSSC exposed to a fluorescent light.

5. Conclusion

DSSCs are broadly regarded as the more promising third generation photovoltaic (solar) technology. These cells are the closest mankind has come to replicating nature's photosynthesis. The work presented in this paper aimed at increasing the efficiency of DSSCs using different natural dyes along with metallic and bimetallic nanoparticles. The data obtained from this research concludes that using metallic and bimetallic nanoparticles as an additive to dye and as cathode increases the efficiency of dye-sensitized solar cells.

Acknowledgements
This paper was done in the project No. 011615-0-000.

References

[1] Parkhomenko Y N and Polisan A A *2014 Physics and technology of Photonics devices: Solar power engineering and nanotechnology* (Moscow: National University of Science and Technology "MISIS")
[2] Ranabhat K, Pylinina A I, Skripkin K S, Sofronova E A, Revina A A, Kasatkin V E, Patrikeev L N and Lapshinsky V A 2016 *IOP Conference Series: Materials Science and Engineering* **151**
[3] O'Regan B and Grätzel M 1991 *Nature* **353** pp 737–740
[4] Lee B and Kim J 2009 *Curr. Appl. Phys.* **9** pp 404–408
[5] Bella F, Gerbaldi C, Barolo C and Gratzel M 2015 *Chemical Society Reviews* **44**
[6] Winans J D, Hungerford C, Shome K, Rothberg L J and Fauchet P M 2015 *Opt. Express* **23**
[7] O'Regan B and Graetzel M 1991 *Nature* **353** pp 737-740
[8] Gratzel M 2005 *Inorg. Chem.* **44** pp 6841-6851
[9] Chernyshova K F and Revina A A 2017 *High-tech nanotechnology* **18** pp 45-50 (Russian)

IOP Conf. Series: Materials Science and Engineering **498** (2019) 012022 doi:10.1088/1757-899X/498/1/012022

Study and production of thin-film memristors based on TiO_2 – TiO_x layers

E V Zhidik [1,2], P E Troyan[1], Y V Sakharov[1], Y S Zhidik[1] and D V Korzhenko[2]

[1] Tomsk State University of Control Systems and Radio-Electronics, pr. Lenina, 40, Tomsk, 634050, Russia
[2] Institute of Physics and Technology, National Research Tomsk Polytechnic University, pr. Lenina, 30, Tomsk, 634050, Russia

Corresponding author's e-mail address: baturina_e.v@mail.ru

Abstract. Results of production of thin-film memristor MIM-structures based on the stoichiometric (TiO_2) and nonstoichiometric (TiO_x) titanium oxides and contacts without precious and rare-earth metals are given. It is shown that such memristor structures without precious metals show its operability only after the process of electrical forming.

1. Introduction

In the sphere of information technology, one of the prior directions is creation of new generation computers, neuronets and artificial intelligence [1]. Resistive random-access memory (RRAM) with a longer time of data storage and higher data density, better in more than dozens of times than any existing type of memory, is considered as the best candidate for computer memory of the future generation.

The structure of switching and memory elements in RRAM represents the system of metal-insulator-metal and the characteristics of this system depend on the composition and structure of the insulating layer. Such structures were predicted in 1971 by L.O. Chua [2] and were called memristors, due to that two stable switching states, high resistance or high conductivity, can be realized in them. The practical realization of memristor-type memory was produced in 2008 by Hewlett-Packard Company [2].

In the Russian Federation, studies concerned with the creation of memristors are conducted in National Research Center "Kurchatov Institute", Southern Federal University, Voronezh State Technical University, National Research Nuclear University MEPhI, as well as National Research University of Electronic Technology MIET.

So, numerous studies of memristors conducted by Hrapovitskaya et al. [4] showed that properties of such elements are determined by their architecture and materials. Eliseev [5] described the method of the production of memristors based on titanium dioxide (TiO_2). Hrapovitskaya et al. studied the insulating layers of TiO_2 – TiO_x. In the paper [6], the insulator was made as a layer of $Ti_{0.85}$–$Al_{0.15}$–O_y and in the paper from Hewlett-Packard Company [7], the insulator was made of 4 kinds of materials: TaO_x–TiO_{2-x}; TaO_x; TiO_{2-x}–TiO_{2-x}; and TiO_{2-x}–TaO_x–TiO_{2-x} with platinum electrodes.

The mentioned studies show that the development of memristors follows the line of the most effective architecture. Besides, the majority of the studies are concerned with the application of the TiO_2 thin films or TiO_2–TiO_x structure, where TiO_x is non-stoichiometric titanium dioxide. At the same time, studies concerned with the impact of the electrode material on the properties of the memristor are very rare.

IOP Conf. Series: Materials Science and Engineering **498** (2019) 012022 doi:10.1088/1757-899X/498/1/012022

This paper is concerned with the impact of the material of the memristor electrode on the characteristics of the memristor.

2. Results and discussion

The experimental memristor Mo–TiO_x–TiO_2–Cu structure was formed on the surface of a coaxial glass-metal kern. The central molybdenum electrode of the Kern was one of the electrodes of the structure (Figure 1).

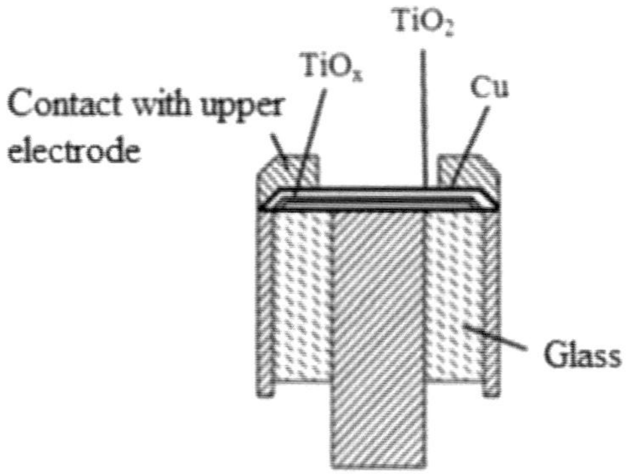

Figure 1. Schematic picture of memristor structures on the kern-type substrates [3].

TiO_2 thin films with the thickness of 50 nm were obtained by means of magnetron sputtering of titanium target in the gas mixture of Ar (65 %) and O_2 (35 %). At the pressure of $7 \cdot 10^{-3}$, the Torr discharge current was 300 mA and the rate of the thin films growth was 16 nm/min. TiO_x thin films were obtained by means of magnetron sputtering of the titanium target in the presence of Ar with O_2 (10%) with sputtering process conditions identical to TiO_2. The rate of growth of TiO_x thin films was approximately 24 nm per minute.

X-ray structural microanalysis was made to control structure of insulating films. It showed that titanium in the obtained thin films had oxidation levels close to that of TiO_2 and Ti_2O_3, which was essential for production of memristor structures.

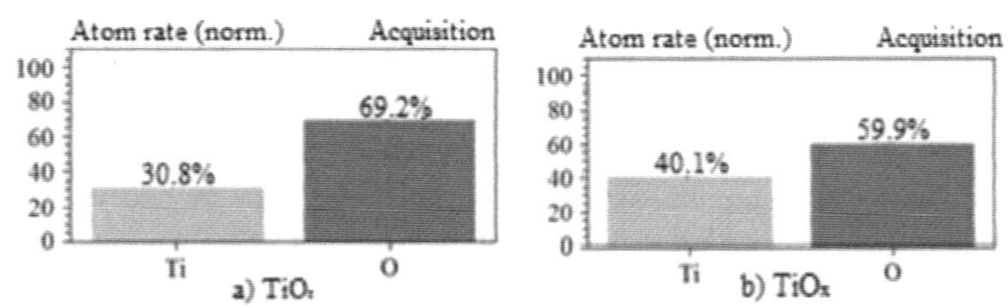

Figure 2. Structure of obtained insulating films.

Films of the upper electrode were obtained by means of thermal evaporation of Cu in vacuum with the pressure of $4 \cdot 10^{-6}$ Torr. The thickness of the film of the upper electrode was 30 nm. The upper electrode was covered by a thick layer of Al outside the center of the kern (Figure 1) for the better electrical contact.

Volt-ampere characteristics were measured in the prepared Mo–TiO_x–TiO_2–Cu structures in normal and reversed polarity at atmospheric pressure. Figure 2 shows the averaged volt-ampere characteristic of Mo–TiO_x–TiO_2–Cu structures.

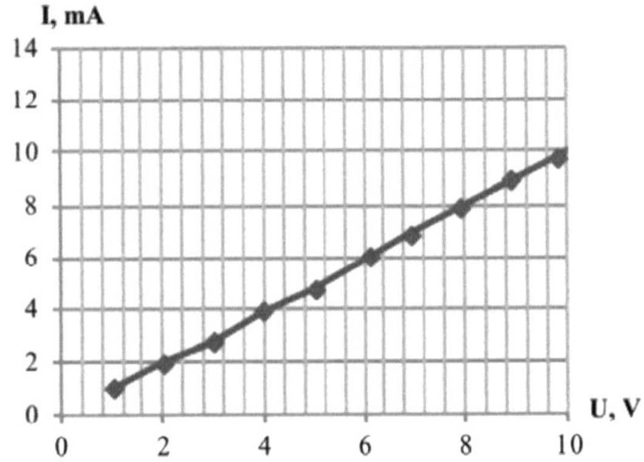

Figure 3. Volt-ampere characteristic of Mo–TiO_x–TiO_2–Cu structure at the atmospheric pressure.

The measured volt-ampere characteristic monotonically increases without a region of negative differential resistance, which means that there are no switching states of low-resistance and high-resistance. The absence of the region of negative differential resistance can be explained by the insufficient work function of electrons from Cu, which is approximately 4.34 eV. At the same time, the work function for materials used for the production of memristors such as TiO_2, Au, and Pt is 4.7 eV, 5.1 eV and 5.32 eV, respectively.

The electrical forming process of the Mo–TiO_x–TiO_2–Cu structure, which was described in detail in [10], allows the compensation of the disadvantage mentioned above.

The samples with Mo–TiO_x–TiO_2–Cu structure were placed in the vacuum at the pressure of 10^{-4} Torr. Electrical forming was conducted by a bias supply of 10 V, the exposition time of 10 minutes at the reversed polarity. Changes in the volt-ampere characteristic after the process of electrical forming are shown in Figure. 4

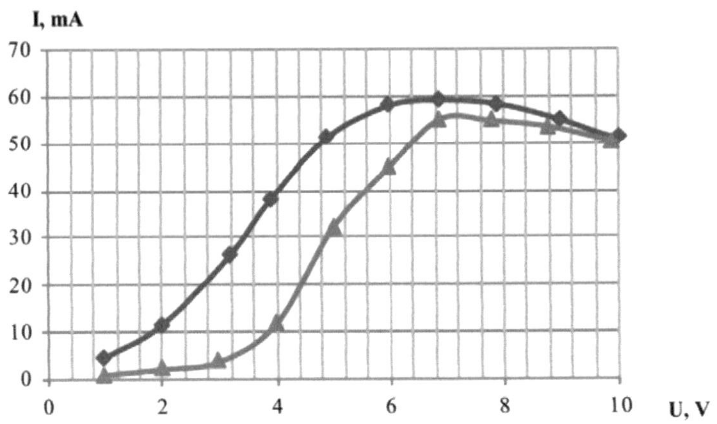

Figure 4. Volt-ampere characteristics of states with different conductivity:
-low-resistance state;
-high-resistance state.

In order to detect the hysteresis of the structure, the voltage was gradually increased on the structure until the value of 10 V. After that, the voltage was gradually decreased, which corresponds to the high-resistance state.

IOP Conf. Series: Materials Science and Engineering **498** (2019) 012022 doi:10.1088/1757-899X/498/1/012022

The measured volt-ampere characteristics point to that initially, the Mo–TiO_x–TiO_2–Cu structure does not show switching properties of memristors, but after the electrical forming of the obtained structures in vacuum its volt-ampere characteristics reveal the region of negative differential resistance.

3. Conclusion

The conducted study of memristor structures shows that structures without precious metals (such as the Mo–TiO_x–TiO_2–Cu structure) are operable only after the electrical forming in vacuum. During the electrical forming, the irreversible changes occur in the MIM-structure, which lead to the appearance of the regions with negative differential resistance in the volt-ampere characteristic.

References

[1] Shchigorev L A and Shagurin I I 2016 *IOP Conf. Ser.: Mater. Sci. Eng.* **151** 1
[2] Chua L O 1971 *IEEE Trans. Circuit Theory* **18** 507
[3] Makrushin M V 2011 *Electronic industry* **2** 62
[4] Hrapovitskaya J V, Maslova N E, Grishenko J V, Demin V A and Zanaveskin M L 2014 *Technical Physics Letters* **40** 87
[5] Eliseev N 2010 *Electronics: science, technology, business* **8** 84
[6] Bobilev A N and Udovichenko S Y 2015 *Reports of TUSUR* **38** 68
[7] Yang J J, Zhang M X, Pickett M D, Miao F, Strachan J P, Li Wen-Di, Yi W, Douglas A A Ohlberg, Choi B J, Wu W, Nickel J H, Ribeiro G M, and Williams R S 2012 *American Institute of Physics* **100** 1
[8] Sakharov Y V, Troyan P E and Zhidik Y S 2015 *Reports of TUSUR* **38** 72
[9] Sidelev D V, Yurjev Y N, Krivobokov V P,. Erofeev E V, Penkova O V and Novikov V A 2016 *Vacuum* **134** 29
[10] Danilina T I and Troyan P E 2010 *Russian Physics Journal* **53** 331

IOP Conf. Series: Materials Science and Engineering **498** (2019) 012023 doi:10.1088/1757-899X/498/1/012023

Gesture recognition system based on Convolutional neural networks

I S Chistyakov* and E V Chepin

National Research Nuclear University MEPhI (Moscow Engineering Physics Institute), Kashirskoe shosse 31, 115409 Moscow, Russian Federation

*Corresponding author's e-mail address: chist-34dml@mail.ru

Abstract. This article tells about gesture recognition system based on Convolutional neural nets. System consists of two parts: the tracking and detection subsystem and gesture recognition subsystem.

1. Introduction.

At the moment, the research of interaction between human and technical devices has received growing attention by the development technologies. Development natural ways for human and machine is actual task. It makes communication easier and more effective and expands the area of problems to solved the computer.

We have several tools for interaction: voice, facial expressions, gestures etc. One of management variants of robotic objects is static and dynamic hand gestures.

Static gestures is some position that can be fixed (Fig. 1).

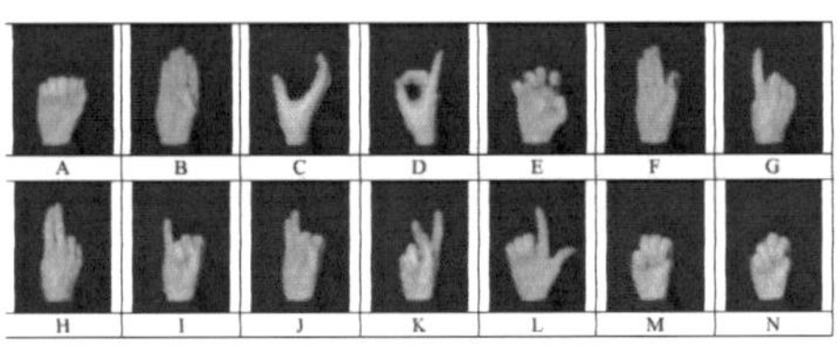

Figure 1. Set of static gestures

Dynamic gesture is gesture that performs for period of time (Fig. 2)

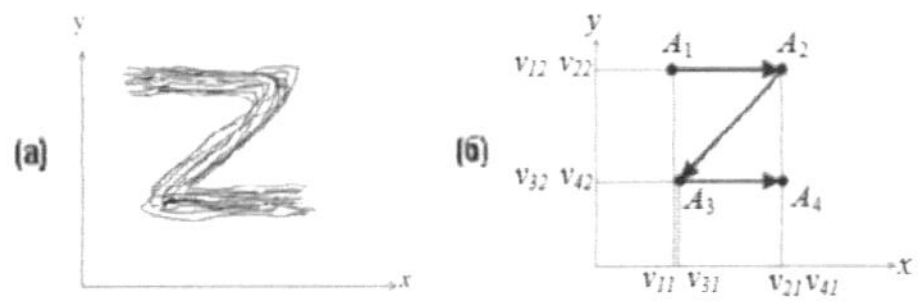

Figure 2. The sample of dynamic gesture

IOP Conf. Series: Materials Science and Engineering **498** (2019) 012023 doi:10.1088/1757-899X/498/1/012023

Gesture - ccontinuous action, so we can't to determine what kind of gesture was performed basing only on current object position. We must store passed object positions and check this array.

Quality system requires reliable performance a several subsystems:

Tracking and detection subsystem. It finds interest object in video stream, watches and sends coordinates to gesture recognition subsystem using TCP/IP protocol.

Gesture recognition subsystem. It receives coordinates, analizes them and notifies user about completed gestures.

2. Analogs

Leap Motion – gesture control system – ***Leap 3D***. This technology based on motion. Architecture of Leap Motion is not very complex. It has two infrared single-action and three powerful infrared light-to diode (Fig. 3).

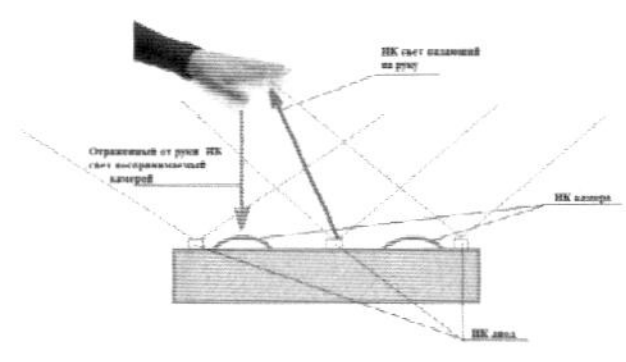

Figure 3. Architecture of Leap Motion

Principle of work is following: infrared diodes illuminate hands, and infrared cameras make hands detection sending data to programming handler of Leap Motion. Mathematical algorithms working on program level allocate contours of hands and detect finger coordinates [6].

Another variant system is Kinect – motion controller from Microsoft. Kinect consists of two deep sensors, multicolor camera and microphone lattice (Fig. 4)

Software makes full 3d-dimension scanning motions of body, facial expressions and voice. Deep sensor consists of infrared projector with monochromatic matrix that allows to get 3d-dimension picture in different lightings.

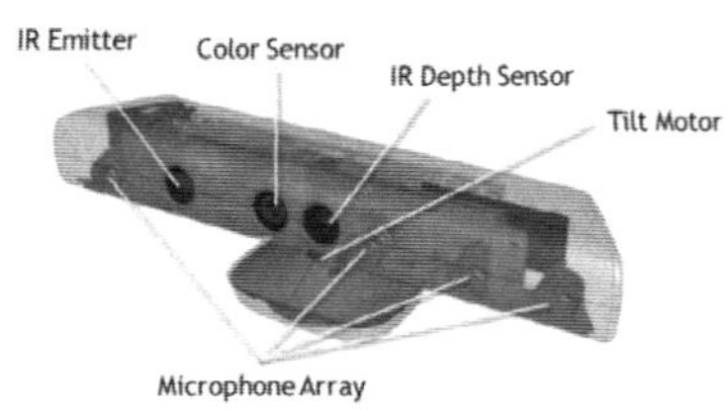

Figure 4. Architecture of Kinect

Computing the human position is two-part process. Firstly, it calculates deep map with help infrared emitter and receiver (deep sensors). After this operation Kinect computes human position (with machine learning algorithms) [7].

Can be useful another variant neural network model of artificial intelligence for handwriting recognition. This neural networks are used for example for handwriting recognition [8] and forensic handwriting examination [9].

3. System Architecture

System consists from several related subsystems. We consider these components separately and together.

3.1. Detection system

We decide to use convolutional neural networks to solve the problem of pattern recognition, since at the moment they are superior in all parameters to other methods of solving this problem. But before you recognize a specific image, you need to find its location on the source image and translate it into a data format that is fed to the input of the convolutional network. With this in mind, the final algorithm looks like this:

1. Find the location of images for recognition on the source image
2. Convert found images to the input format of a convolutional network
3. Run the image recognition and wait for the result

We use the Connected-component labeling algorithm in the version with two full passes through the image to find image locations for recognition. The two-pass version is based on the SNM data structure (the System of Disjoint Sets).

Connected-component labeling works only on a monochrome image, so before you use it, the binarization task of the image arises. Given that the program will work on mobile devices, it was decided to use adaptive binarization, since often the lighting conditions were different in the source image. Before applying binarization, you need to rid the image of random noise. This goal is performed by a Gaussian Filter with a 3-core kernel, which is applied to the source image before binarization. After binarization, a Closing Operator with a core size of 3 elliptic type is also used to remove small tears in the images for recognition.

The single-channel image 28x28 is fed to the input of the convolutional network. But if you bring all the images to this size, then the recognition accuracy will drop, since the neural network was trained on images where the content does not occupy the entire area. Therefore, the found areas for recognition change the size by 24x24 with all the proportions preserved, and then frames with a width of 2 empty pixels on each side are added.

The architecture of the convolutional network was chosen taking into account the computing power of mobile devices and the maximum accuracy available on them. As a result, the following proven architecture was chosen:

1) Convolution layer: kernel size: 5x5, input channels: 1, output channels: 32
2) The maximum pooling layer: kernel size: 2x2, step 2x2
3) Convolution layer: kernel size: 5x5, input channels: 32, output channels: 64
4) The maximum pooling layer: kernel size: 2x2, step 2x2
5) Fully-knit layer: input channels: 64, output channels: 1024
6) Fully-knit layer: input channels: 1024, output channels: 10
7) Softmax filter

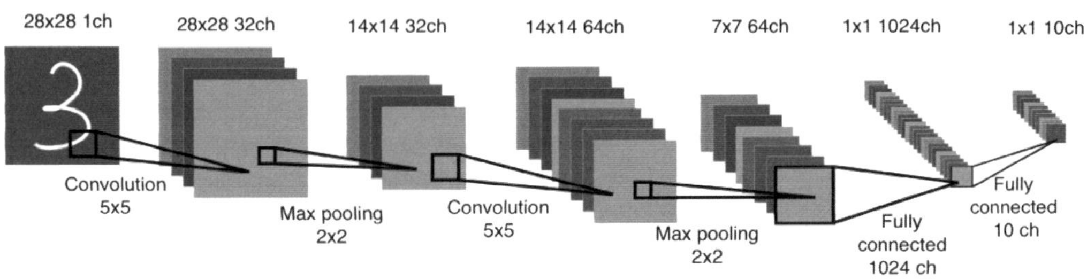

Figure 5. The architecture of the convolutional network

3.2 Gesture recognition system

At the first version this subsystem data was analyzed via analytical methods.

IOP Conf. Series: Materials Science and Engineering **498** (2019) 012023 doi:10.1088/1757-899X/498/1/012023

Later, we decided that the best solution is using neural nets for analyzing trajectories of object.

3.2.1 Analitic algorithms of first system version

We determine gesture such as set of regions that must be visited object for gesture accomplishment.

In the simplest case this regions are circles with particular coordinates and radius in 2-dimension area (in 3D it will be sphere).

Gesture - ccontinuous action, so we can't to determine what kind of gesture was performed basing only on current object position. We must store passed object positions and check this array.

Classical algorithm is a simplest algorithm for gesture detection. Object must visit all regions of gesture in particular order.

Algorithm is very simple for realization but it very much depends from coordinates of regions. We can solve this problem with normalization of trajectory and gesture in library of gestures..

Scanning algorithm - algorithm for gesture detection based on comparing normalizing trajectories with etalon sequence of positions.

Let, solving task is input command to computer with dynamic gestures of hand. We choose 12 different action of hand as etalons (Fig. 6).

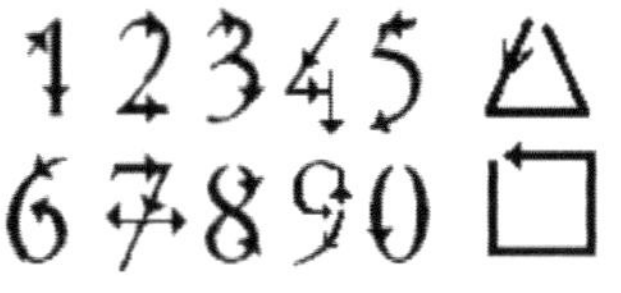

Figure 6. Selected etalon gestures

In case dynamic gestures the task of dentification becomes more complicated in view of indefinite start and end of gesture.

For solving this problem we can analyze multiple trajectories duration from 0.7 to 2.0 seconds.

When system detects probable start and end of gesture we will make next operation:

1. Creating etalon gestures
2. Comparing trajectory with etalons.

Comparing gestures we can make via comparing OX, OY trajectories (Fig. 7)

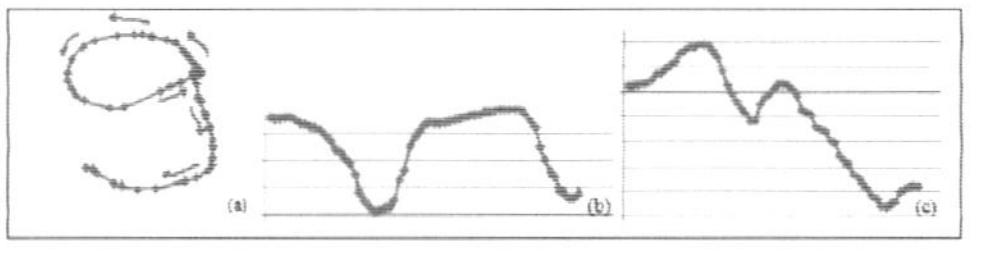

Figure 7. (a) Gesture trajectories. (b) OX trajectory (c) OY trajectory.

It is necessary to normalize the unfolding of the trajectory of the produced gesture and then compare with the standards using the threshold difference-quadratic function (if applicable $\Sigma(y1_i - y2_i)^2 < M\Sigma(y1_i - y2_i)^2 < M$, the gesture can be considered completed) or by using the DTW-technique (date time warping) of comparing [17] time series of different lengths.

3.2.2 The second neural net

Analitycal algorithms is simple for realization but they require many resources (especially DTW-algorithm). For analysis trajectory one of the good methods is neural nets algorithms, because they is oriented for classification images.

Trajectory of object in video stream we may present as bit picture or bit matrix. We mark pixels is "1" where has been object and "0" where object wasn't. Example graphic image of bit matrix is presented on Fig. 8.

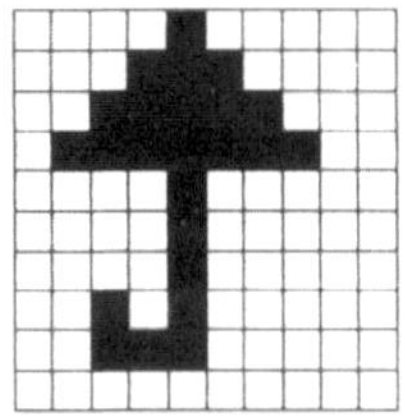

Figure 8. Example of bit matrix

After that we transfer bit matrix to neural net. Previously we must teach neural net using etalons with noise. Neural net analyze input images and notify about completed gestures.

Architecture of neural net is very simple. Simple one-layer perceptron provides logic of detection (Fig.9).

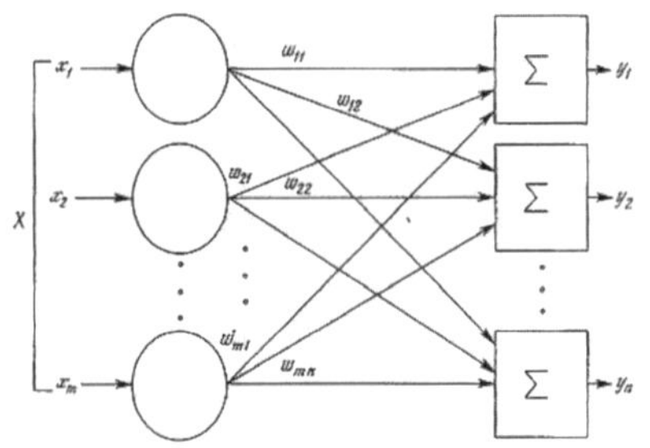

Figure 9. One-layer perceptron

Each neuron of this net present own gesture. If several different neurons will shoot we must choose the neuron with max activation energy (max value).

Training set is presented by etalon gestures with noise.

4. Conclusion

As a result of the work, we developed the system architecture, which has a number of positive qualities, such as cross-platform, low resource consumption and help in improving the current version of the system running on cascading algorithms (TLD-tracking).

Further improvement will increase the efficiency index and expand the possibilities of the system application in the future.

After performing the necessary improvements, we will perform detailed testing of the subsystems (checking the speed and quality of work on various hardware platforms, a comparative analysis of the effectiveness of recognition of dynamic gestures.

References

[1] Dumin A A, Sorocoumov P S, Urvanov G A and Chepin E V 2013 Architechture and prototype the system of BCI with mobile robototechnic device *Messenger of national research univercity MEPHI*

[2] Alomari M H, Samaha A and AlKamha K 2013 *International Journal of Advanced Computer Science and Applications* **2** pp 207-212

[3] Dumin A A, Urvanov G A and Chepin E V 2015 System of relationship with human as agent of mobile robototechnic complex. Program system and computing methods *NB-Media*

[4] Ren Z, Meng J, Yuan J and Zhang Z Autor of translate: A.S. Penkov Link: Robust Hand Gesture Recognition with Kinect Sensor. Web-site: http://www.uran.donetsk.ua/~masters/2013/fknt/penkov/library/article10.htm (date of handling 12.10.2015)

[5] Vasilyev A "Overview: How to arranged the technology of gesture recognition" from 30.03.2014. Web-site: http://www.mobiledevice.ru/69902-raspoznavaniia-zhestov-Technology-realsense-cvs-allsee-kin.asp (date of handling 12.10.2015)

[6] Motions controller "Leapmotion". Web-site:https://www.leapmotion.com/(date of handling 10.10.2015).

[7] Motions controller "Kinect".Web-site: https://msdn.microsoft.com/en-us/library/jj131033(date of handling 12.10.2015).

[8] Kulik S D 2015 *Journal of Theoretical and Applied Information Technology* **73** 2 pp 202-211

[9] Kulik S and Nikonets D 2016 *The Third International Conference on Digital Information Processing, Data Mining, and Wireless Communications* (Moscow Russia) pp 221-226 Doi: 10.1109/DIPDMWC.2016.7529393

[10] Program system "Flutter". Web-site: https://flutterapp.com/(date of handling 12.10.2015).

[11] Приложение "NPointer". Web-site: http://www.neurotechnology.com/npointer.html(date of handling 12.10.2015)

[12] Hardware and software complex «AirGesture» Web-site: http://www.theverge.com/2012/10/3/3447800/pioneer-gesture-control-car-navigation (date of handling 12.10.2015)

[13] System of gestures recognition from company "Atanor". Web-site: http://www.atanor.ru/engineering/systema-raspoznavania-zestov/ (date of handling 12.10.2015)

[14] Gamebox "nTobeBox". Web-site: http://www.innodigital.net (date of handling 12.10.2015)

[15] System of gestures recognition from company "NEC". Web-site: https://www.nec-display-solutions.com/p/ru/ru/company/about.xhtml (date of handling 12.10.2015)

[16] STB (Set-Top-Box)Web-site: http://mediapure.ru/onlajn-servisy/chto-takoe-stb-set-top-box-iptv-

[17] http://jmlda.org/papers/doc/2011/no1/Romanenko2011Dynamic.pdf -DTW algorithm (date of handling 12.10.2015)

IOP Conf. Series: Materials Science and Engineering **498** (2019) 012024 doi:10.1088/1757-899X/498/1/012024

Parametric and topological methods of bulk CMOS IP-blocks yield improvement

Y M Gerasimov, N G Grigoryev, A V Kobylyatskiy*

National Research Nuclear University MEPhI (Moscow Engineering Physics Insitutute), Kashirskoe shosse 31, Moscow, 115409, Russian Federation

*Corresponding author's e-mail address: andreykob91@mail.ru

Abstract. This paper presents the study of the main reasons for the yield loss of modern nanometer VLSIs. The methods of the yield improvement are described on the basis of the impact of parametric and catastrophic reasons for its loss.

1. Introduction

Every step down in the rules of CMOS design (lower than 0.35 um) makes the capabilities of the technological process no longer meet the requirements of the scaling laws. The technological process is complicated by the increase in the number of process steps and stronger requirements thereof. The formation of topological structures by photolithographical means becomes more difficult, leading to lithographical distortions, and as a consequence, increases the influence of different topological variations of the design performance. In this regard, there is a trend of product yield loss.

The modern methodology of designing the digital parts of the IP blocks and VLSI is based on digital standard cells libraries. In modern chips the number of such standard cells reaches millions for the entire area of the crystal, so the main VLSI quality parameters are determined by the quality of standard cells libraries as a whole.

The problem of yield decrease is widely reported in literature. Thus, for example, in [1] a method for the statistical analysis of yield is described. In [2] and [3] are proposed the methods to improve the yield only at the system level of designing. The methods described in [4] are effective; however, they do not allow to reach the maximum yield percent within a project.

In this paper, we tried to give a detailed analysis of the main reasons of yield loss and recommended methods of its improvement.

2. The reasons of design yield loss

Generally, factories provide developers with design rules, verified by the research of the test crystals during the technology characterization. After that, the manufacturer guarantees a 99% level yield of the crystal with limited area and difficulties. Therefore, any occurring reasons of defects are considered by the manufacturer as owing to the poor quality of the design. However, even with the proper level of control of technological steps and cleanness of processes in world practice, there are reasons of the yield decrease that are specifically due to non-ideal technology processes [5].

In general, the reasons for yield loss are divided into the following categories:

1. Catastrophic – these represent functional failures and usually occur due to defects such as small particles falling during the polishing or the etching of metals with very high routing density. In such cases there can occur a short circuit (Figure 1), a break in a topological layer, the inability of the contact to form or the possibility for a hidden defect to be produced. The last one appears either during the failure tests, or during the normal operation of the circuit already as part of the equipment, which is much more critical.

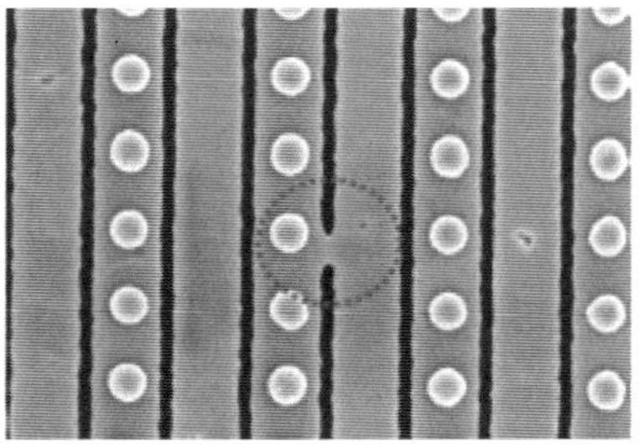

Figure 1. Short circuit between parallel metals [1]

2. Parametric – these represent the parametric mismatch in performance, power consumption or some other parameters of the circuit. The variation of the technological parameters of the transistors, low temperature tolerance and the variation of the supply voltage can have a significant impact on the circuit yield if these effects are not taken into account during the design and if they are the key parameters in the requirements of the circuit. As a rule, modern technological processes guarantee that transistors can work in a wide range of supply voltages and temperatures and that the parameters variation does not exceed $\pm 3\sigma$. However, during manufacturing there are additional variations in topological layers that lead to changes of parasitic structures and hence the whole of the circuit parameters. The variations occur between the crystals on the wafer, between the wafers in one lot and between wafer lots due to the variations of doping profiles, thickness and width of the dielectrics, metal etching features, etc.

It should be noted that both catastrophic and parametric reasons of yield loss may be both random and systematic. For example, the contact defects may occur as a result of the random strikes of particles or the mismatch of photomasks, which lead to systematic defects of the contacts.

Thus, the above-mentioned reasons of yield loss are more relevant for the present development trend of the microelectronics industry. In the development of VLSI for the high reliability applications all possible causes of yield loss should be taken into account and the effects of different variations should be minimized. Otherwise, regardless of the yield loss reasons, the company-developer suffers economic and time costs.

3. Manufacturing yield improvement

One of the most important step in the modern technological process is the use of computational lithography techniques [6]. Such techniques are applied when the minimum size of the structure is less than the wavelength of the exposure light. The most accessible for analysis among them is the correction of optical proximity effect (OPC), as it affects only the image of the topological layer on the photomask. The use of OPC allows to set the right parameters of the structures produced on the wafer, improve the process stability and the crystal yield rate. OPC is based on the use of special masks, which are calculated taking into account the light wave properties in order to achieve higher resolution and lower distortion in the formation of the elements on the wafer at a given wavelength. In Table 1 are presented the types of typical yield loss reasons due to the optical proximity effect.

Each type of the topology defects due to the optical proximity effect in a different rate affects the reliability. Shorts and breaks in the layers have the most negative impact and lead to failures in the circuit functioning. Defective contacts or vias may occur as a result of its insufficient overlay by the corresponding layers, or by the shift of photomasks, or may be struck by microparticles as well as from its burnout by the large current.

IOP Conf. Series: Materials Science and Engineering **498** (2019) 012024 doi:10.1088/1757-899X/498/1/012024

Table 1 The types of typical yield loss reasons due to the optical proximity effect.

Topology defects due to the optical proximity effect		
Name	Description	Illustration
Short	Topological short circuit	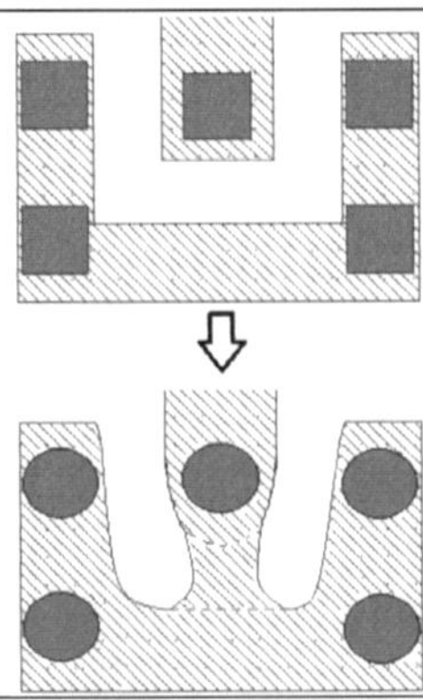
Open	Layer gap	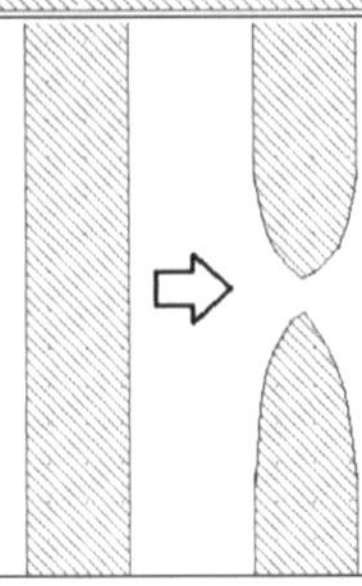
Insufficient contact overlay	Insufficient contact or via overlay by metals or poly	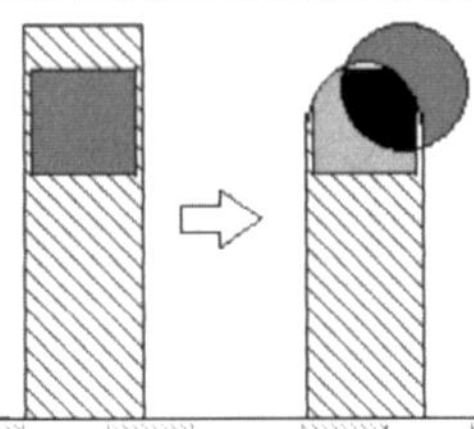
Pinch	Metal narrowing or widening	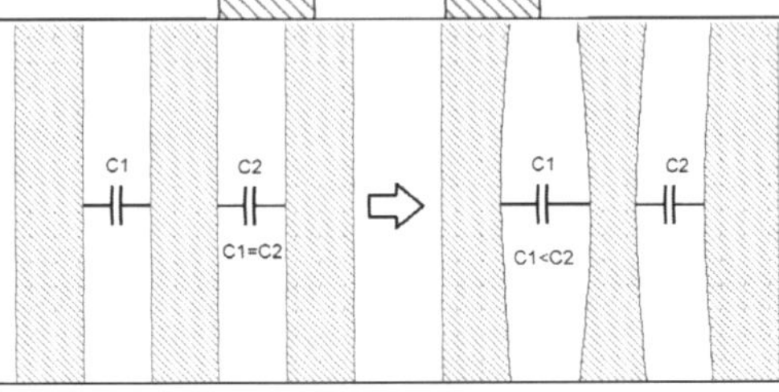
Edge placement error	Rounding of structure corners	

IOP Conf. Series: Materials Science and Engineering **498** (2019) 012024 doi:10.1088/1757-899X/498/1/012024

The last one is the result of bad design of the electromigration rules. The narrowing in the middle of a long wire is a potential place of rupture and the widening can produce shorts when wires are routing parallel in the minimum distance with a high density. Such interconnection defects are caused by the capillary forces acting on resists sides during washing [5]. The yield loss due to the defects in interconnections can also be caused by the unpredictable influence of the parasitic parameters. The variation of the widths of long wires which are parallel can cause a significant change in the side parasitic capacitance which will affect the circuit time parameters. The variations of the time parameters can also be caused by changes in the geometric parameters of the transistors due to the rounding of the topological layers in the U- and L-shaped structures (Table 2).

Thus, the typical distortion of the topological layers can be corrected at the stage of OPC by introducing special topology correction figures [7]. There are three types of correction figures:

1. Change in the nominal layer size by a value of the violation during the photolithography and etching.

2. The introduction of additional lines (assist line) to create the effect of high density structures. The differences in the forming of single and high density structures are the result of the diffraction of light rays on the adjacent layers, so the high density structures are illuminated by larger light density.

3. The introduction of additional figures (serif) or cut-outs on the corners of layers (anti serif) in the rectangular form of to rectify the corners in order to minimize the effects of its rounding.

As a rule, correction figures are introduced in the topology by the manufacturer in automated mode using special software and are aimed at minimizing the causes of catastrophic yield decrease. From the perspective of the influence of the variations, the manufacturer does not know the accuracy requirements of different topology parts. Therefore, the possibility of OPC in improving yield is limited and it is required an additional set of rules and guidelines for the developers to design circuits with taking into account their suitability for mass production. These rules and recommendations are available in some factories. They are based on the analysis of the technology violations during the stages of the manufacturing process. In the world practice this direction of the design is called Design for Manufacturability (DFM).

4. Methods of improvement of yield level

The possibility to achieve the maximum yield value within a given project depends on the quality of the digital and PAD standard cell libraries used in the physical and circuit design synthesis. Therefore, the methods of the yield improvement firstly should be applied on the lowest level of the project abstraction.

During the development of the standard cell libraries and IP blocks we used the following methods to improve the yield:

– the use of the additional set of layout design rules that reduce the risk of short circuits and breaks in the layers, as well as minimize the effects of the structures corners rounding on the geometric parameters of the transistors;

– transistors placement in the same orientation for a better compensation after the masks shifts and other mismatches in the operations of lithography;

– doubling contacts and vias at the place and route step where no hard limitations on the area and traceability, it should be noted that the number of contacts and vias depends on the power of the cell as there are the electromigration requirement;

– the use of methods for calculating precision circuits (like Monte Carlo) with the prediction of the technological parameters variation impact on the timing. Since the calculation of Monte Carlo is rather resource-intensive process, the number of iterations carried out is often limited. In precision circuits we also used transistors with higher width and length to reduce the influence of the relative variation of the transistors geometrical parameters;

– the uniformity of dummies placement to meet the requirements of dummies DRC check. It helps to avoid local layer deformation;

– the power and ground grid optimization – large number of vias, increased metals width and its doubling help to reduce the value of the power and ground buses total resistance and consequently to eliminate its impact on the local supply voltage drops in IP-blocks;

– the antenna effect elimination – during the plasma chemical etching and polishing of the metals the charge is accumulated on the wires, this charge can be the reason of the active elements breakdown or the threshold voltages shifts. When exceeding the specified area of the interconnection limitations in IP-blocks the diodes are placed directly in front of the transistors gates.

Thus, these design methods applied at the level of standard cell libraries and IP-blocks allow to improve the yield of VLSI as a whole achieving a compromise on the basic electrical and geometrical parameters.

At the step of IP-blocks integration as part of VLSI it is also recommended to apply similar methods. To achieve the maximum effect we recommend to provide additional reserves on the setup-hold parameters at the step of the design timing verification. It is necessary because of the inaccuracies occurring during the photomask OPC, and the corresponding errors in the timing characterization of the library cells.

It should be noted that the yield is largely determined by the complexity of the project which depends on the die size, the number of basic cells, IP-blocks and functional content. The simplification of the project can be achieved by the initial HDL-code optimization at the behavioral level. If there are some additional requirements, for example, radiation tolerance, the specialized libraries should be used. It is also required to analyze its structural and topological solutions for the manufacturability and their influence on the design yield.

5. Conclusion

In this work we analyzed the main and the most common reasons of the modern VLSI yield loss as well as the factors that determine the yield, presented the methods of its improve. These methods should be applied at the steps of the circuit and topological synthesis and the project optimization. They are invariant to the level of design rules and can improve not only the yield percentage but the other quality characteristics of the project.

References

[1] Gupta P 2008 Papadopolou E *Yield Analysis and Optimization*
[2] Huang C, Wu C, Li J and Wu C 2003 *IEEE Trans Rel* **52** 4 pp 386-399
[3] Vial J, Virazel A, Bosio A, Dilillo L and Girard P 2010 *International Journal on Advances in Systems and Measurements* **3** 1 pp 1-10
[4] Parahin U N and Sevalynev O S 2012 *Radioelectronics questions* 2012 3
[5] Wong B P, Mittal A, Cao Y and Starr G 2005 *Nano-CMOS Circuit and Physical Design* (USA) p 332
[6] Rodionov I A and Makarchuk V V 2007 *Technology and Engineering in Electronics* **2007** 3 (Russian)

Application of the ion mixing method for doping near surface layers of the silicon single crystals

N Volkov*, B Kalin, Yu Voronov and V Pershenkov

National Research Nuclear University MEPhI (Moscow Engineering Physics Institute), Kashirskoe shosse 31, 115409 Moscow, Russian Federation

*Corresponding author's e-mail address: nvvolkov@mail.ru

Abstract. Penetration of alien atoms (Be, Al, Ni, Mo) into Si, diamond monocrystals substrates was investigated under Ar^+ ion bombardment of samples having thermally evaporated films of 30–50 nm. Sputtering was carried out using a wide energy spectrum beam of Ar^+ ions with mean energy 9.4 keV to dose D=1×10^{16}–10^{19} ion/cm^2. Implanted atom distribution in the targets was measured by Rutherford backscattering spectrometry (RBS) of H^+ and He^+ ions with start energy of 1.6 MeV as well as secondary ion mass-spectrometry (SIMS).

During the bombardment, the penetration depth of Ar atoms increases with dose linearly. This depth is more than 3–20 times deeper than the projected range of bombarding ions and recoil atoms. This is a "ion mixing" process.

The analysis shows that the experimental data for foreign atoms penetration depth are similar to the data calculated for atom migration through the interstitial site in a field of internal (lateral) compressive stresses created in the near-surface layer of the substrate as a result of implantation. Under these experimental conditions atom ratio r_i/r_m (r_i - radius of dopant atom, r_m - radius of substrate atom) can play a principal determining role. Show that maximum penetration depth of the film atoms in the substrates may be determine by "isotropic model" under ion beam (with wide energy spectrum – polyenergy) irradiation of the "film–substrate" systems too.

1. Introduction.

Modern technology of doping materials by ion beams is mainly connected with the use of continuous ion fluxes having narrow energy spectrum and resemblance by their forms to the Gaussian distribution [1-4]. The ratio of the distribution of width on half-height (ΔE) to its mean magnitude (E) is of the order of 1–5%. Doping materials by ions with wide energy spectrum ($\Delta E/E$ = 30–50 %) is of certain scientific and practical interest because in this case the distribution of implanted atoms with the target depth could be changed [5]. However at present time, the data on modification of near-surface layers in structural materials under irradiation by ion fluxes of wide-spectrum energy are very scarce. We attempt to investigate the changes in distribution of dopant (Be, Al, Ni, Mo) with depth- $C(x)$- for some Si (110) and natural diamond – monocrystal substrates) in relation to the fluence (dose) of a polyenergetic argon ion beam.

2. Conditions and materials.

The irradiation was carried out using an instrument wich produces an ion beam with a Gaussian-type energy distribution of argon ion beam within the energy range of E = 4–13.7 keV, at mean energy of $<E>$=9.4 keV [5]. The argon ions were directly implanted into targets made of Si, diamond and atoms of Be, Ni were inserted from sputtered films into substrates made of Si, natural diamond. For that purpose the films were bombarded by a polyenergetic beam of argon ions with mean energy of $<E>$ = 9.4 keV and beam current density of $j = 5$ μA/cm^2.

The films of Be, Ni, Mo were deposited by means of thermal evaporation in vacuum ($p = 4\times10^{-4}$ Pa). Thickness of the wall to (X_t) was chosen to be greater then the penetration depth of the argon ions ($X_t \approx R_p$) and was of X_t = 30–50 nm.

The distribution profile of inserted atoms with depth was determined by two methods: secondary ion (Ar^+ and O^+) mass spectrometry (SIMS) with ion energy E = 4.6 keV using the PHI-600 device and by Rutherford back-scattering (RBS) of fast ions (H^+ and He^+) with initial energy of E = 1.6 MeV. The depth resolution errors of RBS method was not more than ± 5 nm, of SIMS method ±1 nm. The relative concentration errors of RBS method was not more than ± 0.1 at %, of SIMS method – $\pm10^{-2}$ at %.

As parameter of change of conductivity of Si and diamond doped surfaces are investigated to change of spectra of absorption of optical radiation in an interval of lengths of waves 0.1 - 1.4 μm on devices СФ – 46 and Carry – 1000.

3. Experimental results.

3.1. Implantation of Ar atoms

Recovered distribution profiles - $C(x)$ of Ar into Si are presented in Figure 1. The analogous dose relationships for inserted argon distribution were obtained with Al and Si. The results of experiments to determine the most probable (X_p) and maximum (X_m) penetration of argon ions into Be, Al and Si are presented in the Table 1. As can be seen in Figure 1 the accumulation of Ar in Si is changed with the increase of fluence. With low fluencies, the maximum argon concentration is near the projective range (R_p = 12 nm with E = 10 keV) and amounts to X_p =10–20 nm. When the fluence exceed $D = 7\times10^{21}$ m^{-2}, the overall content of Ar drops, gas concentration near the irradiated surface being halved. This probably is connected with gas release through discharge channels created in the process of bombardment. With fluence increase, the location of maximum Ar concentration in Si shifts in depth to X_p= 35 nm. It's important to emphasize that with fluence increase the depth of argon atom maximum penetration (X_m) also increased. As can be seen from the table$_5$ maximum penetrations (X_m) for atoms of Ar are, in the case of penetration into Si, X_m= 82 nm ($D = 1\times10^{22}$ m^{-2}), in Al X_m = 58 nm ($D = 1.5\times10^{22}$ m^{-2}), and in Si X_m = 88 nm ($D = 1.4\times10^{22}$ m^2).

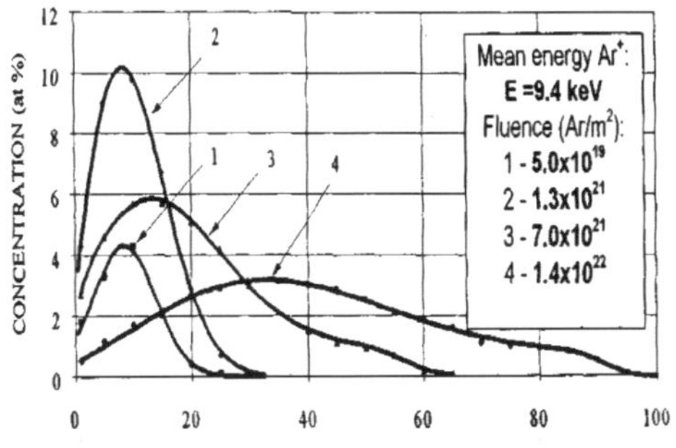

Figure 1. Distribution of implanted Ar^+ ions into Si target with mean energy $<E>$=9.4 keV and fluence D: 1– 5×10^{19} ion/m^2, 2 –1.3×10^{21} ion/m^2, 3– 7.0×10^{21} ion/m^2, 4– 1.4×10^{22} ion/m^2

3.2. Implantation of atoms from films into substrate.

The typical distribution of atoms from surface films in the substrate after bombardment of the film by Ar^+ ions is presented in Figure 2 for the case of Ni implantaion into Si. As can be seen from Figure 2, a considerable penetration of Ni atoms was observed up to depths of X = 5–10 nm, i.e. to depths 3–5 times higher than the range of argon ions (R_p = 3 nm) or the range of primary knock-on atoms of nickel into Si (R_p = 3.5 nm). Maximum penetration of Ni atoms into Si was established at depths exceeding X = 15–20 nm.

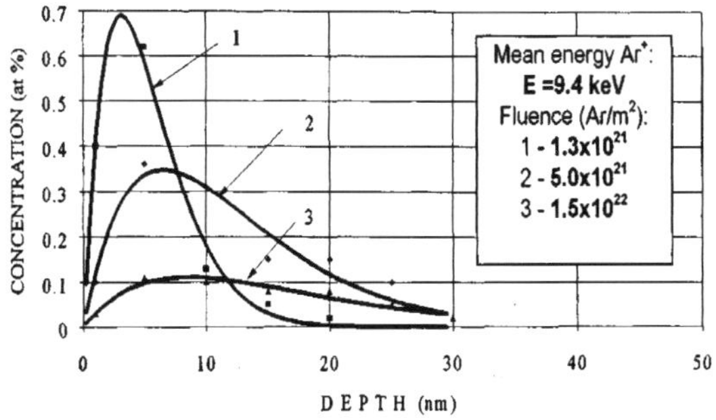

Figure 2. Distribution of implanted Ni atoms into Si substrate under irradiation by Ar^+ ion beam with mean energy <E>=9.4 keV and fluence D: 1- 5×10^{19} ion/m², 2- 1.3×10^{21} ion/m², 3- 5.0×10^{21} ion/m², 4- 1.5×10^{22} ion/m²

As can be seen from data presented in the Table 1, analogous penetration of Be, Ni and Mo atoms was observed from the film into the substrates of Al, Si and diamond. Detailed analysis of Be atom distribution in diamond revealed that with agron ion fluence D = 1×10^{22} m^{-2} Be concentration in the near-surface layer of diamond was $C(X=0)$ = 3 at %. This value is changed very slightly up to X = 20 nm. At the depth of X = 70–90 nm the concentration of Be was 0.5 at%. Then it was falling down and reached the magnitude of 0.1–001 at% by the depth of X> 100 nm.

Table 1. Depth of implanted Ar^+ ions and film atoms

Dopant atoms	Substrate	Fluence, $\times10^{21}$ m^{-2}	Depth, nm	
			X_p	X_m
Ar	Be	0.5	3	33
		1.0	25	48
		5.0	31	63
		10.0	48	82
	Al	1.3	18	45
		5.0	22	40
		15.0	35	58
	Si	0.5	7	23
		1.3	10	26
		7.0	26	42
		14.0	49	88
Be	Al	5.0	26	42
	Si	10.0	31	68
	diamond	10.0	50	113
	Al	7.0	21	45

Ni	Si	10.0	9	25
	Si	7.0	12	26
Mo	diamond	10.0	30	55

3.3. Change of a transmission spectrum. As the diamond in an initial condition has the large specific electro resistance and during it doping under a Ar^{+} ion beam probably accumulation of a electrical charges. Trial doping is carried out on samples from silicon organic polymer of a material with smaller electro resistance and having the optical characteristic close to diamond.

On Figure 3 the results of measurement of transmission spectra of samples, in an initial condition irradiated with Ar^{+} ions with a different doze of an irradiation and doping of the Be atoms with concentration up to 1 at% are submitted. These spectra show decrease of a transparency of samples with increase of a doze of an irradiation and doping of Be atoms in the field of lengths of waves up 400 to 900 nm.

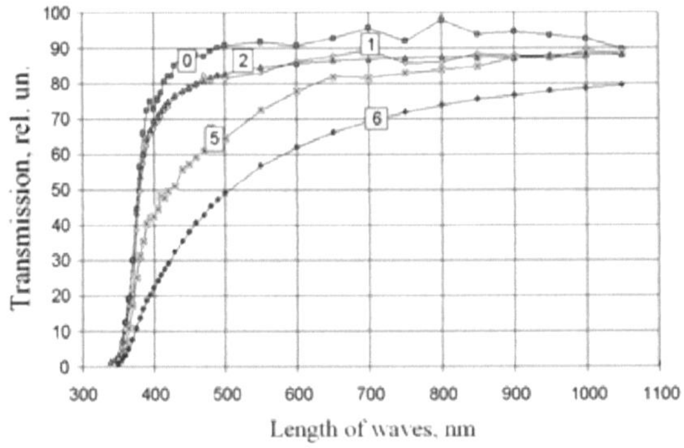

Figure 3. Change of a transmission spectrum of the silicon organic polymer samples at an irradiation by Ar ions and introduction of the Be atoms, irradiation doze of samples D: 0- in an initial condition; 1- $D = 7\times10^{20}$ m^{-2}; 2- $D = 1\times10^{20}$ m^{-2}; 5- $D = 1.3\times10^{21}$ m^{-2}; 6- Be atoms and $D = 1.7\times10^{21}$ m^{-2}

On Figure 4 the transmission spectra of diamond in an initial condition (1) irradiated Ar by ions up to a doze $D = 1{,}3\times10^{21}$ ion/m^{2} (2) and doping by Be atoms (3) are submitted. As visible from figure the irradiation of the sample by Ar^{+} ions is reduces a transparency of diamond sample almost twice in an interval of lengths of waves up 225 to 190 nm. Doping of diamond by Be atoms essentially changes dependence of a transparency of diamond on length of a wave, i.e. the curve of dependence has concave character, on length of a wave 226 nm the abnormal absorption of optical radiation is observed, on a length waves site 265 – 285 nm are observed oscillation in a transparency of the diamond sample.

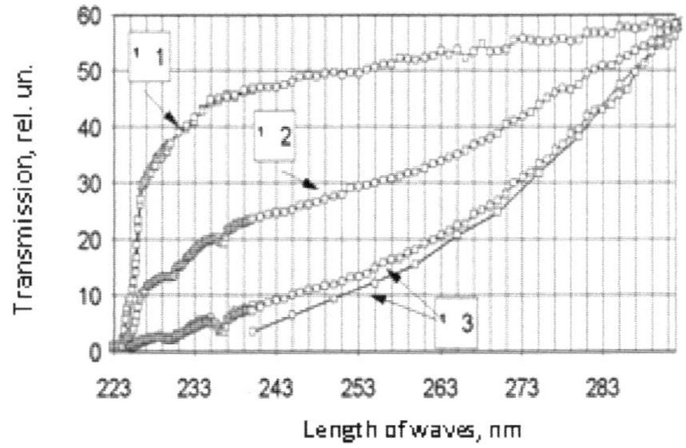

Figure 4. Change of a transmission spectrum of the diamond samples at an irradiation by Ar ions and introduction of the Be atoms, irradiation doze D of samples: 1- D=0 in an initial condition; 2- $D = 1{,}3\times10^{21}$ m^{-2}; 3- Be atoms and $D = 1.6\times10^{21}$ m^{-2}

4. Discussion

The Table 1 presents the basic results of the experiments to investigate the dynamics of Ar, Be, Ni atom penetration Be, Al, Si and diamond under bombardment of samples by polyenergetic Ar^+ ions beam - with different fluences. The comparison of the experimentally gained values concerning Ar^+ ions ranges in Be, Al, and Si with the calculated data on these ranges (R_p) for monoenergetic beams (presented, for instance, in ref. 6) shows that in the case of irradiation by Ar^+ polyenergetic beam, penetration of Ar atoms to depths considerably large then R_p can be observed. This could be connected with the fact that the beam includes Ar ions having energy higher than mean value. As can be seen in Figure 5, the magnitude of Ar^+ ion penetration – $\langle X_p \rangle$ (where $\langle X_p \rangle$ is mean value from $C(x)$ - $C(x)_c$, $C(x)$ is the experimental distribution profile of atoms, $C(x)_c$ is the calculated distribution profile of atoms for a polyenergetic beam) into Al, Si, Be increases linearly with fluence as in [3,4].

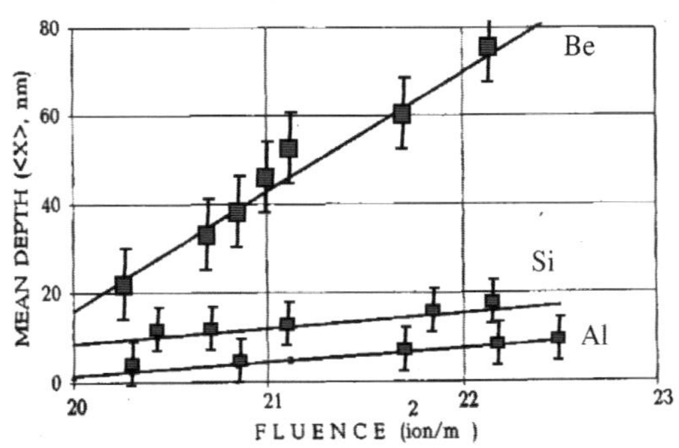

Figure 5. The influence of implanted Ar^+ ions fluence on $\langle X_p \rangle$ range

In this case, X_p increases in the order Al - Si - Be at all fluences. It is important to emphasis that the ratio between the radius of Ar atom and the radius of Al, Si, Be atoms (0.134 : 0.164 : 0.17 nm) as well as Young's moduli of Al, Si, Be (70 : 160 :310, GPa) increase in the same order. The analogous effect of "deep action" can also be observed under implantation into the base layers of atoms films by ions of argon. Atoms of Be and Ni penetrate into Al, Si, and diamond base layers to the depths which exceed the ranges of primary knock-on atoms. Judging by primary knock-on atom energy magnitudes presented in the Table 1 the process of the implantation by Ar ions is accompanied by knock-on atoms of films in the regime of dynamic mixing [2] to considerable depths. Analysis of the film atom maximum penetration (the results ar resented in Figure 6) shows that the case atom penetration depth also increases with the ratio of implanted atom radius ($r_{i;}$) to substrate atom radius (r_m).

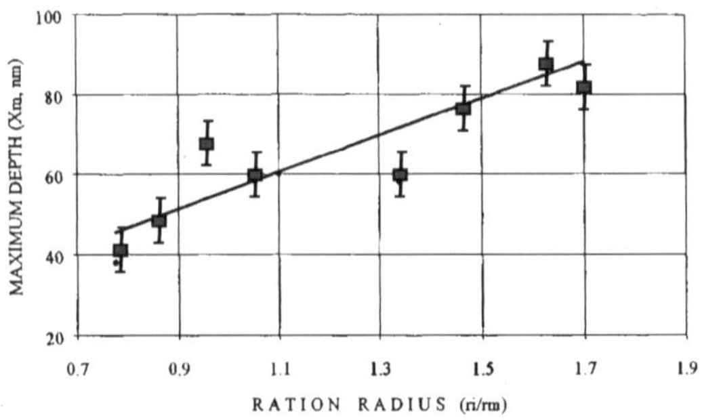

Figure 6. The influence of r_i:r_m on the maxiumum range(X_m) of dopant atoms

To reveal the mechanism of Ar ion and Be, Ni atom deep penetration into the material, the calculation of distribution profiles of inserted atoms was conducted with consideration of the "deep action" mechanisms outlined by Yu. Martynenko [7]. The closest agreement with the experimental data was obtained in the course of assessments concening inserted atom migration through the interstitial site in a field of internal (lateral) compressive stresses created in the irradiated material near-surface layer as a result of implantation [8]. In this case the penetration of the atom being implanted greater than the expected depth of penetration owing to radiation enhanced diffusion. Here $\alpha = (BN_0e/D)^{1/2}$. N_0 is the specific density of the matrix material, B – is a parameter of the radiation enhanced diffusion, D is a coefficient for thermal diffusion, e = 2.718...., α = 10...16 for the case using typical parameters characterizing irradiation, matrix material and inserted atoms. As show experiments doping by film atoms various substrates [9] maximal depths of introduction more precisely can be described within the framework of isotropic ion mixing model [10]. However for systems "film-substrate" it is necessary to take into account an opportunity of physicochemical interaction between their atoms. For example result of a physicochemical analysis of the «Be-film – diamond-substrate» system it has been established that peak mutual penetration of atoms depends on their electronegativities χ, or more precisely on the electronegativity difference for film- and substrate-atoms $\Delta\chi$ [11]. The value of X_m/R_p decreases with strengthening the physicochemical interaction of atoms (with an increase of $\Delta\chi$).

5. Conclusion

1. Irradiation by an Ar ion by polyenergetic beam revealed the penetration of argon atoms to depths 3–10 times greater then the calculated ranges of monoenergetic beams of Ar^+ ions.

2. The ion mixing of Be, Ni and Mo film atoms allows to obtain considerable penetration of these atoms into Al, Si and diamond substrates, the doped layer depth being 20 and more times greater than the culculated projective range of the recoil atoms.

3. The penetration depth of the film atoms into a substrate (X_m/R_p) at a fixed fluence and energy-release in a sample depends on the reduced mass of a beam and a target, the energy reported to recoil atoms, and physicochemical interaction of atoms (for example, electronegativity difference).

References

[1] Brice D K 1971 *J.Rad. Eff* **11** 227.
[2] Sigmund P and Gras-Marti A, 1980 *J. Nucl. Instr. Meth.* **95** 385
[3] Bottiger S J, Nelson S K, Whtelow H J and Wriedl P 1983 *J. Nucl. Instr. Meth.* **218** 684
[4] Averback R S, Thompson L J, Moyle J and Schalif M 1982 *J.Appl. Phys.* **53** 1341

[5] Volkov N V and Kalin B A 1991 *Material science problems of nuclear engineering* pp 64-67
[6] Burenkov A F, Komarov F F, Temkin M M and Schlotzhauer G 1984. *J.Rad. Eff. Lett.* **86**(5) 161
[7] Martynenko Yu V 1993 *Charged particle beams and a solid body* **7** pp 82-112
[8] Eer Nisse E P and Picraux S T 1977 *J. Appl. Phys.* **48**(1) 9
[9] Kalin B A, Gladkov V P and Volkov N V 1996. *J .Nucl. Mater.* **692** 233
[10] Kalin B A, Volkov N V and Atalikova I Kh 1998. *J. Izvestiya RAS, seriya Fizicheskaya* **62** 1477
[11] Darken L S and Gurry R W 1953 *Physical Chemistry of Metals. Mc Granw - Hill Book Company, Inc. New York Toronto London* p 582.

IOP Conf. Series: Materials Science and Engineering **498** (2019) 012026 doi:10.1088/1757-899X/498/1/012026

Factographic information retrieval for semiconductor physics, micro- and nanosystems

S D Kulik

National Research Nuclear University MEPhI (Moscow Engineering Physics Institute), Kashirskoe shosse 31, Moscow 115409, Russian Federation

Corresponding author's e-mail address: sedmik@mail.ru

Abstract. This paper proposes a factographic information retrieval for semiconductor physics, as well as micro- and nanosystems. This factographic information retrieval includes a pattern recognition algorithm and factographic database. The first factographic database includes special keywords from semiconductor physics and micro- and nanosystems. The second factographic database includes document descriptions in the form of Searching Documents Patterns for the fields of semiconductor physics and microelectronics. An analytical model of the factographic information retrieval system is developed. This model is presented by an effectiveness indicator: the average length of the recommendatory list used for semiconductor physics and micro- and nanosystems for space application.

1. Introduction

The Effective Factographic Information Retrieval System is very important for scientists and researchers in semiconductor physics and for space application of micro- and nanosystems [1, 2, 3] to retrieve factographic information with keywords. Next, without loss of generality, we may assume that some keywords are in papers, for example, scientific papers [4-25]. In our research, we used effective information technology.

Professional competencies [18, 19, 27, 28] are very important for scientists and researchers in various areas - for instance, in microelectronics. Factographic information retrieval also improves professional competency at times.

Thus, the paper is organized as follows. A brief introduction and short review of the research field are introduced in Section 1. The research methods are introduced in Section 2. Section 3 contains some examples and keywords (special features of scientific texts) of documents of semiconductor physics and micro- and nanosystem space application. The section contains a brief description of the factographic information retrieval system. Results, discussion and future works are introduced in Section 4. The section contains analytical formulas to evaluate the effectiveness of factographic information retrieval for semiconductor physics and microelectronics. The key defining indicator of factographic information retrieval in the factographic retrieval system is introduced in Section 4. This section presents the results obtained from analytical expressions (analytical models) and brief discussion of the results. Finally, the main results are outlined in Section 5. Some aspects are not discussed in detail because of the article's limited volume.

Many problems were solved and were studied in detail. However, some important problems remain unsolved, for example:

- development of an effective FIRS for application in semiconductor physics and space micro- and nanosystems;
- recognition of new experimental and theoretical results automatically from scientific texts;
- recognition of new scientific ideas from the database for semiconductor physics and micro- and nanosystems for space application;
- recognition of psychological characteristics of a person by handwriting.

Artificial neural networks (ANNs) [11] have been used for Factographic Information Retrieval System (FIRS).

There are important scientific areas of applications for FIRS, artificial neural networks [11], information technology and artificial intelligence methods:

- physics and chemistry [24];
- electric insulators (glass and ceramics) [25];
- semiconductor physics, micro- and nanosystems;
- medicine [22, 23];
- police, criminology [20] and forensic experts;
- retrieval system;
- ill-posed approximation problems [12, 13];
- Data Analysis [14];
- housing market [21];
- Gaze-Based Control [15];
- Eye-Brain-Computer Interface [16];
- scientific recommender systems [17];
- knowledge management system [19];
- Big Data, Fast Data [26];
- Information Security [27, 28];
- competences forming [18];

and many other areas.

The effective factographic information retrieval problem has been investigated in the paper for semiconductor physics and micro- and nanosystems for space application.

It is necessary to give a short review of the research field. Some important problems of semiconductor physics and micro- and nanosystems for space application were considered in the indicated papers [1, 2, 3, 4, 5, 6]. Paper [1] explores a conversion model of radiation-induced interface-trap buildup. Paper [2] deals with conventional silicon bipolar transistors. Paper [3] deals with the formation of dielectric silicon compounds. Paper [4] deals with an analytical model of plasma-chemical etching in a planar reactor. Reactive ion etching of silicon using a low-power plasma etcher is introduced in [5]. Sensitive elements are introduced in [6]. Numerous papers and monographs are concerned with very different important topics, for example, neural networks theory in [12, 13], data analysis in [14], Gaze-Based Control in [15], Eye-Brain-Computer Interface in [16], recommender systems in [17], software simulators for the evaluation of socio-personal competence in [18] and cognitive competence of graduates in [19], artificial intelligence methods in [20] and Big Data in [26]. Monograph [11] is concerned with the problem of artificial neural networks. The problem of factographic information retrieval for semiconductor physics and microelectronics is very important. Despite a vast amount of literature, this problem has never been studied from a new viewpoint.

2. Methods

The research was conducted on the basis of the methods of semiconductor physics, micro- and nanosystems [1, 3, 5, 7], information retrieval, operations research, Big Data [26], artificial intelligence methods [14, 20], neural networks [11, 12, 22, 23], mathematical modeling [13, 21], competence forming [18, 19, 27, 28] and pattern recognition.

3. Factographic information retrieval system

The factographic information retrieval system includes FDB I and FDB II. FDB I includes keywords of documents of semiconductor physics and micro- and nanosystems for space application. FDB II includes factographic information of documents.

Special features of scientific texts, for example, keywords, are very important for the factographic retrieval system. Some examples of these keywords (special features of scientific texts) are shown in Tables 1, 2, 3 and 4.

Table 1. Keywords for the factographic database from [1].

Special keywords (from paper)	Identification code
bipolar device	B273d1078
radiation degradation	R105d978
averaged dose rate	A580d57R8
silicon forbidden gap	S700f1G02
thermoactivated tunneling mechanism	T400t2M01
space application	S007a567
radiation-induced degradation	R105i04D4
transition time of the electron	T207t30E07
hydrogen-electron concept	H005e07C1
hydrogen-electron mechanism	H005e07M01
averaged temperature	A058t800
conversion model	C057m054

Table 2. Keywords for the factographic database from [2].

Special keywords (from paper)	Identification code
bipolar transistor	B273t105
SiGe-transistor	SG10t105
silicon transistor	S700t105
bipolar device	B273d17
SiGe device	SG10 d17
electron-hole	E007h03
whole oxide layer	W702o5L04
emitter junction	E201j310

IOP Conf. Series: Materials Science and Engineering **498** (2019) 012026 doi:10.1088/1757-899X/498/1/012026

We can assume that the Ω is a set of special keywords (see Table 1 and Table 2) and Z_j is the element of Ω, i.e. $Z_j \in \Omega$ and

$$\Omega = \{Z_1, Z_2, Z_3 ..., Z_j, ..., Z_k\}.$$

Table 3. Keywords for the factographic database from [3, 4, 5, 6, 7, 8, 9, 10].

Special keywords (from title)	Identification code
dielectric silicon compounds	D148s60C98
plasma-chemical etching in planar reactor	P885c9e16p08R203
low-power plasma etcher	L015p84p77E55
dielectric membrane structures	D148m888S500
reactive magnetron sputtering	R445m22S054
high temperature annealing	H021t800A404
dielectric membrane	D148m888

Table 4. Keywords for the factographic database from [19, 20, 21, 22, 23, 24, 25].

Special keywords (from title)	Identification code
micro-arc oxidation	M61a43O91
electric insulators	E002i28
neural networks	N622n15
new knowledge	N222k34
knowledge management system	K34m34S505
intelligence methods	I111m324
mass appraisal	M0017a008

Without loss of generality, we can assume that the $\hat{\Omega}$ is set of special keywords from title only (see Table 3 and Table 4) and $\hat{Z}_j$ is the element of $\hat{\Omega}$, i.e. $\hat{Z}_j \in \hat{\Omega}$ and

$$\hat{\Omega} = \{\hat{Z}_1, \hat{Z}_2, \hat{Z}_3 ..., \hat{Z}_j, ..., \hat{Z}_m\}.$$

The factographic information retrieval system (see Figure 1) for semiconductor physics, micro- and nanosystems (i.e. MN-systems) consists of different blocks.

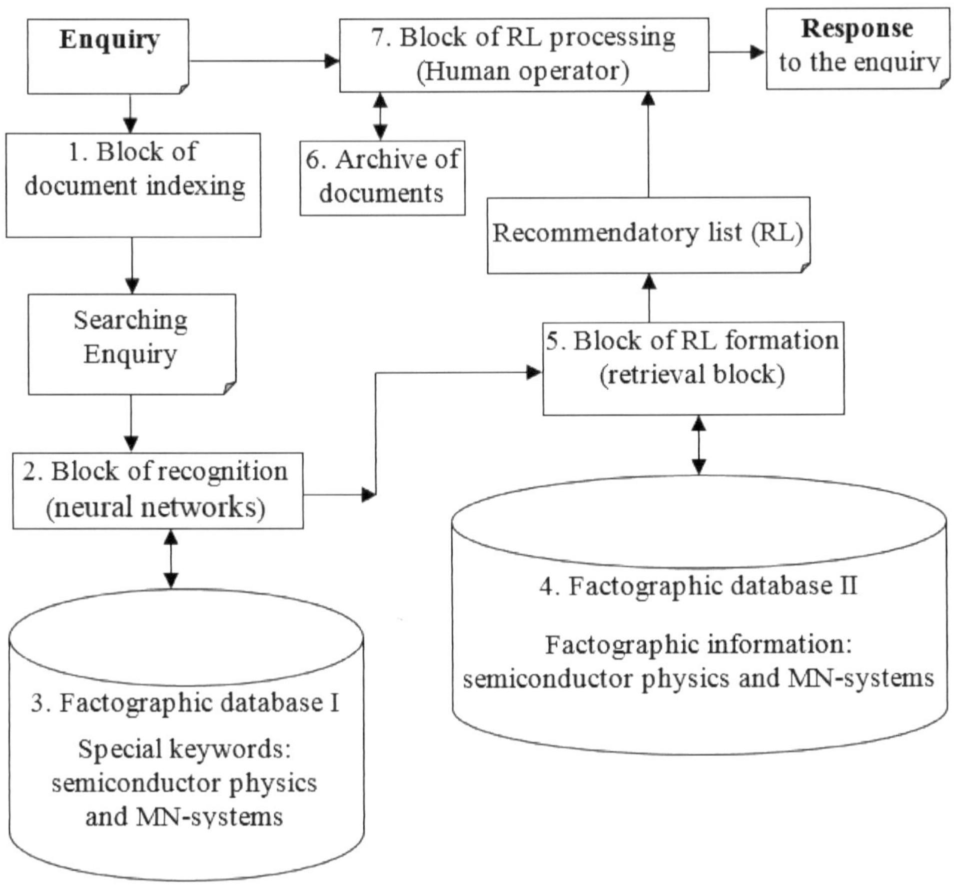

Figure 1. Factographic information retrieval system for semiconductor physics (MN-systems).

The factographic information retrieval system includes seven blocks: the block 1 of document indexing; the block 2 of recognition (neural networks); the block 3 of Factographic database I (Special keywords: Ω and $\hat{\Omega}$); the block 4 of Factographic database II (factographic information) which includes document descriptions in the form of Searching Documents Patterns (SDP); the block 5 of RL formation (retrieval block); the block 6 archive of documents; the block 7 of Recommendatory List (RL) processing. It is supposed that there are N documents which are stored in the archive and every enquiry has Searching Enquiry Pattern (SEP). Factographic database II has N SDPs.

4. Results, discussion and future works

Experiments and analytical results for estimation of error recognition and evaluation of the effectiveness of retrieval are very important for us.

It is supposed that L_x is average length of the RL. Analytical formula was obtained to evaluate the factographic information retrieval effectiveness with the help of indicator (see (1) and (2)):

$$\bar{L} \approx F\left(N, L, P_2\right). \tag{1}$$

It is supposed that:

P1 – probability of the correct comparison of two identical documents based on their descriptions (determines the target mission probability);

P2 – probability of the correct comparison of two non-identical documents based on their descriptions (determines the false alarm probability);

$$E_i^N = \binom{N}{i} = \frac{N!}{i!(N-i)!}, \tag{2}$$

were $N \gg i$.

The following analytical expressions were obtained to evaluate the effectiveness:

$$F(N,L,P_2) = \sum_{i=0}^{L-1}\left[E_i^N \psi_i i\right] + L\sum_{i=L}^{N}\left[E_i^N \psi_i\right], \quad \psi_i = P_2^{N-i}(1-P_2)^i. \tag{3}$$

As a result of researches (1) and (2), it was set that $\bar{L}$ in (1) is changed during the changes of L, N, and P_2. A small part of this research for different L we can see in Table 5.

Table 5. Example of the effectiveness, if N=50000 and P_2=0.995

L	$\bar{L}$
5	5
7	7
15	15
50	50
140	140
240	237
245	241
850	250
$2 \cdot 10^3$	250
$7 \cdot 10^3$	250

According to Table 5, if P_2=0.995, N=50000 (N – the number of SDP) and L=2000 (L – maximum length of the RL), the length of the RL of the FIRS is $\bar{L} \approx 250$.

Future steps of our work include getting the answers to the following important questions:

1. How do parameters of the algorithm of pattern recognition influence information retrieval for semiconductor physics and micro- and nanosystems for space application?
2. Could one recognize new experimental and theoretical results automatically?
3. Which structure of neural network is most suitable for recognizing scientific texts from semiconductor physics and micro- and nanosystems space application?
4. Could one recognize thought from scientific texts for semiconductor physics?
5. Could one recognize new scientific ideas from database for semiconductor physics and micro- and nanosystems for space application?
6. How do parameters of neural network model influence error discrimination (classification) scientific texts from semiconductor physics, micro- and nanosystems for space application?
7. Could one extract features of special of scientific texts automatically?
8. What activation function is most suitable for recognizing scientific texts for semiconductor physics and microsystem electronics?

The special neural network algorithm will be the subject of our future work in which we are aiming to improve factographic retrieval tools for superconducting microelectronics, solid-state microelectronics and microsystem electronics. That is why our plans are:

- Developing the special neural network algorithm for recognizing special keywords for superconducting microelectronics, solid-state microelectronics and microsystem electronics.
- Experiment with structures of special neural networks.
- Developing the special neural network algorithm for recognizing the scientific theoretical facts from scientific texts for other space applications.

5. Conclusion

Thus, as a result of research, the analytical formula (2) allowing the evaluation of the RL average length was developed. Set of keywords were explored for microelectronics. Factographic database I (special keywords: set Ω and $\hat{\Omega}$) was created for semiconductor physics and MN-systems.

Properties of the important indicator of effectiveness – the RL average length for the human operator were explored. Necessary software, allowing the evaluation of the retrieval's effectiveness, was created for the developer of the Factographic Information Retrieval System for semiconductor physics and micro- and nanosystems for space application. The knowledge of Factographic information retrieval can be used as illustrative material in teaching modern information technology for microelectronics in the university.

We believe that our results are useful for scientists and researches in the fields of semiconductor physics and micro- and nanosystems for space application to retrieve factographic information.

Acknowledgements
This work was supported by Competitiveness Growth Program of the Federal Autonomous Educational Institution of Higher Education National Research Nuclear University MEPhI (Moscow Engineering Physics Institute).

We would like to express our gratitude to students of NRNU MEPHI for their valuable discussion and suggestions.

References
[1] Pershenkov V S, Sogoyan A V and Telets V A 2016 *IOP Conf. Series: Materials Science and Engineering* **151** 012001
[2] Bakerenkov A S, Felitsyn V A and Rodin A S 2016 *IOP Conf. Series: Materials Science and Engineering* **151** 012013
[3] Veselov D S and Voronov Yu A 2016 *Journal of Physics: Conference Series.* **747** Art. No. 012022
[4] Veselov D S, Bakun A D, Voronov Yu A, Kireev V Yu and Vasileva O V 2016 *Journal of Physics: Conference Series* **748** 012015
[5] Veselov D S, Bakun A D and Voronov 2016 *Journal of Physics: Conference Series.* **748** 012017
[6] Veselov D S 2015 *Physics Procedia* **72** pp 499-502
[7] Veselov D S, Voronov Yu A and Vanyukhin K D 2015 *Physics Procedia* **72** pp 494-498
[8] Veselov D S and Voronov Yu A 2015 *Physics Procedia* **71** pp 423-427
[9] Veselov D S 2015 *Applied Mechanics and Materials* **799-800** pp 994-997
[10] Veselov D S and Voronov Yu 2015 *Applied Mechanics and Materials* **799-800** pp 919-922
[11] Haykin S 1999 *Neural Networks. A Comprehensive Foundation* (Second Edition, Pearson Education, Inc.)
[12] Kruglov I A, Mishulina O A and Bakirov M B 2012 *Neurocomputing* **96** pp 74-82
[13] Kruglov I A and Mishulina O A 2013 *Journal of Computer and Systems Sciences International* **52** 4 pp 503–518
[14] Skiteva L, Trofimov A, Ushakov V, Malakhov D and Velichkovsky B 2016 *Biologically Inspired Cognitive Architectures for Young Scientists Proceedings of the First International Early Research Career Enhancement School* **449** pp 135-140
[15] Shishkin S, Nuzhdin Y, Svirin E, Trofimov A, Fedorova A, Kozyrskiy B and Velichkovsky B 2016*Frontiers in Neuroscience* **10** pp 1-20

[16] Nuzhdin Y O, Shishkin S L, Fedorova A A, Trofimov A G, Svirin E P, Kozyrskiy B L, Medyntsev A A, Dubynin I A and Velichkovsky B M *Proceedings of the 2017 ACM Workshop on An Application-oriented Approach to BCI out of the laboratory* pp 39-42.
[17] Guseva A I, Kireev V S, Bochkarev P V, Kuznetsov I A and Philippov S A 2017 *AIP Conference Proceedings* **1797** 020002
[18] Guseva A I, Kireev V S, Silenko A N, Tikhomirov G V and Sheina E A 2017 *AIP Conference Proceedings* **1797** 020003
[19] Kireev V, Silenko A and Guseva A 2017 *IOP Conf. Series: Journal of Physics: Conf. Series* **781** 012060
[20] Yasnitsky L N, Vauleva S V, Safonova D N and Cherepanov F M 2015 *Criminology Journal of Baikal National University of Economics and Law* **9** (3) pp 423-430 DOI: 10.17150/1996-7756.2015.9(3) pp 423-430.
[21] Yasnitsky L N and Yasnitsky V 2016 *Journal of Applied Economic Sciences* **XI** 8 46 pp 1519-1530
[22] Yasnitsky L N, Dumler A A, Poleshchuk A N, Bogdanov C V and Cherepanov F M 2015 *Biology and Medicine* (Aligarh) **7** 2 BM-095-15 8
[23] Yasnitsky L N, Dumler A A, Bogdanov K V, Poleschuk A N, Cherepanov F M, Makurina T V and Chugaynov S V 2013 *Biomedical Engineering* **47** 3 pp 160-163
[24] Gorchakov A I, Semenova A V, Syrovatskaya Yu V, Shcherbakov Yu V and Yasnitskij L N 2004 *Fizika i Khimiya Obrabotki Materialov.* 1 pp 43-47
[25] Dobrynin G F and Yasnitskii L N 1995 *Glass and Ceramics* **51** 7-8 pp 275-278
[26] Miloslavskaya N and Tolstoy 2016 *Procedia Computer Science* **88** pp 300-305
[27] Miloslavskaya N and Tolstoy A 2016 *IEEE 4th International Conference on Future Internet of Things and Cloud Workshops* (Vienna) pp 83-90
[28] Miloslavskaya N and Tolstoy A 2015 *Information and Communication Technology* **453** pp 135-145

IOP Conf. Series: Materials Science and Engineering **498** (2019) 012027 doi:10.1088/1757-899X/498/1/012027

Investigation of electrochemical sensors with thin film nanocomposite Pt/C electrodes obtained by magnetron sputtering

E Medvedeva[1,*], **A Baranov**[1], **N Samotaev**[2]

[1]FSUE SPA 'Analitpribor', Smolensk, Russia
[2]National Research Nuclear University MEPhI (Moscow Engineering Physics Institute), Moscow, Russia

*Corresponding author's e-mail address: elenamedvedeva78@mail.ru

Abstract. The electrochemical gas sensors with the film nanocomposite electrodes are produced and tested in several gas atmospheres. The experimental results demonstrate that a-C/Pt has high sensitivity and stability for continuous monitoring of carbon monoxide (CO) and hydrogen sulfide (H_2S) gases compare to with main counterparts are mass-produced.

1. Introduction

In connection with environmental pollution, the need for real-time air monitoring is constantly growing. For these purposes the different monitoring systems are used which include electrochemical [1], semiconductor [2], catalytic [3] and IR sensors [4].

The world trend aimed at solving problems of air monitoring over large areas associated with leaks toxic gases has been the development of sensory networks of various sizes that collect, analyze and make decisions on the management of executive devices to prevent emergencies and alert appropriate services when exceeding the maximum permissible concentration (MPC) of toxic gases.

Depending on the way data is transferred from the sensors to the data reception devices, wire and wireless sensor networks are allocated [5]. The drawback of wired systems is their dependence on power supply systems, which limits the use of wire-based monitoring systems for air composition, in particular, in places with no network power supply. Overcoming of existing limitations is possible due to the creation of autonomous wireless systems for continuous monitoring of toxic gases that provide air composition measurements on the territory of industrial enterprises.

To create autonomous wireless monitoring systems, in addition to low-power electronic components and energy efficient algorithms [6], it is required to use gas sensors with minimal power consumption. From the whole variety of gas sensors, minimal energy consumption and, at the same time, sufficient sensitivity and selectivity have only electrochemical sensors. The wide use of electrochemical sensors in gas analyzers is due to their ease of maintenance, price and a wide range of measured components.

In the electrochemical sensor, the detectable component reacts with the sensitive layer directly on the working electrode. Therefore, the working electrode must simultaneously perform several functions that ensure the operation of sensors: to have the ability to pass gas, play the role of a catalyst, have high electrical conductivity, and prevent the liquid electrolyte from flowing out.

The electrochemical sensor electrode in the general case is a gas permeable membrane with a catalytically active layer deposited on it (Fig. 1). The active layer can be applied by using well-known deposition techniques such as screen printing [7], chemical and electrochemical deposition [8], ultrasonic spraying [9] and laser technics [10].

Recently, the technology of magnetron deposition of thin-film nanocomposite catalysts based on platinum and amorphous carbon (graphite) has been developed to create working electrode of electrochemical sensor, which improves the reproducibility of electrode parameters [11]. Amorphous carbon is a highly inert material with good stability in many acids, solvents and electrolytes and has a high electrical conductivity. The co-sputtering of graphite and platinum provides the increasing platinum surface and its porosity. Note that platinum and graphite do not participate in the oxidation-reduction process, but are only carriers of atomic oxygen.

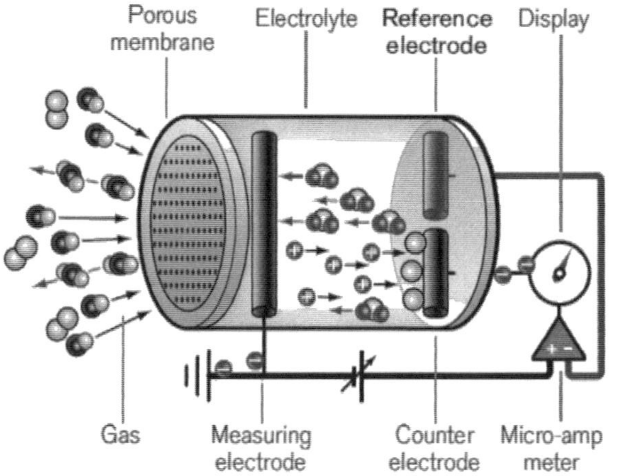

Figure 1. Scheme of the electrochemical sensor.

2. Experimental

Thin a-C/Pt films of the composite were produced on a "Elato 600-R" (Izovak) unit equipped with a DC magnetron and a rotation system for the substrate holder. The deposition of the films was carried out in an argon atmosphere at a pressure of ~ 10^{-1} Pa. Spray mode: U = 500V, I = 200mA, t = 20 min. Substrates for the application of the catalyst were Teflon FM-400 (PTFE) with a porosity of 50% and a pore diameter of 1 μm without any additional preparation (Fig. 2a). Photos of the working and auxiliary electrodes are shown in Fig. 2b and 2c. The composition of a-C/Pt catalytic layers was investigated by EDX spectra (Fig. 3). The developed thin film electrodes are embedded in the sensor nodes with a liquid electrolyte and tested in the presence of CO and H_2S. We evaluate the sensor performance in terms of sensitivity, selectivity and response.

a b c

Figure 2. Teflon surface (a) and magnetron sputtering electrodes: (b) working, (c) reference and auxiliary.

IOP Conf. Series: Materials Science and Engineering **498** (2019) 012027 doi:10.1088/1757-899X/498/1/012027

Table 1 demonstrates the sensor sensitivity (response) in the presence of two gases CO and H_2S in the environment. The sensors improve the results of commercial sensors by Nemoto (NAP-505 for CO and Ne4 for H_2S) while guarantying low background current which is lower than 1%.

The evaluation of the electrochemical sensor cross sensitivity to different acid gases is highly important when considering the long-term operation of the sensor. The results of this evaluation are presented in Table 2. As a matter of fact, H_2S sensor has no sensitivity to other gases listed in the column 'Tested Gas' in Table 2. In contrast, Nemoto's NE4 sensor for H_2S has significant response to SO_2 and NO_2 . As for the CO sensor fabricated in this work, it has negligible sensitivity (at the level of an error rate) to H_2S and NO_2.

Comparison of the characteristics – sensitivity (K), lower background currents (I_0), response times (t_{09}), temperature dependence of background current (T_w) and detection limits for gases (C_{min}) are presented in Fig.4. It is seen that developed electrochemical sensors have better characteristics.

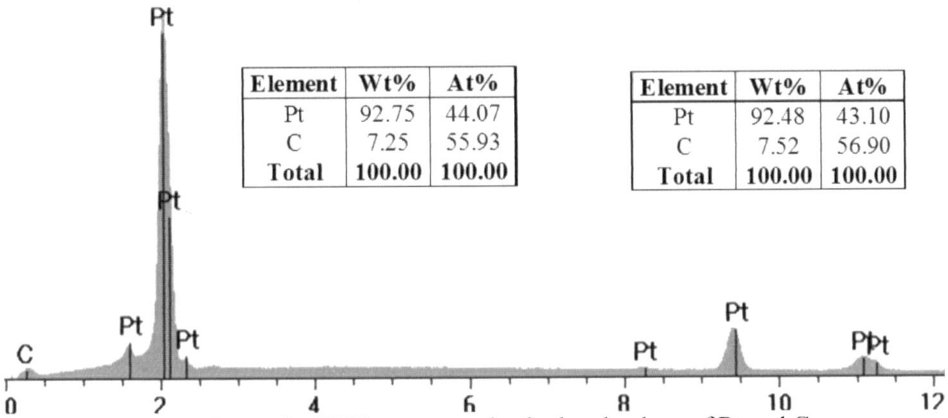

Element	Wt%	At%
Pt	92.75	44.07
C	7.25	55.93
Total	**100.00**	**100.00**

Element	Wt%	At%
Pt	92.48	43.10
C	7.52	56.90
Total	**100.00**	**100.00**

Figure 3. EDX spectra and calculated values of Pt and C.

Table 1. Sensitivity of electrochemical sensors.

Gas	Sensitivity of electrochemical cell	
	Background current, μA	Sensitivity, μA
CO	0,05	48,6 (100 ppm)
H_2S	0,01	30,1 (30 ppm)

Table 2. Cross sensitivity of carbon monoxide and hydrogen sulfide sensors

Tested Gas	Concentration (ppm)	Sensors' readings (ppm)	
		CO	H_2S
Chlorine (*Cl*)	100	0	0
Carbon monoxide (*CO*)	100	100	0
Hydrogen sulfide (H_2S)	30	0.5	30
Nitrogen dioxide (NO_2)	30	0.3	0
Hydrogen chloride (*HCl*)	100	0	0
Sulfur dioxide (SO_2)	25	0	0
Formaldehyde (CH_2O)	10	0	0
Ethylene oxide (C_2H_4O)	100	0	0
Ammonia (NH_3)	200	0	0

3. Conclusions

It was shown that the developed electrochemical sensors with Ptl/C electrodes have greater sensitivity, lower background currents, comparable response times and detection limits for gases.

To better understand the relationship between the structure of a-C/Pt catalyst layers and their electrochemical properties, it is planned to use the in-situ X-ray monitoring method [12].

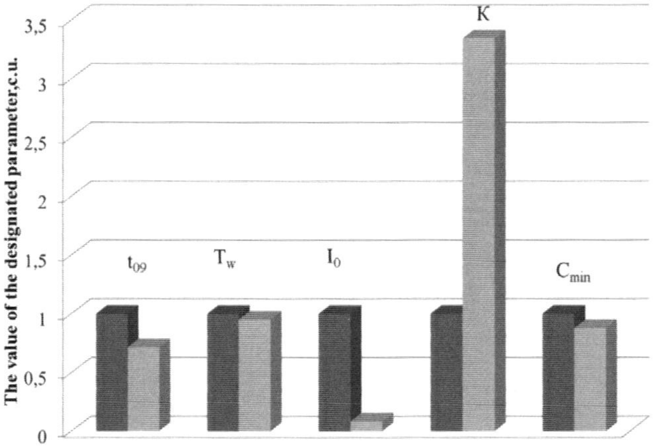

Figure 4. Comparison of the characteristics of the developed sensors (columns on the right) in comparison with the sensors of the main manufacturers: Alphasense (Great Britain), City Technology (Great Britain), Drager (Germany), Nemoto (Japan); MST (Germany).

References

[1] Spilker B, Randhahn J, Grabow H, H Beikirch, Jeroschewski P. 2008 New electrochemical sensor for the detection of hydrogen sulfide and other redox active species. *Journal of Electroanalytical Chemistry* 612(1) pp 121-30

[2] Vasiliev A A, Pisliakov A V,. Sokolov A.V., Samotaev N N et. al. 2016 Non-silicon MEMS platforms for gas sensors *Sensors and Actuators, B: Chemical* vol. 224 pp 700-713

[3] Somov A, Karpov E F, Karpova E, Suchkov A, Mironov S, Karelin A, Baranov A *et al* 2015 Compact low power wireless gas sensor node with thermo compensation for ubiquitous deployment. . *IEEE Transactions on Industrial Informatics* 11 (6), pp 1660-1670

[4] Makeenkov A, Lapitskiy I, Somov A and Baranov A. 2015 Flammable gases and vapors of flammable liquids: Monitoring with infrared sensor node. *Sensors and Actuators B: Chemical.* 209 pp 1102-07

[5] Samotaev N, Ivanova A, Oblov K, Soloviev S, Vasiliev A. 2014 Wi-Fi wireless digital sensor matrix for environmental gas monitoring. *Procedia Engineering*,vol. 87 pp 1294-1297

[6] Baranov A, Spirjakin D, Akbari S and Somov A. 2015 Optimization of power consumption for gas sensor nodes: A survey. *Sensors and Actuators A: Physical.* 233 pp 279-289

[7] Eggins Brian R, Chemical Sensors and Biosensors, John Wiley & Sons Ltd.,London, 2002.

[8] Hart J, Crew A, Crouch E, Honeychurch K, Pemberton R M, 2004 Some recent designs and developments of screen-printed carbon electrochemical sensors/biosensors for biomedical environmental, and industrial analyses, *Anal. Lett.* 37 pp 789–830

[9] Vork F, Janssen L, Barendrecht E. 1986 Oxidation of hydrogen at platinum-polypyrrole electrodes, *Electrochim. Acta* 31 pp 1569–1575

[10] Litvinov A V, Nikolaev I N and Samotaev N N. 2010 MIS - Field effect sensors for low concentration of H2S for environmental monitoring. *Procedia Engineering.* 5 pp 1216-1219

[11] Medvedeva E, Baranov A and Somov A. 2016 Design and investigation of thin film nanocomposite electrodes for electrochemical sensors, *Sensors and Actuators B* 236 pp 858–864
[12] Baranov A M. 1998 Planarization of substrate surface by means of ultrathin diamond-like carbon film *Surface and Coatings Technology* 102 pp 154-158

Radiation hardness estimation in various temperature conditions under radiation impact

A S Bakerenkov·, A S Rodin, V A Felitsyn, V S Pershenkov
National Research Nuclear University MEPhI (MEPhI), Kashirskoe shosse 31, Moscow, 115409, Russian Federation

·Corresponding author's e-mail address: ASBakerenkov@mephi.ru

Abstract. The application of the ELDRS conversion model for numerical simulation of the radiation degradation of input bias current in widely used voltage comparator is presented. The recurrence relations for numerical simulations were obtained and used calculations of the estimation of the radiation response of LM111 input bias current in real operation temperature conditions of TechEdSat satellite.

1. Introduction

In space environment the radiation dose rate and temperature can vary drastically and unexpectedly. As it was shown earlier [1], in order to estimate the operation time of electronic equipment under radiation impact with variable dose rate, it is possible to use the averaged value of the dose rate. To predict the operation time correctly it is necessary to take into account only the change in temperature over time for a specified orbit, which the satellite equipment is located on. Using a set of fitting parameters of the conversion model [1] for a particular integrated circuit and a plot of the actual temperature dependence on time for those conditions which the spacecraft will operate in, it is possible to calculate the dependence of the radiation-sensitive parameter degradation on total dose.

In this work we apply the conversion model of enhanced low dose rate sensitivity effect to estimate the dependence of input bias current of LM111 voltage comparator on total dose in real temperature conditions of space environment.

2. Mathematical expressions for numerical simulation

According to basic assumptions of the conversion model we can describe the kinetics of radiation induced positive charge generation in the oxide (Q_{ot}) and the rate of radiation degradation of a radiation sensitive parameter (I_{in}) by following differential equations:

$$\frac{dQ'_{ot}}{dt} = q \cdot K_D \cdot \gamma - \frac{Q'_{ot}}{\tau_D}, \tag{1a}$$

$$\frac{dI_{in}}{dt} = K_s \cdot \gamma + \frac{1}{q}\frac{Q'_{ot}}{\tau_D}, \tag{1b}$$

$$\tau_D(T) = \tau_{D0} \exp\left(\frac{E_A}{kT}\right), \tag{1c}$$

where K_D, K_S, τ_0 and E_A are fitting parameters, γ is dose rate, q is electron charge and $Q'_{ot} = AQ_{ot}$. Expression (1c) represents the Arrhenius law, which we use to describe the dependence of the conversion process rate on temperature.

For numerical calculations expressions (1a) and (1b) can be presented as recurrence relations:

$$Q'_{ot}(t+\Delta t) = \left(qK_D\gamma(t) - \frac{Q'_{ot}(t)}{\tau_D(T)}\right)\cdot \Delta t + Q'_{ot}(t), \quad (2a)$$

$$I_{in}(t+\Delta t) = (K_s\gamma(t) + \frac{1}{q}\frac{Q'_{ot}(t)}{\tau_D(T)})\cdot \Delta t + I_{in}(t), \quad (2b)$$

The recurrence relations (2a) and (2b) can be used to calculate values of $Q'_{ot}(t+\Delta t)$ and $I_{in}(t+\Delta t)$ after very short time interval Δt using corresponding values of $Q'_{ot}(t)$ and $I_{in}(t)$ in the beginning of the interval. The value of $\tau_D(T)$ must be calculated for time moment t using expression (1c) and substituted to relations (2a) and (2b). Corresponding temperature value T must be taken at the same time moment. Consequently if we know initial values of Q'_{ot} and I_{in} for $t = 0$ we can calculate this values for any following time moment step-by-step. For correct simulations the duration of Δt must be short enough to consider the temperature and dose rate as constants.

3. Application of the simulation technique for real temperature profile.

In [1] the fitting parameters of the conversion model extracted for LM111 voltage comparator are presented. Using the parameters and information from the on-board temperature sensor of the TechEdSat satellite (Fig. 1, information from the site https://directory.eoportal.org), which has low Earth orbit, the numerical simulation of the dependence of input current degradation of bipolar voltage comparator LM111 on time was performed for constant dose rate 1 mrad(SiO_2)/s and a variable temperature.

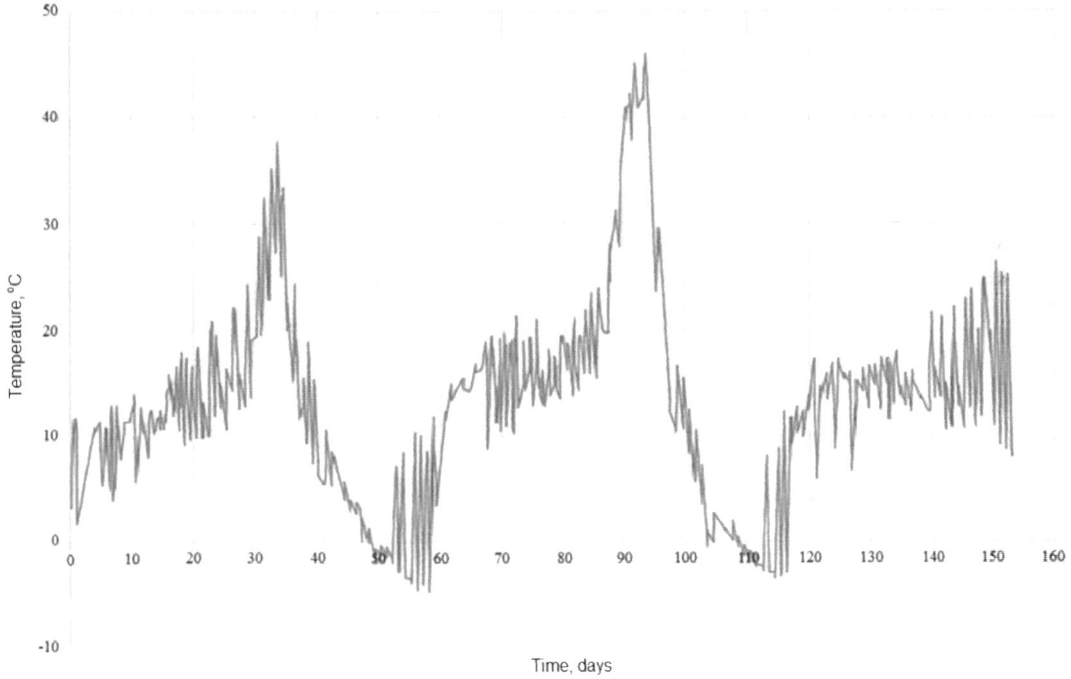

Figure 1. The dependence of the TechEdSat satellite temperature on time for low Earth orbit

For automation of numerical simulation a computer program was developed. The developed program allows to perform a numerical estimation of radiation hardness of integrated circuits for any temperate transient processes. Moreover it enables to perform the updating of the life time estimation for a satellite electronics using real time temperature data.

The numerical estimation of the radiation degradation of the voltage comparator input bias current based on the real time temperature data was performed using relations (2a), (2b) and (1c). The results of performed simulation are presented in Fig. 2. In Fig. 3 the total dose response of the input bias current is presented for averaged temperature value (22.6°C) and for 25.0°C, which is typical for hardness assurance test procedures, together with real time simulation data from Fig. 2 for comparison.

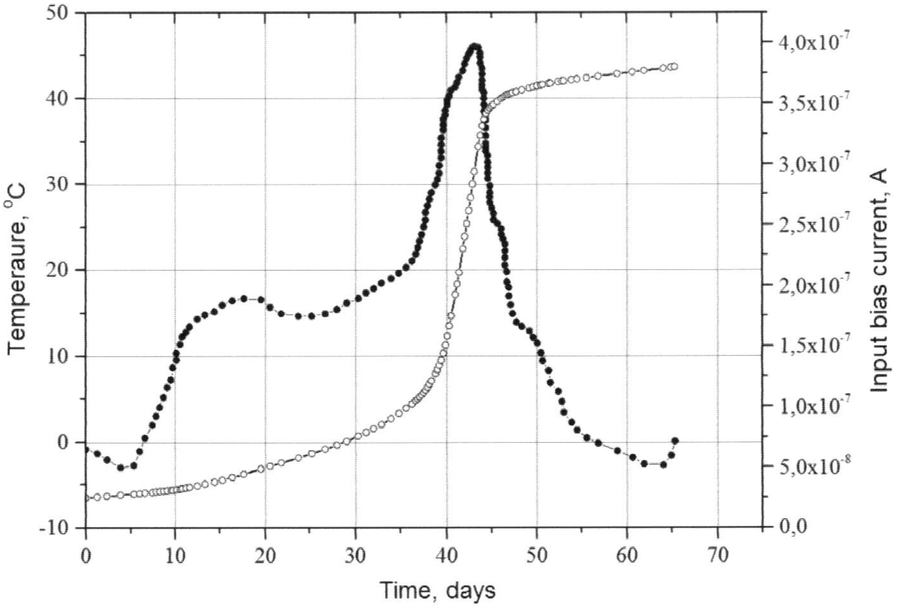

Figure 2. The calculated dependence of the input current degradation of the voltage comparator LM111 on the irradiation time (●), based on the real time dependence of temperature (○) in low-earth orbit for the TechEdSat satellite

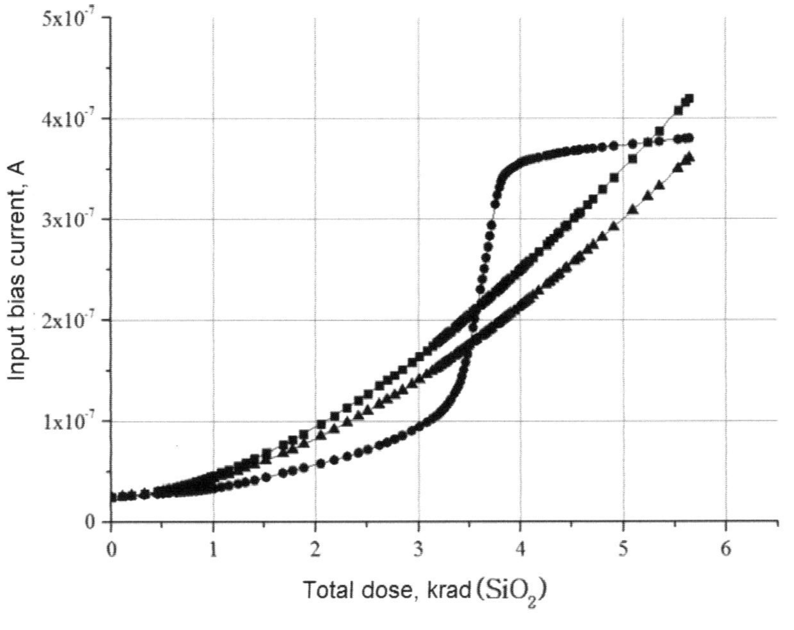

Figure 3. The calculated dependence of the degradation of the input current of the voltage comparator LM111 on total dose at 25°C (■), for complex temperature profile (●) and at an average temperature value (22.6°C) (▲)

The results of the numerical simulation presented in Fig. 2 shows a significant difference between the total dose responses for a constant temperature value and complex temperature profile. It is extremely important to take into account real temperature conditions during space mission.

4. Conclusion

From the simulation results we can see the significant impact of the temperature profile on the degradation rate of the input current of the LM111 voltage comparator. It demonstrates the importance of taking into account the dynamics of the temperature variation when estimating the operation time of the satellite electronic equipment. When using the presented calculation and experimental technique for estimating of the radiation hardness of semiconductor devices, it becomes possible to adjust the prediction of the operation time of electronic devices in the space environment using actual data on the dose rate and operating temperature.

Acknowledgements

The authors of the paper would like to thank for the support from National Research Nuclear University MEPhI in the framework of the Russian Academic Excellence Project (contract No. 02.a03.21.0005, 27.08.2013).

References

[1] V.S. Pershenkov, A.S. Bakerenkov, A.S. Solomatin, V.V. Belyakov 2013 *Proc. Conf RADECS*. article number 6937456

Investigation of memristor effect on basis of thin-film oxide dielectrics modified by carbon

Yu V Sakharov

Tomsk State University of Control Systems and Radioelectronics (TUSUR), 40, Lenin Str., Tomsk, Russian Federation

Corresponding author's e-mail: iurii.v.sakharov@tusur.ru

Abstract. The regularities of the change of electrophysical properties of memristor structures based on porous films of titanium dioxide are considered at their modification by carbon. It is shown that the introduction of impurity results in the significant change of the electrophysical properties of memristor structures based on porous films of titanium dioxide. The purpose of this paper is to reveal the regularities between the electrophysical properties of elements of non-volatile memristor memory and the phenomena of switching and memory at atmospheric pressure.

1. Introduction

To date, the main problem of electronics is to find an alternative to FLASH memory. It relates to electrically reprogrammed semiconductor memory and stores bits in the form of electric charges. The main disadvantage of FLASH memory consists in electrostatic leakages that must be addressed at the transition to microcircuits of the nanometer scale [1]. As a result, all attempts to increase the density of information storage result in the significant shortening of the time of its storage.

The implementation of new perspective memristor elements will allow us to create effective computing systems to fulfil the goals set. A memristor is a passive element of electronics, the key feature of which is the ability to change electrical resistance depending on the value of the electric field applied to the active layer. They possess resistive switching and can remember the conductivity state for a long time because it is stored not in the form of a charge and is consequently completely non-volatile. Resistive switching is the control of the value of electrical resistance of metal-dielectric-metal structures by means of short-term influence by strong electric fields. Non-volatile memory possesses a number of advantages, such as longer period of storage of information and high-density of data recording, significantly exceeding that of existing types of memory.

A memristor has a simple enough structure: a layer of metal oxide, embedded between two metal electrodes, i.e. metal-dielectric-metal structures (MDM). Also, this effect is present in the structures based on semiconductor substrates, i.e. metal-dielectric-semiconductor structures (MDS). The titanium oxide Ti_nO_{2n-1} with the thickness of 5-40 nm is the most widely investigated. This memristor effect was observed for the first time on the mentioned type of titanium oxide. Widely investigated are also other metal oxides: ZrO_x, HfO_x, VO_2, Nb_2O_5, TaO_x, MoO_2, WO_3, Fe_2O_3, NiO, Co_2O_3, Al_2O_3 [2]. Nevertheless, the switching effect differs for different materials. On the basis of the current-voltage characteristics (CVC), the switching effect can be divided into two types: unipolar switching and bipolar switching. The typical CVC for the two types is shown in Fig. 1 [3].

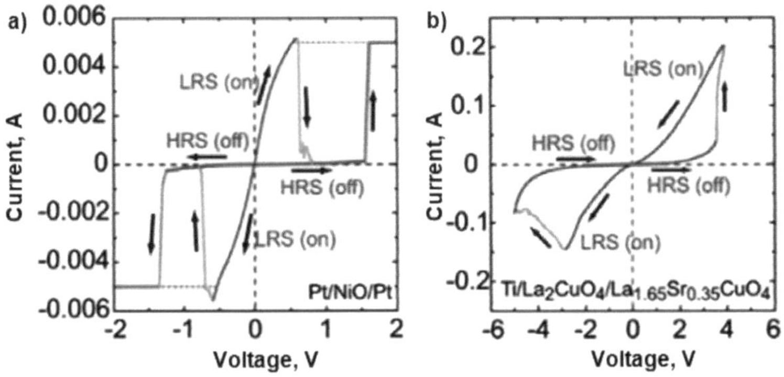

Figure 1. Current-voltage characteristics of a memristor for unipolar (a) and bipolar (b) switching [3]

At unipolar switching, the switching depends on the amplitude of applied voltage rather than on polarity (Fig. 1a). The structures are exposed to the forming process, after which they go into a low-resistance state (LRS). After the voltage reset, structures go into a high-resistance state (HRS). The switching from HRS to LRS occurs by the feeding of the threshold voltage that exceeds the voltage of the reset. Such a type of switching is observed in many insulation oxides, for example, in TiO_2 [4].

At bipolar switching, the switching depends on the direction of the applied voltage (Fig. 1b). This type of switching is observed in such complex oxides such as $Hf_xAl_{(1-x)}O_y$ and TiO_2-TiO_x [4, 5].

2. The process of electrical forming

The forming process consists in the following: an MDM structure is placed in a vacuum of no more than 1 Pa and the voltage of some critical value is applied. After a while, some nonreversible changes of electrical properties will occur in the structure; in particular, the conductivity will significantly increase, the structure will obtain a switching effect and a memory effect, and also electroluminescence in visible range, there will occur the electron emission in vacuum, and the CVC will obtain the N-type negative differential resistance. This process is referred to as forming, and the voltage under which the forming occurs is named as a voltage of forming [6].

These changes are connected to the presence of formed channels in the structure. The distinguishing characteristic of the formed channels is the absence of the upper electrode. Immediately after forming, the formed channels have an average diameter of 0.1-1 microns. The channel body is referred to as high conductivity channel; it is filled with microcrystals of the formed dielectric. The upper electrode over the channel is missing. It melts and detaches itself due to local heating and intensive gas emission, and a gap of 10-20 nm (a nanogap) is formed between the electrode edge and the high conductivity channel [6].

3. Experiment and measuring techniques

The experiment was carried out on Al-TiO_2-Al and Al-TiO_2-Ag structures. The Al films of 100 nm were applied by thermal evaporation in vacuum for use as a bottom electrode. For the counter electrode were used 30 nm films of Al and Ag, applied by thermal evaporation in vacuum. The dielectric was applied by means of magnetron sputtering of the titanium target in an atmosphere of oxygen at the pressure of 4×10^{-2} Pa in a vacuum chamber [7]. The thickness of the dielectric was 30 nm. The modification of the films by carbon was carried out by the sputtering of composite target Ti:C with various area ratios. At the same time, the magnetron construction allowed us to change the area ratio of Ti and carbon on the composite target. The amount of carbon introduced into the TiO_2 film was determined using the carbon area on the composite target and expressed in percentages of S_c. The modification of the TiO_2 film by carbon will permit to increase its porosity and to create additional oxygen vacancies, which will accelerate the process of electrical forming and will improve the resistive switching [8].

4. Experiment results and analysis

When applying the voltage of 12 V to an Al-TiO_2-Al structure, some nonreversible changes of electrical properties occur, which can be observed in the fact that the steady leakage current increases sharply, and the electroluminescence in the visible range and emissive current appear [9]. The forming time is counted from the moment of the voltage application to the moment of the relative stabilisation of steady leakage current. The dependence of steady leakage current from time is presented in Fig. 2a. The analysis of this dependence allows us to note that the modification of TiO_2 film by carbon considerably accelerates the forming process, thus increasing the steady leakage current passing through the structure. The main cause of the steady leakage current growth is the increase in density of the formed channels, which are mainly localised near pores. It is indicated by the electroluminescence pattern and the analysis of the structure surface before and after forming.

After the termination of the forming process, the CVC of both structures have considerably changed: steady leakage currents have significantly increased, and also there appeared a site of the negative differential resistance of N-type (Fig. 2b). Both structures have a hysteresis of current. However, the structures with the modified dielectric have higher I_{max}/I_{min} relation.

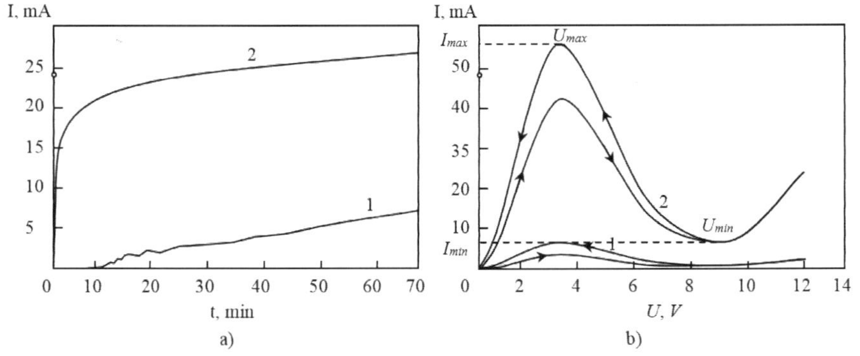

Figure 2. The change of steady leakage current in time (a) and CVC after forming (b) for Al-TiO_2-Al structures without the modification of the dielectric by carbon S_c=0% (1) and with the modified dielectric S_c=60% (2)

A rapid disconnection of voltage from the MDM structure on a site results in the appearance of memory effect. As a result, the structure "remembers" the electrical conductivity at the moment of the disconnection, and a new CVC will have the given electrical conductivity from 0 V to some threshold voltage Uth (3-4 V) or reset voltage (Fig. 3) on the site. This state can be erased by applying a voltage higher than threshold Uth or can be recorded again by applying a voltage U>Umin with subsequent rapid disconnection. Thus, the system takes on two stable states: LRS and HRS. Also, the given states are completely non-volatile and are resistant to moderate heat, light, as well as magnetic fields.

A prominent feature of the forming process is the necessity of a vacuum to fulfil it. Also, all above-mentioned effects can be observed only in a vacuum.

A distinct feature of the Al-TiO_2-Al structure with a carbon modified dielectric is the possibility to observe all of these effects in argon at atmospheric pressure when Ag films are used as an upper electrode. Besides, steady leakage currents decrease by a factor of 10-20.

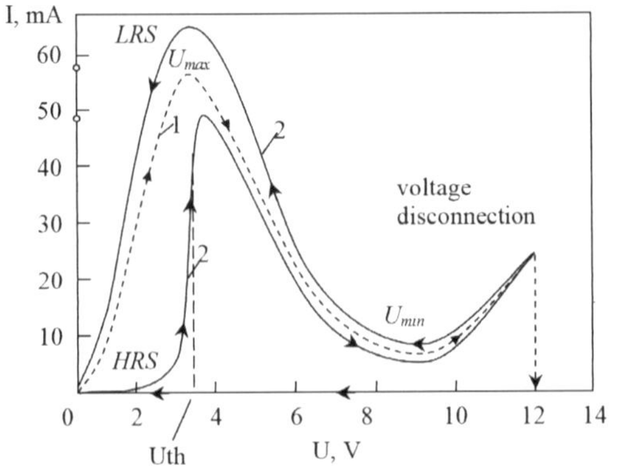

Figure 3. Initial CVC (1) and a memory effect (2) for an Al-TiO_2-Al structure with a modified dielectric S_c=60%

5. Conclusion

The modification of TiO_2 films by carbon results in the formation of a loose structure of the dielectric, which contains a considerable quantity of pores and gas inclusions [10]. It is supposed that it is the pores that are centres for the localisation of the formed channels, defining all properties of the formed structures. Accordingly, the increase in density of the formed channels results in the increase of steady leakage currents passing through the dielectric.

The reason of these changes consists in the chemical reaction between oxygen and carbon, resulting in the formation of a volatile compound of CO or CO_2. The evaporation of this volatile compound from the dielectric film leads to the formation of see-through pores and gas inclusions, and also to the formation of areas of partial titan oxidation. It is supposed that the kinetics of the forming process development can vary in the case of use of a porous dielectric with a higher number of defects.

References

[1] Shchigorev L A and Shagurin I I 2016 *IOP Conference Series: Materials Science and Engineering* **151** 1 012004
[2] Ma H, Feng J, Lv H, Gao T, Xu X, Luo Q, Gong T фтв Yuan P 2017 *Nanoscale Research Letters* **12** 1,118
[3] Sawa A 2008 *Materials Today* **11** 6 pp 28-36
[4] Magyari-Köpe B, Tendulkar M, Park S-G, Lee H D and Nishi Y 2011 *Nanotechnology* **22** 25 254029
[5] Kundozerova T and Stefanovich G 2013 *Applied Mechanics and Materials* **346** pp 29-34
[6] Troyan P E *Russian Physics Journal* **39** 10 pp 948-995
[7] Burman R 2010 *Proceedings - 2010 International Conference on the Origins and Evolution of the Cavity Magnetron* 5565572 pp 3-4
[8] Troyan P E et al 2006 *Russian Physics Journal* **49** 2 pp 219-220
[9] Troyan P E 1996 *Russian Microelectronics* **25** 2 pp 136-137
[10] Liu C, Li Y, Xu P, Li M and Zeng M 2015 *Materials Chemistry and Physics* **149** pp 69-76

IOP Conf. Series: Materials Science and Engineering **498** (2019) 012030 doi:10.1088/1757-899X/498/1/012030

Design of AIS for space applications on the basis of dynamic current cascades providing formation of synergetic effect

A A Lebedev*, N M Klokov and I I Lyakhchilina

National Research Nuclear University MEPhI (Moscow Engineering Physics Institute), Kashirskoe shosse 31, 115409 Moscow, Russian Federation

*Corresponding author's e-mail address: lebedev741@mail.ru

Abstract. - Two approaches to the design of analog ICs (AIS) for space applications are considered. The first approach is related to the modernization of industrial AIS by including dynamic current cascades (DCCs) in their structures. The second approach is based on the development of an input differential cascade (preamplifier) built on dynamic cascades (RH-CM). In the design approaches under consideration, the improvement of AIS quality is achieved by using the synergistic effect (SE).

1. Introduction

Article [1] is devoted to the analysis of the synergistic effect between the accumulated dose (TID) and the SET reaction. However, in the articles [2, 3, 4] it is emphasized that the effect of radiation always leads to deterioration in the characteristics of AIS. Moreover, the study [5] of OpAmp LM124 of three manufacturers showed that the support of the scheme, including its schematic and topological components, has a great influence on the formation of the SE. It should be noted that works in which architectural, structural and circuit engineering methods are used, including the inclusion of dynamic current stages, are practically absent in literature. The purpose of this paper (based on the results obtained by Rosche) is to improve the performance (increase in slew rate) in the LM124 op-amp, as well as to reduce the vulnerability to the effects of a heavy charged particle.

2. The principle of radiation-resistant RH-CM

Fig. 1 and Fig. 2 show the Widlar current reflector and the proposed radiation-resistant current reflector, as well as the combination of the input DC circuit with the intermediate stage and the proposed RH-CM. CM Widlar has a low input resistance, so it is capable to work from a current source. In the RH-CM circuit, an additional transistor (Q6), which is connected by a common-base circuit, is introduced, thereby providing a large input resistance and the ability to operate from the source of the voltage signal.

IOP Conf. Series: Materials Science and Engineering **498** (2019) 012030 doi:10.1088/1757-899X/498/1/012030

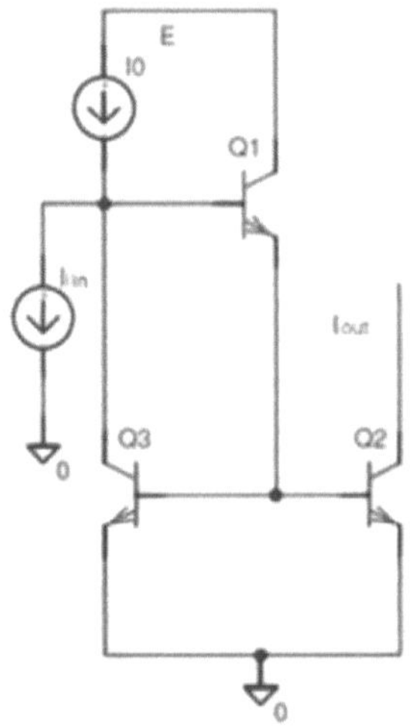

Figure 1. Widlar CM

$K_I = 1 - \frac{2}{\beta^2+\beta+2}$ Works for β>20

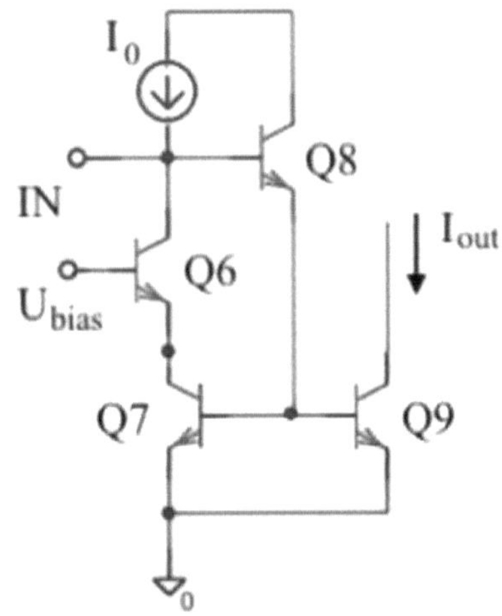

Figure 2. proposed CM

$K_I = 1 + \frac{\beta-2}{\beta^2+2}$ Works for β>3

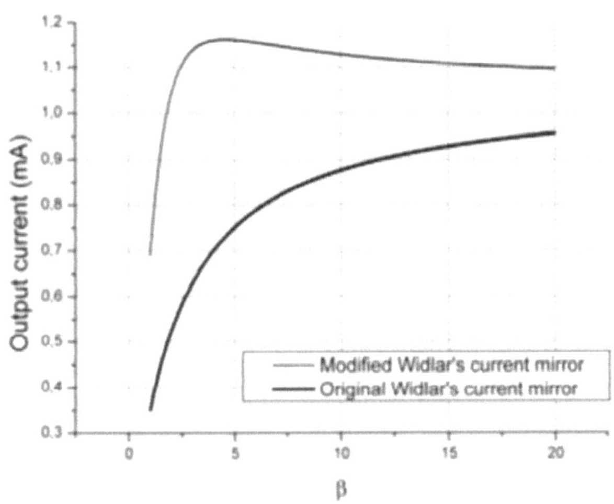

Figure 3. Current reflector with current "flush" (upper curve)

Note that the synergistic effect generates a current boost due to the introduction of local positive feedback (Fig. 3). In this case, the loop gain of the op amp is generally less than that of unity, so the op amp operates steadily when it is covered by the common negative feedback. In [5], it is shown that a combination of 2-3 current reflectors into a common ring allows them to be used as independent cells (CC-II, CC-III) that emulate certain mathematical functions.

3. Decreasing the SET time

In order to reduce the SET time under the action of HCP, it is proposed to use a CM with a current throw in operation. Such a current reflector has the property of increasing or generating an additional current in dynamic mode. Modernization of the structure of the OpAmp is achieved with the inclusion of DCC - CM with increased radiation characteristics.

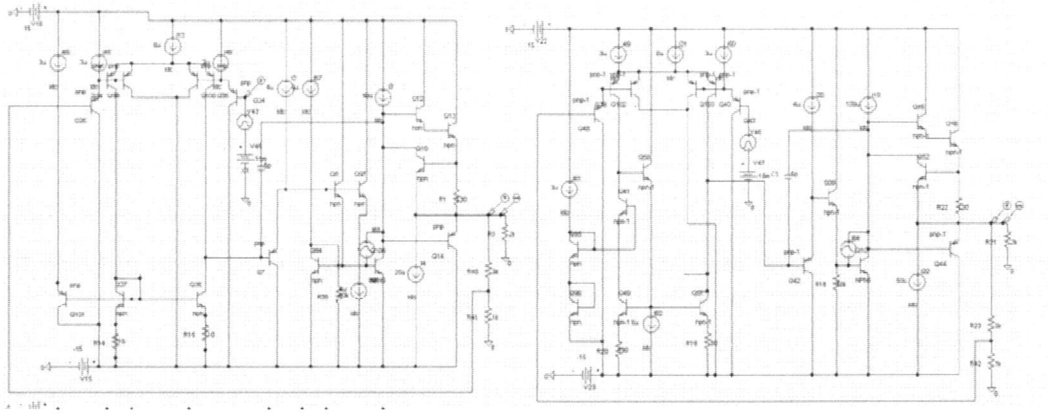

Figure 4. The structural diagram of the LM124 op-amp and the block diagram of the modified CM.

Fig. 4 shows that the difference lies in that between the output of the first and the input of the second cascade, a radiation-resistant CM is included. Simulations of both schemes were carried out for their response to HCP at the accumulated dose of 0, 20 and 50 krad, which corresponds to the degradation of β from 100 to 50 and 5. The inclusion of CM sharply reduced the transition time, which can be explained by that the charge accumulation and resorption occur practically instantly.

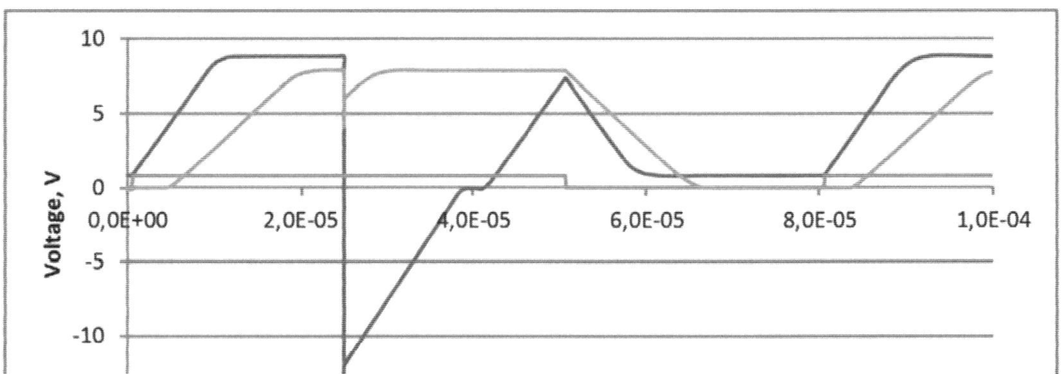

Figure 5. HPC with beta 100 in non-invertor with a gain of 10

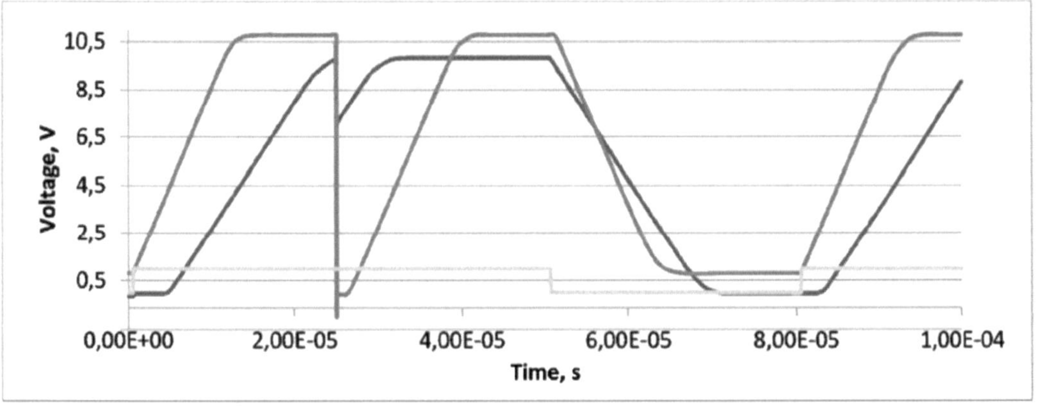

Figure 6. HPC with beta 5 in non-invertor with a gain of 10

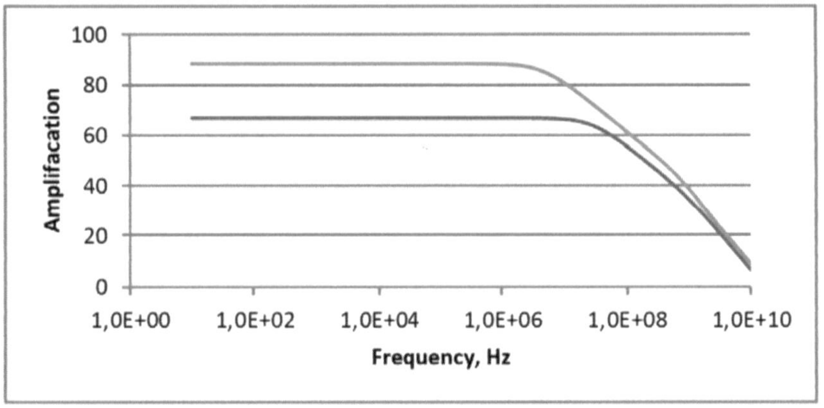

Figure 7. K degradation with least and most accumulated dose

Fig. 5, 6 and 7 shows SET reaction with HPC at time 25 μs and degradation of beta with least and most accumulated doses. Blue (Fig. 5 and Fig. 6) line represents our modification of OpAmp. Another line represents a typical model of LM124.

4. Conclusion

A new dynamic current reflector has been proposed, which tends to automatically increase the mode current when the gain β is degraded, the operating principle of such a current reflector is considered, and one of the methods of the coordinated switching of the current reflector into the electric circuit of an operational amplifier is shown. Considering the factor that the circuit of the developed operational amplifier practically repeats the prototype circuit, except for the connection of the current reflector with a throw, which allows to upgrade the LM124 OpAmp circuit with simultaneous increase of the life cycle of the spacecraft by 2-3 years.

5. References

[1] Roche N J-H, Dusseau L, Boch J, Gonzalez Velo Y, Vaillé J-R, Saigné F, Auriel G, Azais B, Buchner S P, Marec R, Calvel P and Bezerra F 2010 *IEEE T. Nucl. Sci* **57** 1861

[2] Roche N J-H, Dusseau L, Vaillé J-R, Mekki J, Gonzalez Velo Y, Perez S, Boch J, Saigné F, Marec R, Calvel P, Bezerra F, Auriel G and Azaïs B 2010 *IEEE T. Nucl.* Sci **57** 3392

[3] Roche N J-H, Perez S, Mekki J, Velo Y G, Dusseau L, Boch J, Vaillé J-R, Saigné F, Marec R, Calvel F, Bezerra F, Auriel G, Azaïs B and Buchner S P 2011 *IEEE T. Nucl. Sci.* **58** 2980

IOP Conf. Series: Materials Science and Engineering **498** (2019) 012030 doi:10.1088/1757-899X/498/1/012030

[4] Zebrev G I, Petrov A S, Useinov R G, Ikhsanov R S, Ulimov V N, Anashin V S, Elushov I V, Drosdetsky M G and Galimov A M 2014 *IEEE T. Nucl. Sci.* **61** pp 1785-1790
[5] Roig F, Dusseau L, Khachatrian A, Roche N J-H, Privat A, Vaillé J-R, Boch J, Warner J H, Saigné F, Buchner S P, McMorrow D, Ribeiro P, Auriel G, Azais B, Marec R, Calvel P, Bezerra F and Ecoffet R 2013 *IEEE T. Nucl. Sci.* **60** 4430
[6] Toumazou C, Lidgey F J and Haigh D G 1990 *Analogue IC design: the current-mode approach* (London: Six Hills Way) pp 20-68

IOP Conf. Series: Materials Science and Engineering **498** (2019) 012031 doi:10.1088/1757-899X/498/1/012031

Donor ionization tuning in AlGaAs/InGaAs/GaAs PHEMT quantum wells with AlAs nanolayers in spacer

D A Safonov, A N Vinichenko, Yu D Sibirmovsky, N I Kargin, I S Vasil'evskii

National Research Nuclear University MEPhI (Moscow Engineering Physics Institute), Kashirskoe shosse 31, 115409 Moscow, Russian Federation

Corresponding author's e-mail address: ivasilevskii@mail.ru

Abstract. A comparison of the electron transport properties of pseudomorphic quantum well $Al_xGa_{1-x}As/In_{0.2}Ga_{0.8}As/GaAs$ with those of conventional donor layer (x=0.15 and x=0.25) and with AlAs nanoinserts around the delta-Si layer or AlAs: delta-Si donor layer is presented. The structures with added AlAs layers exhibit electron concentration decrease, combined with increased electron mobility. This effect is related to the suppression of remote ionized impurity electron scattering, change of band structure and decreasing efficiency of silicon atoms doping when incorporating in pure AlAs.

1. Introduction

Due to high electron velocity and two-dimensional electron gas density, pseudomorphic high electron mobility transistors (PHEMT) are one of the most widely used materials for microwave semiconductor electronics. Increasing electron mobility while a high electron concentration is maintained is still an important challenge. Delta doping through the spacer layer allows high electron concentration, yet remote ionized impurity scattering becomes a significant factor, as it vastly decreases electron mobility. Also, some portion of conduction electrons begin to move in the region of the doping layer [1]. Thus, parallel conduction appears in two layers: in the $_xGa_{1-x}As$ quantum well layer with high electron mobility and the wide-bandgap barrier layer with low mobility [2-4]. This second parallel conducting channel leads to nonlinear transistor gate characteristics and increasing noise [5]. This parasitic effect can be avoided by using two-sided doping or by increasing the height of the $Al_xGa_{1-x}As$ barrier due to a larger AlAs mole fraction. The first drawback is that the wide-bandgap AlGaAs layer in HEMT heterostructures can increase ohmic contact resistivity between drain and source. The second is that the properties of $Al_xGa_{1-x}As$ become ineligible due to the increase of ionization energy and reduction of doping efficiency if AlAs content is high enough. DX-centers appear at x>20%, leading to [6, 7]. However, in contrast to bulk layers, the position of the donor level energy can be influenced by the electrostatic potential distribution in heterostructure. The ionization of deep levels can be high enough if they are located considerably higher than Fermi level in heterostructures if the donor layer is located close to the quantum well. Some papers show increased electron mobility in AlGaAs/GaAs HEMT with AlAs spacer and barrier layer [8]. Thus, the problem of eliminating parallel conduction without using thick AlGaAs barrier layers with increased AlAs content is of great importance.

In this paper, we study the electron properties of PHEMT quantum wells with two types of barrier/donor layer: one containing AlAs nanoinserts around the delta doping layer and another with

AlAs:δ-Si doping layer. A comparison with standard PHEMT structures with a smooth $Al_xGa_{1-x}As$ barrier with different Al mole fraction is carried out (x=0.15 and x=0.25).

2. Samples growth and preparation

A series of PHEMT samples with one-sided silicon delta doping was grown using molecular-beam epitaxy on semi-insulating GaAs (100) substrates. To improve buffer quality, an AlGaAs/GaAs superlattice was grown on the substrate. GaAs buffer layer is also a barrier for the $In_{0.2}Ga_{0.8}As$ quantum well from one side. From the other side, the $Al_xGa_{1-x}As$ layer serves as a barrier and spacer with aluminium mole fraction x = 0.16 for samples 1-3 and x = 0.25 for samples 4 and 5. The distance between the quantum well and delta-layer was 8 nm for samples 1-3 and 5 nm for samples 4 and 5. After the delta-layer $Al_xGa_{1-x}As$, a Schottky barrier was grown: 26 nm wide for samples 1-3 and 20 nm for samples 4 and 5. The top of the heterostructure was capped with a slightly doped 13 nm GaAs layer. In sample 2, the delta-layer was surrounded by two 1 nm-wide AlAs nanolayers separated by 4.2 nm. In sample 3, an 8 nm layer AlAs with the delta-layer in the middle was grown at the distance of 4 nm from the quantum well heterojunction.

For transport measurements, Hall-bar mesa structures were made using photolithography and wet etching, with Ni/AuGe/Ni/Au ohmic contacts, which were annealed at 390 °C [9]. Electron mobility and concentration measurements were carried out at magnetic fields up to 6 $kgs^{-2}A^{-1}$ in a temperature interval of 2.1 K ÷ 300 K with a Cryo-free 404 system.

3. Band structure and electron states modeling

The band structure, the potential profile, the energies of subbands and the profiles of the electron wave functions were calculated for samples 1, 2 and 3. Due to heavy doping in sample 1, the conduction band potential around the delta-layer is decreasing, which leads to the change in the wave functions profile: the ground state wave function ψ_1 is mainly localized in the quantum well, yet partially permeates into the barrier. Wave functions of excited states become hybrid: ψ_2 and ψ_3 has the same amplitude around donors and in the quantum well. In this case, electrons in these subbands have low mobility because of ionized impurity scattering. Moreover, these subbands are close in energy ($E_3 - E_2 \leq 15$ meV) and thus the wave functions ψ_2 and ψ_3 have a large spatial overlap. The addition of two thin AlAs barriers around the Si delta-layer significantly changes the shape of the wave functions and the energies of the upper subbands. The AlAs inserts do not noticeably change the energy of the ground state; however, the amplitude of wave function ψ_1 around the donors is 4.5 times lower in the second sample and 200 times lower in the third sample. The excited states ψ_2 and ψ_3 are mostly localized in separate parts of the structure – ψ_2 inside the quantum well, while ψ_3 is predominantly around the delta-layer. The reason for these changes is that adding AlAs nanobarriers mainly increases electron state energy around the donor delta-layer. At the same time, the energy gap $E_3 - E_1$ is increased by 20 meV and the ψ_2 and ψ_3 states overlap is greatly reduced. Confinement increase in the quantum well for the state ψ_2 also leads to a slight increment in the intersubband gap $E_2 - E_1$ by ~ 14 meV.

In sample 3, an 8 nm AlAs layer greatly increases the energy of the bottom of the conduction band around the region of donors, thus forcing out electron states from the donors region. As Figure 1 shows, in that case the wave functions of the first and second subband become almost completely confined in the InGaAs quantum well.

With increasing barrier height caused by adding AlAs, the difference between first and second subband energies increases, and the bottom energy of conduction band around the delta-layer does not influence them because these two subbands are associated with the quantum well. Energy of the third subband, associated with the delta layer, increases by 20 meV with the addition of AlAs inserts around the donor layer and considerably more so in the case of an AlAs donor layer.

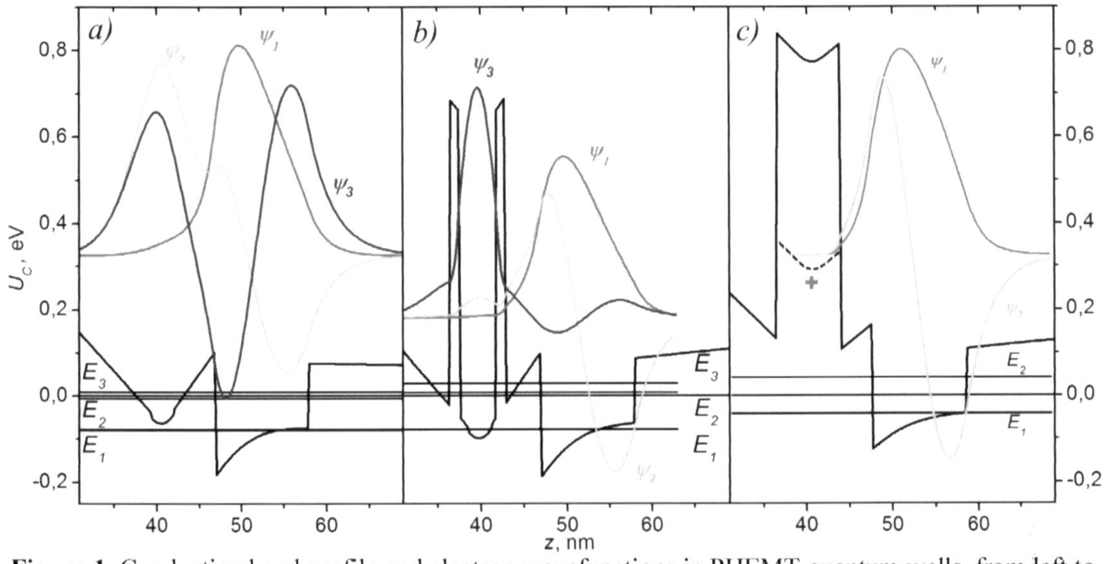

Figure 1. Conduction band profile and electron wavefunctions in PHEMT quantum wells, from left to right: a) conventional structure, b) structure with the two AlAs nanolayers surrounding δ-Si area, c) structure with the AlAs:δ-Si donor layer.

Furthermore, band structure around the donor delta-layer affects the donor ionization energy. With adding AlAs nanobarriers, Si donor states, that in hydrogen-like approximation have an effective width of wave function of about 5 nm, suffer a level shifting proportionate to the increase of average Al mole fraction x in the wide-gap barrier layer. In case of a solid AlAs(Si) layer in sample 3, the ionization donor energy decreases relatively to the conduction band bottom, however the donor level turns out to be higher than the conduction band bottom in the surrounding ternary layer AlGaAs and about ~ 260 meV higher than Fermi level.

4. Results and discussion

Specific resistance of samples, Hall mobility and concentration of electrons were measured in the temperature range from 2.1 K to 300 K. Electron mobility increases with decreasing temperature, and along the whole range of measurements, the third sample has the highest mobility.

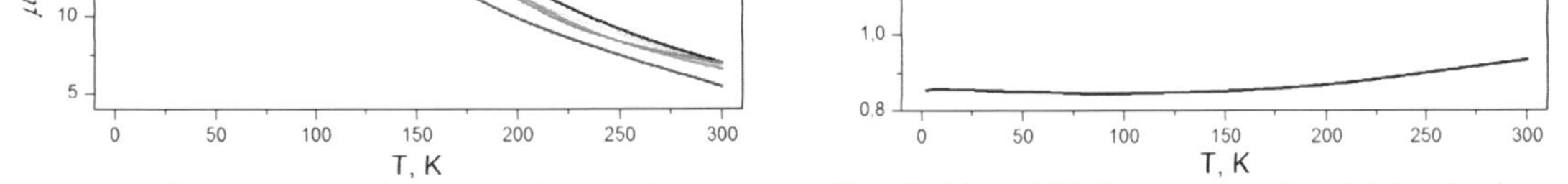

Figure 2. Temperature dependencies of electron mobility (left) and Hall concentration (right) in the samples.

The AlAs mole fraction in the barrier layer $Al_{0.16}Ga_{0.84}As$ of samples 1 and 2 was chosen to be comparatively small, so as to keep ionization donor energy sufficiently low relative to the edge of conduction band. On the other hand, conduction band energy in the barrier turns out to be small. On the contrary, in samples 4 and 5, AlAs content is high, which provides higher conduction band energy in the barrier.

In sample 1, electron concentration temperature dependence shows the largest increase at temperatures T > 100 K.

Adding two AlAs barriers slightly decreases Hall concentration, while temperature dependence becomes weaker. An even fainter change of electron concentration is observed in samples 4 and 5, despite the fact that sample 5 has the greatest value of electron concentration.

This can be explained by the fact that the ionization effectiveness of Si donors depends on the position of donor levels relatively to Fermi level, which in turn depends on both aluminum content in $Al_xGa_{1-x}As$ and donor concentration. Adding nanobarriers surrounding Si delta-layer increases donor ionization energy level since Bohr radius for ionized impurity of Si in the hydrogen-like model is an order of about ~ 10 nm [10], so in sample 2, donor level rises relative to the edge of the conduction band. Donor ionization is better at low temperatures; electron concentration growth begins at higher temperature and has a lesser value. Nevertheless, despites identical doping of samples 1 and 2, sample 2 with AlAs nanoinserts shows less electron concentration value. Also, in sample 3 with the AlAs donor layer, donor impurity concentration was 1.3 times higher; however, Hall electron concentration turned out to be ~ 2 times less than in sample 1. Band structure calculation shows that in the AlAs donor layer, the impurity level associated with X-valley is located ~ 0.52 eV lower than the G-valley conduction band bottom, and in the case of sample 3 is ~ 300 meV higher than Fermi level. So in such a way, the decrease of the electron concentration effect in the discussed PHEMT heterostructures is not related to ionization since Si donor level is located high enough, however with the decrease of silicon incorporation, effectiveness as donor impurity with increasing AlAs content. Besides, since donor level in AlAs is associated with X-valley and the effective mass of X-electrons is substantially larger than G-electrons, effective wave function width of the donor state turns out to be 4 times less. Thus, impurity ionization energy in the delta-doped AlAs nanolayer must correspond to its value in bulk AlAs. This explains the donor ionization peculiarity of increasing temperature in different samples: in sample 1, $n(T)$ has maximum growth, whereas in sample 2, concentration change is less and the temperature of change-over to increasing dependence is higher. In sample 3, increasing $n(T)$ turns out to be the weakest. Similar behavior of concentration temperature dependence shows samples 4 and 5 with uniform barrier layer with higher AlAs content.

Hall electron mobility behavior differs in samples 1-3 and 4, 5. With increase of silicon donors concentration by 2.1 times in sample 5 in comparison with sample 4, Hall concentration increases by ~ 1.5 times. According to data obtained from band structure modeling and Shubnikov-de Haas effect analysis, in samples 4 and 5 with a higher aluminum mole fraction, only one subband is populated. At the same time, due to better screening and increase in Fermi-momentum electron mobility is also increasing with electron concentration. A different situation is observed in samples 1-3. Electron mobility in these samples is increasing with decreasing Hall concentration. This can be explained with the decrease of remote ionized impurity electron scattering due to suppression of electrons tunneling from the quantum well to wideband barrier. Also, the electron mobility changes with the addition of AlAs is related to the population of higher subbands in samples 1 and 2. In sample 2, the electron concentration in the second subband is about 2% of the concentration in the first subband. In sample 1, where mobility is the lowest, electron concentration in the second subband is 10% of the first one. In sample 3, which has the highest mobility, the second subband population is neglected because of lower Si doping efficiency in AlAs. Thus, mobility in samples 1-3 increases with the addition of AlAs layers in the barrier because electrons localize predominantly in the ground subband with the highest electron mobility.

5. Conclusion

A comparison of electron transport properties of the pseudomorphic quantum well $Al_xGa_{1-x}As/In_{0.2}Ga_{0.8}As/GaAs$ with a conventional donor layer (x=0.15 and x=0.25) and with AlAs nanoinserts around the delta-Si layer or the AlAs doped barrier layer is presented. Structures with AlAs layers show an increase of electron mobility combined with slightly decreased electron concentration. This effect is related to the suppression of remote ionized impurity electron scattering, change of band structure and decreasing efficiency of silicon atom doping in pure AlAs.

Acknowledgments
The research was supported by the State assignment program № 8.3887.2017/PCH.

References

[1] Vasil'evskii, I S, Galiev G B, Klimov E A, Mokerov V G, Shirokov S S, Imamov R M and Subbotin I A 2008 *Semiconductors* **42** 9
[2] Van der Burgt M, Karavolas V C, Peeters F M, Singleton J, Nicholas R J, Herlach F, Harris J J, Van Hove M and Borghs G 1995 *Phys. Rev. B.* **52** 16
[3] Lisesivdin S B, Yildiz A, Balkan N, Kasap M, Ozcelik S and Ozbay E 2010 *J. Appl. Phys.* **108**
[4] Masselink W T, Braslau N, Wang W I and Wright S L 1987 *Appl. Phys. Lett.* **51** 19
[5] Chen Y W, Hsu W C, Shieh H M, Chen Y J, Lin Y S, Li Y J and Wang T B 2002 *IEEE Trans. Elec. Dev.* **49** 2 pp 221-225
[6] Schubert E F and Ploog K 1984 *Phys. Rev. B.* **30** 12
[7] Godts, P, Constant E, Depreeuw D, Zimmerman J and Danneville F 1988 *Electr. Lett.* **24** 15
[8] Holland M C, Skuras E, Davies J H, Larkin I A, Long A R and Stanley C R 1995 *J. Cryst.Growth* **150**
[9] Vanyukhin K D, Zakharchenko R V, Kargin N I, Pashkov M V and Seidman L A 2015 *Russian Microelectronics* **44** 8 pp 122-127
[10] Bastard G 1990 *Wave mechanics applied to semiconductor heterostructures.*

IOP Conf. Series: Materials Science and Engineering **498** (2019) 012032 doi:10.1088/1757-899X/498/1/012032

Investigation of the deposition of metal nanoclusters on the surface of porous electrode materials by electrophoresis

V Sleptsov[1], A Savkin[1,*], D Kukushkin[1], A Diteleva[1], N Samotaev[2]

[1]Moscow Aviation Institute (National Research University), Moscow, Russia
[2]National Research Nuclear University MEPhI (Moscow Engineering Physics Institute), Moscow, Russia

[*]Corresponding author's e-mail address: savkin@mati.ru

Abstract. The problems of coating on the entire surface of a highly porous electrode material based on activated carbon from a colloidal metal solution are considered. Experimental results demonstrated the possibility of metallization of porous carbon materials with nanoclusters of metals 2-10 nm in size by electrophoresis without reducing their initial porosity.

1. Introduction

The miniaturization of electronics has led to the fact that the main volume and weight of many devices is occupied by a power source, which is subject to ever-increasing demands on the duration of battery life, the amount of stored energy, the number of recharging cycles and the service life, which must be at least the lifetime of the electronic device itself. Thus, today there was a need for a sharp increase in the energy characteristics of current sources, which can be achieved by using microelectronic technologies and nanotechnologies in their production [1, 2].

Priority areas for improving miniature sources of electrical energy have become chemical sources of current, primarily lithium batteries and supercapacitor structures. Inferior to chemical sources of current in specific energy, supercapacitors considerably exceed them in terms of specific power and stability of charge-discharge characteristics over a wide temperature range.

Microelectroncs manufacturing technologies has been studied a lot recently [13-15]. Microelectronic technologies and nanotechnologies are in demand for improving the parameters of electrode materials, as the main elements of chemical current sources and supercapacitor structures [3, 4].

Prospective electrode materials must meet several criteria: developed surface area, high electrical conductivity, availability of porous structure to wetting, chemical inertness to structural materials and so on. The most fully listed requirements are met by carbon materials of varying degrees of porosity [5-10]. One of the promising materials in this group are carbon cloths from activated carbon fibers – "busofit" (Figure. 1) [11].

The capacitance of the capacitor depends primarily on the surface area, which in carbon materials can increase in comparison with the geometric surface by 10^3-10^5 times. In addition to a highly developed surface, electrode materials must have a low electrical resistance, both along the layer and across, as this determines the internal resistance of the source, which limits its output current. To reduce the electrical resistance of busofit and create an electrode material on its basis, two problems must be solved. First, apply a continuous layer of metal to the surface of the busofit, which will be a

IOP Conf. Series: Materials Science and Engineering **498** (2019) 012032 doi:10.1088/1757-899X/498/1/012032

current collector and reduce the internal resistance, which can be done using, for example, magnetron sputtering. Secondly, it is necessary to deposit a layer of metal on each filament of the busofit, which, on the one hand, will lower their electrical resistance, and on the other hand will increase the surface.

To solve this problem, it is proposed to use the technology of depositing nanoclusters of metals on the surface of porous materials by electrophoresis. As is known, nanometer-sized particles have a number of unique properties. They carry a positive charge, and their electrophysical and physicochemical properties are determined by the size of the nanoclusters. Therefore, the main task is to obtain metal particles with the desired size range and then position them in the porous space by electrophoresis. In this case, nanoparticles should have a uniform distribution over the surface.

a b

Figure 1. Initial busofit (a) and a snapshot with increasing (b)

2. Experimental

In the studies, the serial carbon cloth Busofit T-40 and similar carbon materials Busofit T-1, Busofit-T, Busofit-TM-4 were used. On the busofit threads, nanoclusters of nickel and silver were applied from the hydrosol. The preparation of metal nanoclusters was carried out by the method of electropulse dispersion due to an electric discharge between electrodes made of the metal immersed in distilled water [12]. The interelectrode distance was 100 µm, air gap of 1 mm, capacitor voltage up to 3 kV, capacitance of 1000 pF. Produced nanoclusters of metals are shown in Figure. 2.

Under such a synthesis process, colloidal solutions of silver and nickel were obtained in which the initial sizes of the nanoclusters were 25-40 nm and 2-10 nm, respectively.

To metallize busofit with metal nanoclusters, an experimental laboratory reactor were developed using electrophoresis, its 3D model and a photograph of which is shown in Figure. 3.

Busofit (3) was placed between the electrodes (2 and 5), after which the reactor was filled with a colloidal solution with silver or nickel nanoclusters and connected to an AC power source with a voltage of 40 V. Colloid particles have a charge, so they can move in an electric field. Reaching the destination electrode, the particles lose their charge and stick together - the colloidal solution coagulates on the surface of the material.

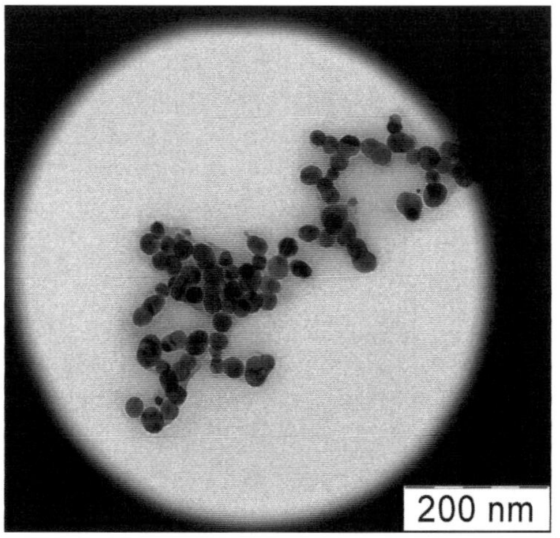

Figure 2. Produced silver nanoclusters of various sizes

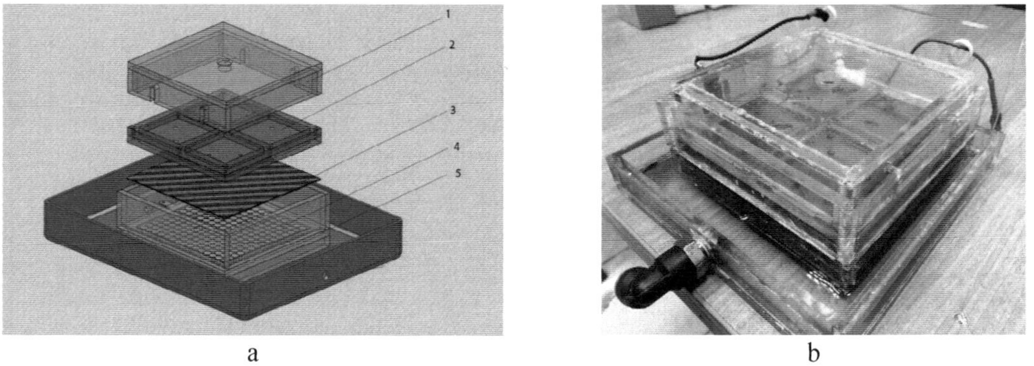

a b

Figure 3. 3D model (a) and photo of the laboratory reactor for the application of metal nanoclusters (b) 1 - upper clamping cover, 2 - upper electrode, 3 - carbon cloth "busofit", 4 - base, 5 - lower electrode

Figure 4 shows shots of metallized busofit fibers obtained with an electron beam microscope. As can be seen, nickel is deposited on the busofit in a thin layer, and silver on the surface of the busofit fibers is formed in the form of large nanoparticles and crystallites, developing the surface of the fiber. The experiment of silver deposition on pre-nickel-plated busofit fibers also did not allow a single layer of silver to be deposited on the nickel sublayer as well as the initial busofit fibers. Apparently, this metallization result is associated with the initial size of metal nanoclusters, thin-film coatings are obtained from nanoclusters with sizes of 2-10 nm, and large nanoparticles and crystallites are formed from nanoclusters of a larger fraction.

IOP Conf. Series: Materials Science and Engineering **498** (2019) 012032 doi:10.1088/1757-899X/498/1/012032

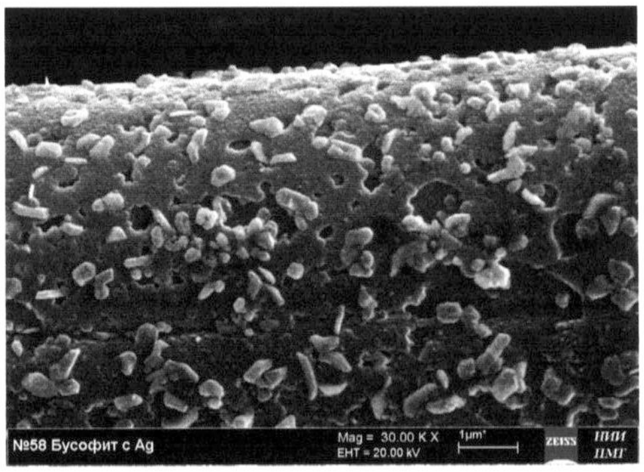

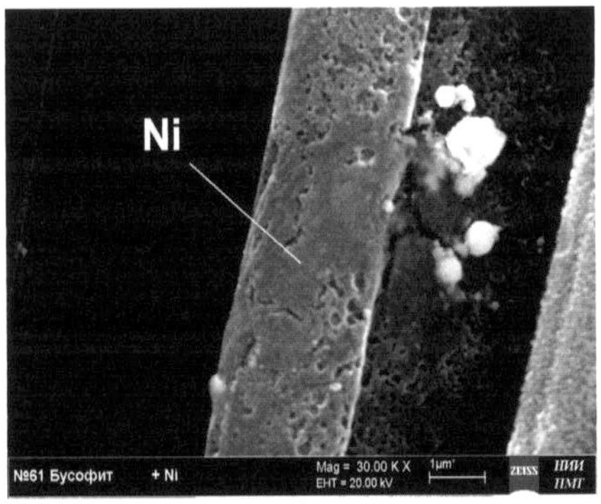

a b

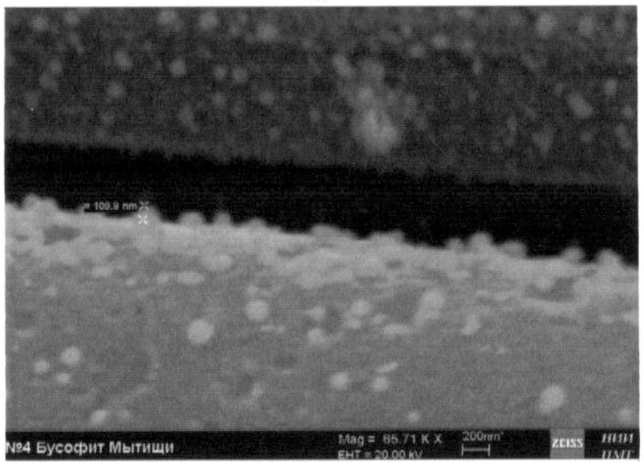

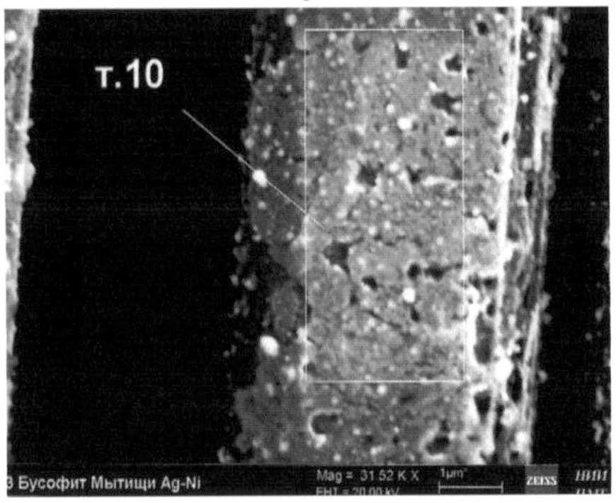

c d

Figure 4. The busofit fibers metallized with silver (a), nickel (b), silver over nickel (c, d)

3. Conclusions

It was shown the possibility of metallization of porous materials with metal nanoclusters by electrophoresis without reducing their initial porosity. This treatment allows the formation of a coating on the entire surface of a highly porous material, which reduces the electrical resistance and stabilizes the parameters of the electrode material.

To further study the metallization of highly porous materials, it is planned to use graduated colloidal solutions with a smaller spread of nanoclusters from various metals and alloys.

Acknowledgement

This work was supported by the grant № 8.7552.2017/BP from the Ministry of Education and Science of Russian Federation

Reference

[1] Arico A S, Bruce P, Scrosati B, Tarascon J-M and van Schalkwijk W 2005 *Nature Materials* **4** pp 366 – 377
[2] Huang X, Tan C, Yin Z and Zhang H 2014 *Advanced Materials* **26** 14 pp 2185–2204
[3] Scrosati B 2007 *Nature Nanotechnology* **2** pp 598 - 599
[4] Lee S W, Yabuuchi N, Gallant B M, Chen S, Kim B S, Hammond P T and Shao-Horn Y 2010 *Nature Nanotechnology* 5 pp 531–537
[5] Sleptsov V V, Elinson V M, Baranov A M and Tereshin S A 1995 *Wide Band Gap Electronic Materials NATO ASI Series. 3 High Technology* 1 pp 257-264
[6] Li X and Bingqing W 2013 *Nano Energy* **2** 2 pp 159-173
[7] Pandolfo A G and Hollenkamp A F 2006 *Journal of Power Sources* **157** 1 pp 11-27

IOP Conf. Series: Materials Science and Engineering **498** (2019) 012032 doi:10.1088/1757-899X/498/1/012032

[8] Blomquist N, Wells T, Andres B, Bäckström J, Forsberg S and Olin H 2017 *Scientific Reports* **7** 39836
[9] Weinstein L, Dash R 2013 *Materials today* **16** 10 pp 356-357
[10] Dai L, Chang D W, Baek JB and Lu W 2012 *Small* **8** 8 pp 1130–1166
[11] Astakhov M V, Galimzyanov R R, Klimont A A, Krechetov I S, Kundu M and Stakhanova S V 2016 *Current Nanoscience* **12** 1 pp 83-89
[12] Ostroukhov N N, Tyanginskii A Y, Sleptsov V V and Tserulev M V 2014 *Inorganic Materials: Applied Research* **5** 3 pp 284-288
[13] Vanyukhin K D, Zakharchenko R V, Kargin N I, Pashkov M V and Seidman L A 2015 *Russian Microelectronics* **44** 8 pp 564-568
[14] Shostachenko S A, Zakharchenko R V, Ryzhuk R V, Kulyamina D A and Kargin N I 2016 *Proceedings of SPIE - The International Society for Optical Engineering* **10224**
[15] Veselov D S, Bakun A D and Voronov Y A 2016 *Journal of Physics: Conference Series* **748** 1

IOP Conf. Series: Materials Science and Engineering **498** (2019) 012033 doi:10.1088/1757-899X/498/1/012033

Investigation of the synthesis of nanoparticles by the method of spark erosion with overvoltage of the discharge gap

V Sleptsov, A Savkin*, V Berdnik, D Kukushkin, A Diteleva and N Samotaev

National Research Nuclear University MEPhI (Moscow Engineering Physics Institute), Kashirskoe shosse 31, 115409 Moscow, Russian Federation

*Corresponding author's e-mail address: savkin@mati.ru

Abstract. The silver nanoparticles were studied, which were obtained by spark erosion with an overvoltage of the discharge gap. Experimental results show the possibility of producing nanoparticles of conductive materials with a narrow size distribution.

1. Introduction

The unique properties of nanoparticles have been the subject of intense research for several decades. Modern science has shown the possibility of using them in many applications, including for storage devices, creating new energy storage devices, MEMS, etc. [1-3]. A number of methods for obtaining nanoparticles have been developed, which can be conditionally divided into two groups: chemical methods and physical methods (dispersion) [4-6].

To obtain submicron and nanoscale particles of conductive materials (metals, alloys and semiconductors), a method of synthesis based on the electrospark erosion of raw materials in a liquid dielectric using the Electrical Discharge Machine (EDM), which is very attractive in its simplicity and accessibility, was proposed [7]. However, this method and a number of other known methods for synthesizing nanoparticles have the disadvantage of a large dispersion in the sizes (from 1 to 100 nm) of produced nanoparticles, which makes it extremely difficult to obtain or isolate nanoparticles with the required characteristics.

Further studies aimed at studying the method of electrospark erosion confirmed the possibility of obtaining nanoparticles of various conductive materials, including complex composition [8, 9]. Investigations of the influence of dielectric liquid media used in the synthesis of nanoparticles [10], including cryogenic liquids [11], have been carried out. A decrease in the size of the resulting metallic nanoparticles upon exposure to a liquid dielectric in the discharge zone by ultrasound due to cavitation and mixing has been established [12].

The results obtained by different groups of researchers made it possible to better understand the process of electrospark synthesis of nanoparticles, but could not eliminate the drawback associated with a large scatter of nanoparticle sizes. This can be due to several factors. In [10], it is proposed to use pulses with minimal energy for the synthesis of nanoparticles, which is difficult with the use of standard pulse generators for EDM. In addition, as shown by the results of mathematical modeling of the behavior of vapor-gas bubbles formed in a liquid dielectric in breakdown [13], their collapse is in most cases accompanied by division into several parts whose dimensions depend on the condition of their formations, which subsequently leads to the formation of several nanoparticles.

On the basis of these results and arguments, we made the assumption that a decrease in the spread of the sizes of the produced nanoparticles is possible due to a change in the conditions for the occurrence of a spark breakdown, so that the process proceeds much faster. This can be achieved if a significant overvoltage is created on the EDM electrodes, which is not possible in any previously described scheme. To create an overvoltage, another switching element must be added to the circuit, which would commute the charged capacitor and the EDM electrodes to a high voltage.

2. Experimental

An experimental EDM was developed with the possibility of creating an overvoltage at the electrodes. The functional diagram of the experimental EDM is shown in Fig. 1.

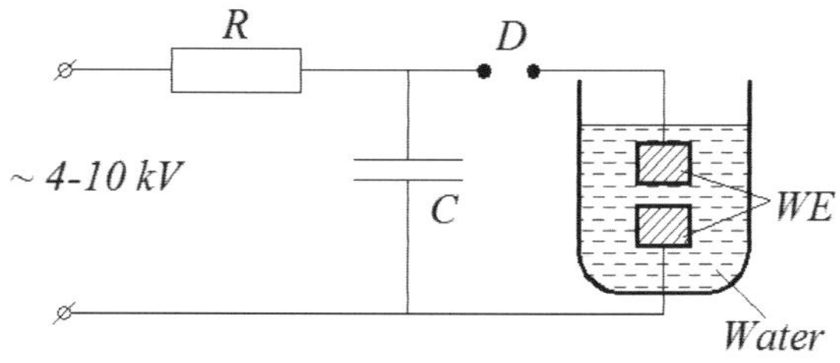

Figure 1. Functional diagram of the experimental EDM. C - high-voltage capacitor, R - limiting resistor, D - spark gap, WE - working electrodes

A high-voltage capacitor (C) with a capacity of 1000 pF was used in the experimental EDM. The capacitor was charged from a high-voltage AC transformer (50 Hz) to a voltage of 1-10 kV. For capacitor switching (C), an air electric spark gap (D) is used, which is an air gap of 1 mm width between two 50 mm2 electrodes. The photo of the air electric spark gap is shown in Fig. 2a. The charge rate of the capacitor is determined by the resistor (R). The capacitor charged to a voltage sufficient to breakdown the air gap (D) and the electrode system (WE) is discharged through the air gap D to the working electrodes (WE) immersed in the liquid dielectric.

Figure 2. Air electric spark gap (a), immersed silver electrodes (b), experimental EDM (c)

IOP Conf. Series: Materials Science and Engineering **498** (2019) 012033 doi:10.1088/1757-899X/498/1/012033

As a liquid dielectric, distilled water was used in the experiment. The working electrodes were made of silver, their photograph is shown in Fig. 2b. For the experiment, a gap of 70 μm was established between them. For uniform consumption of electrode material during erosion, rotation of working electrodes relative to each other was provided. The general view of the collected experimental EDM is presented in Fig. 2c.

The experiment showed that at the indicated values of the parameters of the experimental setup, the breakdown voltage was of the order of 4 kV, and the discharge current in the pulses reached values of 1 ÷ 2 kA. The obtained samples of silver nanoparticles were subjected to the investigation of the size distribution by the method of dynamic light scattering (DLS), the measurement was carried out with the Photocor Compact-Z. The resulting particle size distribution is shown in Fig. 3. As can be seen from the figure, the distribution of the nanoparticles obtained in size is rather narrow. The diameter of more than 80% of nanoparticles lies in the range of 20-45 nm.

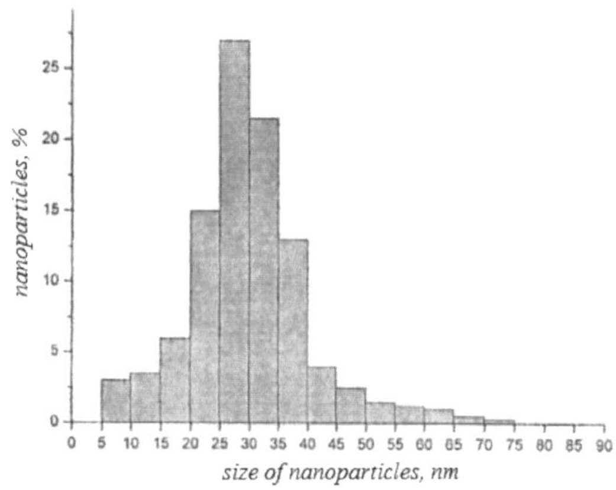

Figure 3. Distribution of silver nanoparticles by size

In addition to the DLS method, images obtained with the transmission electron microscope LEO-912 AB OMEGA were used to confirm the nanoparticle size distribution. In the pictures, it is seen (Fig. 4) that most particles are spherical and have a size of 20 to 40 nm.

The electron diffraction pattern of the obtained nanoparticle samples and its comparison with a sample of crystalline silver showed that the nanoparticles obtained consist of crystalline silver, without a noticeable admixture of oxides or salts (Fig. 5).

Figure 4. Photographs of the silver nanoparticles on an electron microscope

3. Conclusions

The possibility of producing nanoparticles by electrospark erosion under conditions of considerable overvoltage of the discharge gap was demonstrated. The resulting silver nanoparticles had a spherical shape and a rather narrow size distribution.

For a better understanding of the effect of overvoltage of the discharge gap on the production process of nanoparticles, it is planned to investigate this process for different ratios of discharge gaps and various conductive materials.

a b

Figure 5. Electronogram of silver nanoparticles (a) and solid silver (b)

Acknowledgement
This work was supported by the grant № 8.7552.2017/BP from the Ministry of Education and Science of Russian Federation

Reference

[1] Yan Y, Warren S C, Fuller P and Grzybowski B A 2016 *Nature Nanotechnology* **11** pp 603–608
[2] Karmakar S, Kumar S, Rinaldi R and Maruccio G *Journal of Physics: Conference Series* **292** 012002
[3] Theodorakos I, Zacharatos F, Geremia R, Karnakis D and Zergioti I 2015 *Applied Surface Science* **336** pp 157-162
[4] Gubin S P, Koksharov Yu A, Khomutov G B and Yurkov G Yu *Russian Chemical Reviews* **74** 6 pp 489-520
[5] Lu A H, Salabas E L and Schüth F 2007 *Angewandte Chemie* **46** 8 pp 1222–1244
[6] Rajput N 2015 *International Journal of Advances in Engineering & Technology* **7** 4 1806-1811
[7] Berkowitz A E and Walter J L 1987 *Journal of Materials Research* **2** 2 pp 277-288
[8] Hong J I, Parker F T, Solomon V C, Madras P, Smith D J and Berkowitz A E *J. Mater. Res.* **23** pp 1758-1763
[9] Nguyen P K, Lee K H, Moon J, Kim S I, Ahn K A, Chen L H, Lee S M, Chen R K, Jin S and Berkowitz A E 2012 *Nanotechnology* **23** 415604
[10] Carrey J, Radousky H B and Berkowitz A E 2004 *Journal of applied physics* **95** 3 pp 823-829
[11] Monastyrsky G 2015 *Nanoscale Research Letters* 10 503
[12] Liu Y, Li X, Bai F, Chen J, Wang Y and Liu N 2014 *Particuology* **17** pp 36–41
[13] Shervani-Tabar M T, Mobadersany N 2013 *Theoretical and Computational Fluid Dynamics* **27** pp 701–719

IOP Conf. Series: Materials Science and Engineering **498** (2019) 012034 doi:10.1088/1757-899X/498/1/012034

Perspective technology for low-scale production of SMD MOX gas sensors

N N Samotaev, **K Y Oblov**, **A V Ivanova**

National Research Nuclear University MEPhI (Moscow Engineering Physics Institute), Kashirskoe shosse 31, Moscow, 115409, Russian Federation

Corresponding author's e-mail address: NNSamotaev@mephi.ru

Abstract. The article discusses technological and economic aspects of small-scale production of gas MOX sensors. The main technical factors affecting the use of sensors in various applications are described. The perspective laser micro-milling technology for ceramic MEMS producing of microhotplate and SMD package for the MOX sensor is discussed.

1. Introduction

The modern MOX sensor is a combination of several factors - a microhotplate (responsible for power consumption and temperature operation mode), MOX gas sensitive layer (responsible for sensitivity to target gases) and a package (responsible for the possible application in which the sensor is used). Last decades MOX gas sensors development there is a concentration of efforts, in the field of technologies allowing obtaining a cheap product for mass applications in the manner of indoor air quality (IAQ)monitoring [3-5]. Similar solutions were found by the leading manufacturers of MOX sensors (Figaro, Sensirion, Bosch, AMS, SGX and etc). Technological solutions that make it possible to obtain a sensor with satisfactory characteristics for such application consist in the use of a combination of silicon MEMS microhotplate and plastic, metal-plastic or metal-ceramic SMD package. But what to do if for scientific groups which, at the moment there is no market or is it too small for the return of investment in silicon MEMS technology and the production of specialized plastic or ceramic SMD package? In that case need technology process close to 3D prototype philosophy –fast, simple and cheap with ratio to a single sample to total costs.

2. Experimental

As a technology for manufacturing ceramic MEMS microhotplate and SMD packages, it is proposed to use laser micro-milling of monolithic Al_2O_3 [1,2] and screen-print, jet (aerosol) printing technologies to form platinum metallization on MEMS and package. The equipment involved in the described technology does not need clean rooms and is widely represented on the market. The software for technology is simple and accessible at the student level. The only specialized product of the proposed technology is the adaptation of machine vision to obtain the minimum possible size of the MEMS structure of the microheater and the deposition of the MOX gas sensitive layer and platinum metallization, as well as the translation program of the 3D models of MEMS microhotplate and 2D topology of metallization to a 4-axis laser micro-milling facilities for automatic production. Technological steps are follows:

- 3D modeling of MEMS microheater (Fig.1), top and bottom part of the package (Fig. 2) in * .stl format (Autodesk Inventor or SolidWorks program) and 2D modeling topology of MEMS metallization in * .dxf format (Autodesk AutoCAD program).
- Simulation of the MEMS microheater parameters in the ANSYS program for prediction approximate representations about the MOX sensor thermal characteristics (if it need)
- Laser micro-milling of monolithic ceramics on a 4-axis laser facility by using 3D model of MEMS microheater (Fig 3), top and bottom part of the package.
- Deposition of platinum metallization (Fig. 6) using 2D model topology and technical annealing of metallization on MEMS according to specification on screen-print, jet or aerosol platinum materials.
- Laser micro-milling of metallization according 2D model (if it need).
- Deposition and annealing of the MOX gas sensitive layer on the MEMS microhotplate.
- Assembling and conglutination by special glass single parts of sensor into a monolith package, like schematically represented for SOT-23 package on Fig. 2.

Figure 1. 3D model of microhotplate as a task for laser micro-milling. Right scale is MEMS thickness.

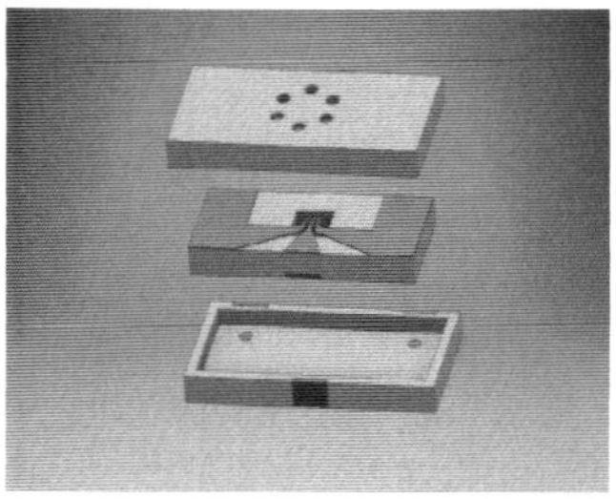

Figure 2. Assembling single parts of sensor into one SOT-23 packages (demension 1,6 x 2,8 x 1,1 mm).

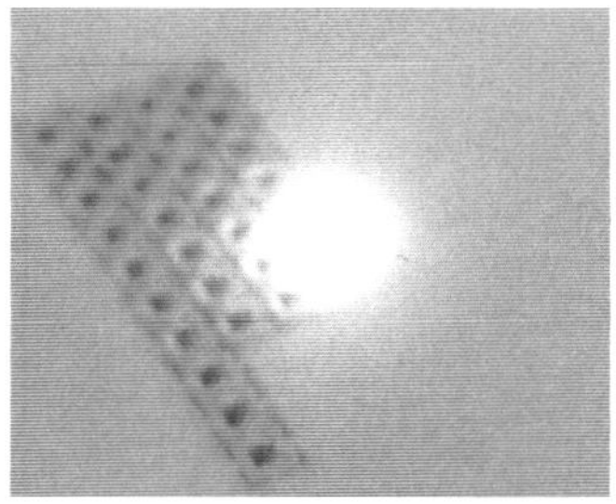

Figure 3. MEMS laser micro-milling on 250 μm thick Al_2O_3substrateby 4-x axis CNC facility.

Figure 4. Bottom view of MEMS after laser micro-milling. Chip size 1x1 mm by border of internal frame.

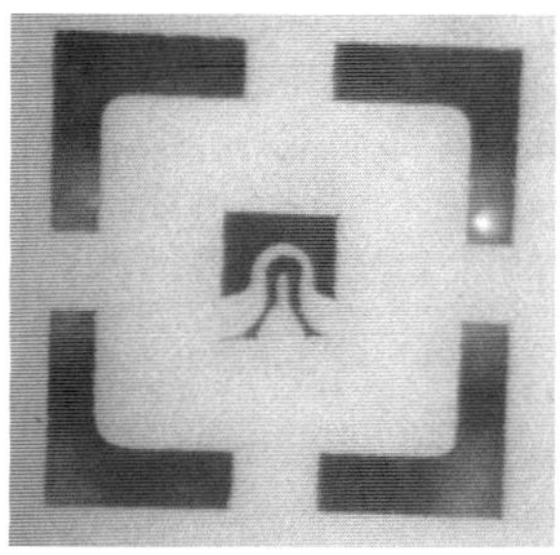

Figure 5. Top view of MEMS. Internal window size 0,3x0,3 mm. Central track width is 30 μm.

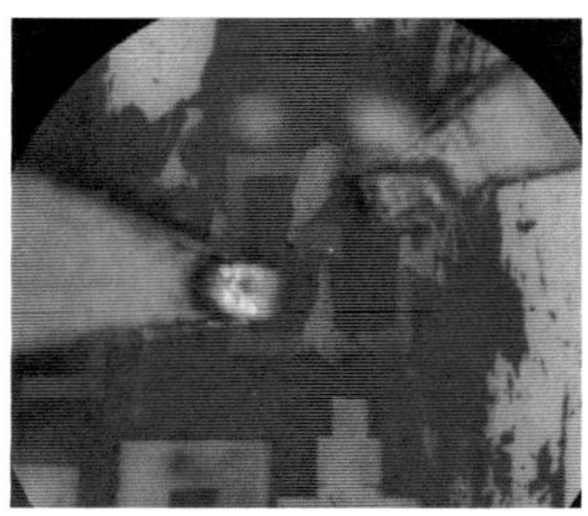

Figure 6. Testing MEMS microhotplate under voltage load (pins). Hot area at ≈ 600°C.

The quality of the final MEMS structure strongly depends on its surface roughness. The minimal roughness will make the MEMS structure more durable, as well as allow for more qualitative metallization deposition processes on the chip surface. To obtain a minimum surface roughness, it is required to use a material whose thickness does not exceed the height of the obtained structure in order to avoid ablation of the volume of the material only in order to reduce its thickness. However, the cost of the final product depends heavily on the cost of the material, in our case, ceramic substrates Al_2O_3. The minimum cost have the workpieces of standard size. The standard sizes of the ceramic substrates being manufactured start at a thickness of 500 μm or higher, which, at a chip height of 250 μm, requires excess ablation of the material at a thickness of 250 μm. Operation of laser evaporation of excess material can be justified in the case of a high cost of manufacturing substrates of non-standard thickness and only in the case of a qualitatively selected laser micro-milling mode that does not lead to a strong increase in the surface roughness after removal of excess material thickness.

Laser radiation allows to carry out the process of Al_2O_3 ceramics ablation to a depth of 0.5 to 50 microns in one pass. High radiation power and low speed of the laser beam accelerate the process of micro-milling, but lead to a higher surface roughness, the values of which can be more than 50 μm. Roughness is formed due to the fact that the intensity of laser radiation has a Gaussian distribution, which requires a very accurate selection of the laser beam overlapping region, when the entire material is engraved. On average, the surface roughness is 50 to 300 percent of the ablation depth per one pass of the laser beam. The minimum roughness was obtained by evaporation of the material in one pass at 1 μm and amounted to 2.3 μm (Fig. 7).

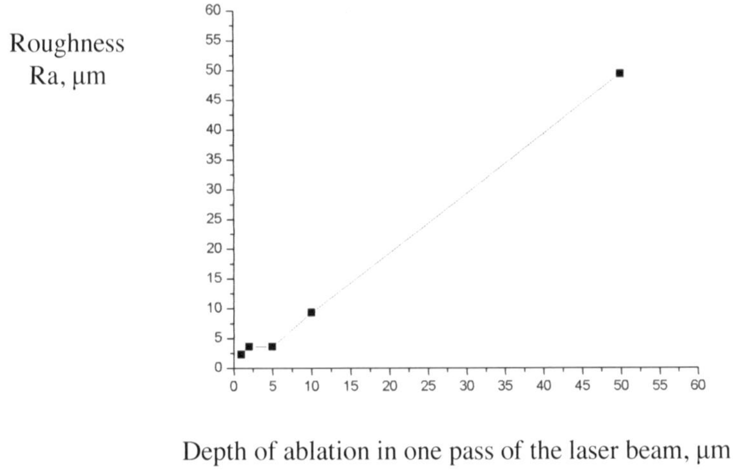

Figure 7. Dependence of the roughness on the depth of ablation in a single pass of the laser beam.

The laser micro-milling process is a complex multifactor physical process, whose behavior, depending on the type of material being processed, has a different character, which requires an empirical selection of the optimal surface treatment regime. Acceleration of the debugging process can be achieved by integrating the measuring equipment, able to produce a visual analysis of the surface obtained. After the ablation of each layer of material, such integration allows to analyze the resulting surface and correct the laser radiation regime by adjusting the radiation parameters or introducing additional cleaning passages aimed only at the surface roughness reducing (Fig. 8). In the process of laser micro-milling, the surface roughness increases with the increase in the number of engraved material layers.

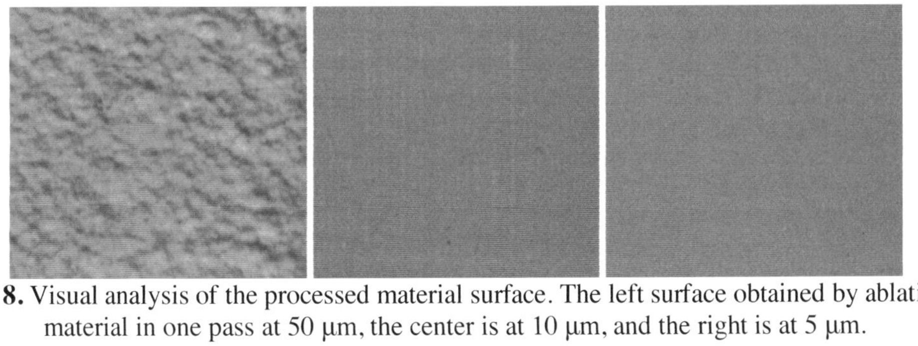

Figure 8. Visual analysis of the processed material surface. The left surface obtained by ablating the material in one pass at 50 μm, the center is at 10 μm, and the right is at 5 μm.

Other than the laser radiation modes, the quality of the final structure depends on the sequence of technological operations performed in one production cycle. Four coordinate laser microprocessing allows to process two sides of the substrate with accuracy of alignment up to 5 μm, which is permissible at the minimum MEMS structure (the central track) size of 30 μm. Two-sided laser microprocessing makes it possible to obtain a more accurate chip geometry, which is close to the 3D model, by reducing the conicity of the side faces of the chips, which is about 9 degrees when microprocessing Al_2O_3 ceramics, or about 80 microns with a substrate thickness of 500 microns.

When working with MEMS structures, one of the problems is the manipulation and moving of chips, the sizes of which do not exceed several millimeters. The application of metallization, a gas sensitive layer, the assembly of small-size chips is a very labor-intensive process, coupled with a large percentage of defects. Two-sided laser processing of the material makes it possible to locate bridges that do not allow the chip to fall out of the substrate after processing, symmetrically relative to the upper and lower planes of the substrate surfaces, which makes possible to fabricate an array of chips that do not require separation during metallization operations, a gas sensitive layer deposition, and gluing of the SOT-23 housing, in other words, until the final assembly (Fig. 9).

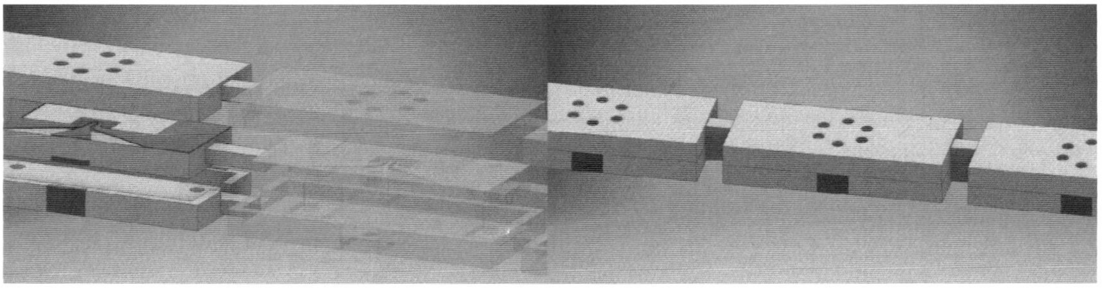

Figure 9. The process of assembling the chip into the SOT-23 housing, without dividing the array to separate chips.

This technology allows the group chips production, which greatly facilitates and speeds up the production process, and most importantly reduces the amount of rejects to a minimum.

Using this approach, experiments were carried out to fabricate a MEMS microhotplate, whose track width was 30 μm and a thickness of 20 μm. The power consumption of MEMS microhotplate at 450°C was approximately 250 mW (350 mW at ≈ 600°C - “burning” platinum temperature, see Fig.6).

3. Conclusion

Current results gives the prospect that manufacture MOX sensor in the SMD SOT-23 package type (max dissipating power by form-factor of package is 350 mW) for surface mounting in a tape is available by using described technology.

Acknowledgements

This work was supported by the Ministry of Education and Science of the Russian Federation (Grant No. 14.584.21.0030, unique identifier RFMEFI58417X0030

References

[1] Vasiliev A A, Pisliakov A V, Sokolov A V, Samotaev N N, Soloviev S A, Oblov K Y, Guarnieri V, Lorenzelli L, Brunelli J, Maglione A, Lipilin A S, Mozalev A and Legin A V 2016 *Sensors and Actuators, B: Chemical* **224** pp 700-713

[2] Ivanova A V, Oblov K Y, Soloviev S A, Samotaev N N, Gurkovskiy B V and Mironov V D 2016 *IFMBE. Proceedings* **55** pp 285-288

IOP Conf. Series: Materials Science and Engineering **498** (2019) 012034 doi:10.1088/1757-899X/498/1/012034

[3] Veselov D S, Bakun A D, Veselov D S, Bakun A D and Voronov Y A 2016 *Journal of Physics: Conference Series* **748** 1
[4] Podlepetsky B 2017 *Sensors and Actuators, B: Chemical* **238** pp 1207-1213.
[5] Podlepetsky B I, Kovalenko A V and Nikiforova M Y 2016 *IOP Conference Series: Materials Science and Engineering* **151** 1

IOP Conf. Series: Materials Science and Engineering **498** (2019) 012035 doi:10.1088/1757-899X/498/1/012035

Simulation of low energy ion implantation in silicon

D S Veselov*, Y A Voronov and Yu V Metel'

National Research Nuclear University MEPhI (Moscow Engineering Physics Institute), Kashirskoe shosse 31, 115409 Moscow, Russian Federation

*Corresponding author's e-mail address: dsveselov@mephi.ru

Abstract. This report is devoted to simulation of distributions of boron and indium ions implanted into the silicon substrates with energies up to 10 keV. The simulation was conducted immediately after implantation and after subsequent annealing in inert and oxidizing ambient. Post-implant and post-anneal distributions of impurities concentration versus depth were obtained. It was demonstrated, that the super-steep retrograde distribution of indium in silicon is provided exactly by the annealing in an oxidizing ambient. Also, it was demonstrated, that the annealing does not provide the super-steep retrograde distribution of boron in silicon. Thus, it was shown, that indium doping of gate region contributes to suppression of the short-channel effects in nanoscale MOSFETs, unlike boron doping.

1. Introduction

Doping of the gate region applies to the threshold voltage control of MOSFETs. For this purpose, boron doping is widely used at manufacturing of microscale transistors. But this impurity does not suppress the short-channel effects of nanoscale transistors. Recently, indium doping of the gate region is increasingly used at manufacturing of the nanoscale MOSFETs to this purpose. Indium atoms provide a super-steep retrograde well (SSRW) doping profile in the channel. The SSRW distribution of impurities implies a sufficiently large ratio of the maximum concentration to the surface one [1,2].

In several articles are presented the experimental results of impurities distribution of boron and indium in the channel of MOSFETs. These results were obtained by secondary ion mass spectroscopy. These results provide a comprehensive properties description of already developed devices but do not contribute to the appropriate choice of processing modes on the stage of devices development. Thus, unfortunately, an absence of physical process models complicates an appropriate choice of processing modes for nanoscale transistors manufacturing [3].

This paper presents the simulation results of the impurities distribution of boron and indium in silicon. The results were obtained for ions with initial energy up to 10 keV. The ion implantation with such ion energies is typically used at manufacturing of nanoscale MOSFETs. The impurities distribution was conducted immediately after implantation and after subsequent annealing in inert and oxidizing ambient. At such technological processes, after ion implantation, for impurity activation and crystal lattice defect removal, the high-temperature annealing is needed [4].

2. Simulation

The one-dimensional approximation was used at the simulation. The impurities distribution was simulated by the Gaussian function. The true distribution of impurities in most cases is much more precisely described by the Pearson function. At the depths exceeding the ions mean range, the impurity concentration values, calculated by the Pearson function, practically coincide with the values, calculated by the Gaussian function. At the depths, smaller then the ions mean range, the experimental values of the impurities concentration coincide with the values, calculated by the Pearson function, and greatly exceed the values, calculated by the Gaussian function. Moreover, these differences are increased with the increasing of initial ions energy. To obtain the small doping depth, ion implantation is carried out at low initial ions energy. At that the differences between the values of impurity concentration, calculated by the Pearson function and the Gaussian function, becomes insignificant. Therefore, the use of the Gaussian function, in this case, is acceptable. This approach provides quite an accurate description of the low energy ion implantation process.

Post-implant distribution of impurities was calculated using the following expression:

$$N(x) = \frac{Q}{\sqrt{2\pi}\Delta Rp} \cdot exp\left[-\frac{1}{2}\left(\frac{x-R_p}{\Delta R_p}\right)^2\right], \qquad (1)$$

where x is the coordinate, counted from the substrate surface into depth; Q is the radiation dose, equal to the number of ions, bombarding a surface unit of the substrate during the implementation; R_p is the ion mean range in silicon; ΔR_p is the standard deviation of ion mean range in silicon. The values of ions mean range and standard deviation were calculated according to the stopping Lindhard-Scharff-Schiott theory [5,6]. Obtained results are shown in table 1.

Table 1. Mean range and standard deviation of boron and indium ions in silicon.

Ions type	Boron	Boron	Boron	Indium	Indium	Indium
Ion energy, keV	1	5	10	1	5	10
R_p, nm	8,586	18,250	33,274	1,135	5,840	8,924
ΔR_p, nm	3,142	6,681	12,182	0,368	1,893	2,892

Using expression 1 and values of table 1 were obtained post-implant distributions of boron and indium in silicon, shown in Fig. 1. The simulation was conducted for ions with low initial energies (1, 5 and 10 keV). The radiation dose was assumed to be equal $3 \cdot 10^{13}$ $см^{-2}$. Smooth curves were obtained with interpolation by the method of the cubic spline using software Origin Pro 9.1 (Origin Lab corp.). As can be seen from Fig. 1, with the decreasing of the initial energy of ions, the concentration gradient of both impurities increases sharply. At the same modes of implementation (dose and ions energy), indium ions provide a steeper distribution of impurity, than boron ions. At that, the maximum concentration of indium ions is located closer to the substrate surface than it is typical for boron ions. However, post-implant impurities concentration on the substrate surface is too high. It does not allow providing SSRW impurities distribution, which is necessary to the formation of the channel of the nanoscale transistor with a low threshold voltage.

For the reduction of crystal lattice defects and impurity activation after ion implantation, it is necessary to carry out the high-temperature annealing of silicon substrates. During the annealing, due to the diffusion, the impurity distributes into the depth of the substrate. This leads to the reduction of impurity concentration on the substrate surface, as well as to decrease of the maximum impurity concentration. The distribution of impurities after implantation and subsequent annealing in an inert ambient at the temperature of 1000 degrees of Celsius for 10 seconds was calculated using the following expression:

$$N(x,t) = \frac{Q}{\sqrt{2\pi(\Delta Rp^2+2Dt)}} exp\left[-\frac{1}{2}\frac{(x-R_p)^2}{(\Delta Rp^2+2Dt)}\right], \qquad (2)$$

where D is the diffusion coefficient of impurity in silicon. The values of the diffusion coefficients were calculated according to [5], and amounted for indium $5{,}81 \cdot 10^{-15}$ см^2/с and for boron $3{,}09 \cdot 10^{-13}$ см^2/с. The distributions of boron and indium ions in silicon were obtained and shown in Fig. 2. As can be seen from Fig. 2, both for indium and boron the ratio of the maximum to surface concentration is decreases. For boron this ratio decreases by one order of magnitude. For indium this ratio decreases by two orders of magnitude. This indicates that the annealing in inert ambient contributes to the violation of SSRW distribution for both impurities.

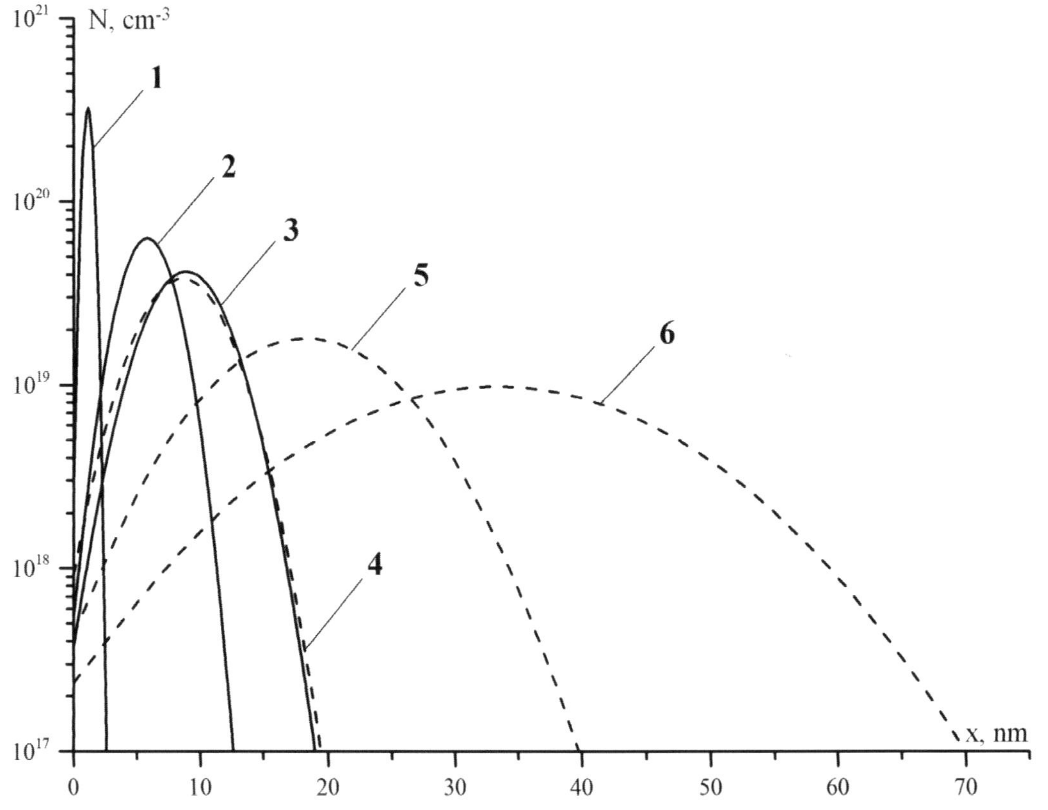

Figure 1. Post-implant distribution of boron and indium ions in silicon: solid curves for indium, dashed curves for boron. Initial energy of ions: 1,4 – E= 1 keV; 2,5 – E= 5 keV; 3,6 – E= 10 keV.

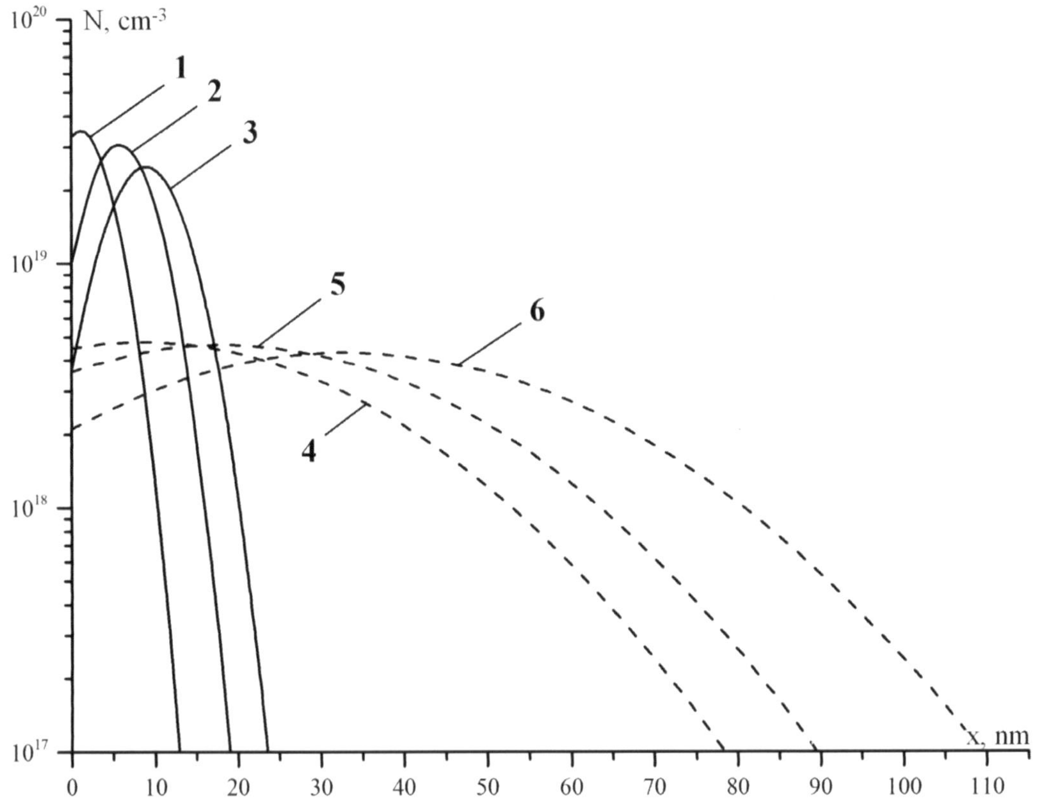

Figure 2. Distribution of boron and indium ions in silicon after implantation and subsequent annealing in inert ambient: solid curves for indium, dashed curves for boron. Initial energy of ions: 1,4 – E= 1 keV; 2,5 – E= 5 keV; 3,6 – E= 10 keV.

During the annealing in oxidizing ambient except for the diffusion processes have a significant impact on the processes of impurities redistribution due to the segregation at the interface Si - SiO_2. The essence of the segregation process is consisting of the redistribution of impurity between Si and SiO_2. This process is determined by the segregation coefficient k, equal to the ratio of the equilibrium concentration of impurity in silicon oxide to concentration in silicon. The segregation coefficients of the impurities have been taken equal 3 for boron and 103 for indium [5,7,8]. The distribution of impurities after implantation and subsequent annealing in an oxidizing ambient at the temperature of 1000 degrees of Celsius for 60 seconds was calculated using the following expression:

$$N(x,t) = \frac{N_0 \Delta R_p}{2\sqrt{\Delta R_p^2 + 2Dt}} \left\{ exp\left[-\frac{(x-R_p)^2}{2(\Delta R_p^2+2Dt)}\right] \times \left(1 + erf\left[\frac{R_p\sqrt{Dt}}{\Delta R_p\sqrt{(\Delta R_p^2+2Dt)}} + \frac{x\Delta R_p}{2\sqrt{Dt(\Delta R_p^2+2Dt)}}\right]\right) + \right.$$

$$\left. \frac{1-\alpha}{1+\alpha} exp\left[-\frac{(x+R_p)^2}{2(\Delta R_p^2+2Dt)}\right] \times \left(1 + erf\left[\frac{R_p\sqrt{Dt}}{\Delta R_p\sqrt{(\Delta R_p^2+2Dt)}} - \frac{x\Delta R_p}{2\sqrt{Dt(\Delta R_p^2+2Dt)}}\right]\right)\right\}, \quad (3)$$

where $N_0 = \frac{Q}{\sqrt{2\pi}\Delta Rp}$ is the maximum concentration of impurities at ion implantation; $\alpha = k \times \sqrt{\frac{D_1}{D}}$, where D_1 is the diffusion coefficient of impurity in silicon dioxide [8].

Fig. 3 presents the distributions of impurities, implanted into silicon with an initial energy of 10 keV, with subsequent annealing in an oxidizing ambient. Also, for comparison, Fig. 3 shows post-implant distributions of boron and indium ions in silicon. At the figure can be seen, that only indium ion characterized by SSRW impurity distribution. At that, the concentration and depth of indium implantation meet the requirements for nanoscale MOSFETs.

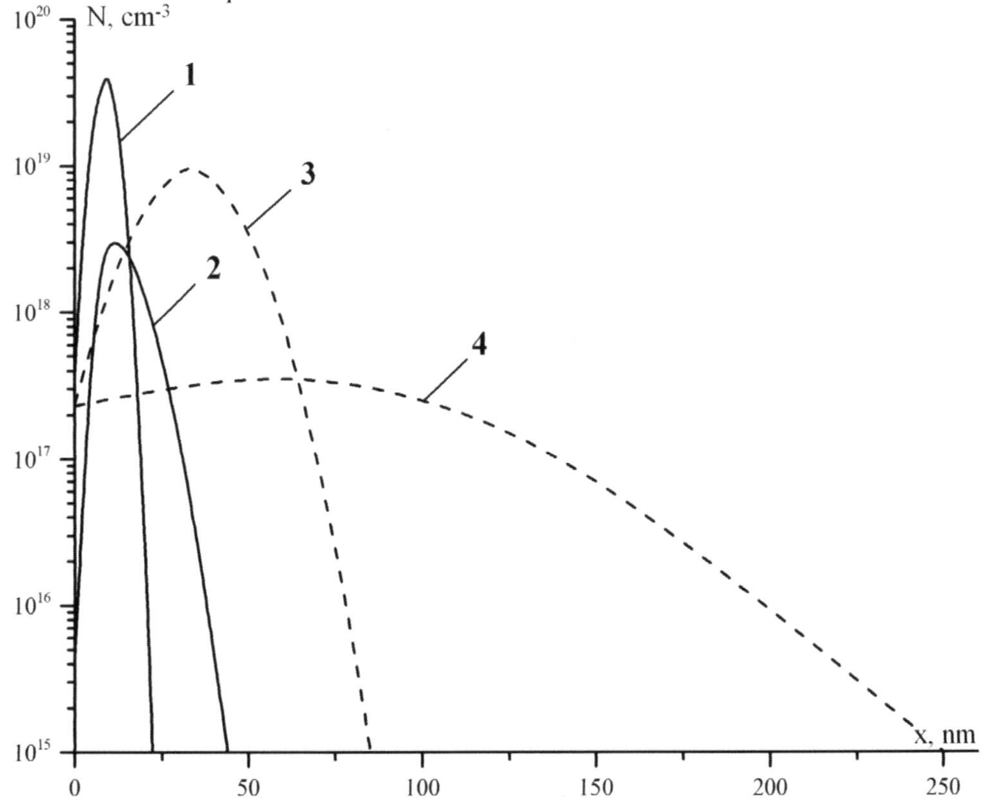

Figure 3. Distribution of impurities implanted into silicon with energy 10 keV: solid curves for indium dashed curves for boron. 1,3 – post-implant distributions; 2,4 – post-anneal distributions.

3. Conclusion

At the simulation on the basis of calculating stopping coefficients were obtained distributions of indium and boron ions in silicon before and after annealing. It was revealed, that indium ions provide steeper impurity distribution, in comparison with boron. The post-implant ratio of the maximum to a surface concentration of an impurity for boron equal 40, and for indium equal 100. But such values of the ratio are insufficient for SSRW for MOSFETs. It was shown, that subsequent annealing in inert ambient leads to decrease of this ratio for both impurities. This annealing mode does not provide the required steepness of the retrograde distribution in the gate region of nanoscale MOSFETs for both impurities. It was also shown, that annealing for indium ions in oxidizing ambient leads to increase of this ratio by one order of magnitude. This ensures SSRW distribution of impurity. Thus, it was demonstrated, that the SSRW distribution of indium atoms is provided exactly by the annealing in an oxidizing ambient. At the same time was revealed, that the annealing of boron ions in the oxidizing ambient leads to decrease of this ratio by one order of magnitude, as well as the annealing in the inert ambient. This is because indium characterized by a much higher value of the impurity segregation coefficient. Thus, it was shown, that indium doping of the gate region contributes to suppression of the short-channel effects in nanoscale transistors, unlike boron doping. The results of this report allow to correctly approach the selection of the processing modes at nanoscale MOSFETs manufacturing.

Acknowledgements

This work was supported by the Competitiveness Program of NRNU "MEPhI" (M.H.U.).

Reference

[1] Solmi S, Parisini A, Bersani M, Giubertoni D, Soncini V, Carnevale G, Benvenuti A and Marmiroli A 2002 *J. of Appl. Phys.* **92** 3 pp1361-1366
[2] Shahidi G G, Davari B, Bucelot T J, Ronsheim P A, Coane P J, Pollack S, Blair C R, Clark B, and Hansen H H 1993 *IEEE Transactions on Electron Device Letters* **14** 8 pp 409-411
[3] Damrongplasit N, Xu N, Takeuchi H, Stephenson R J, Cody N W, Yiptong A, Huang X, Hytha M, Mears R J and Liu T J K 2013 *IEEE Transactions on Electron Devices* **60** 5 pp 1790-1793
[4] Gorai P, Kondratenko Y V and Seebauer E G 2012 *J. of Appl. Phys.* **111** 9 094510
[5] Sze S M 1988 *VLSI Technology* **1** 2-nd Edition
[6] Lindhard J, Scharff M and Schiøtt H 1963 *Mat.-Fys. Med. Dan. Vid. Selsk* **33** 14 pp 1- 42
[7] Kizilyalli I C, Rich T L, Stevie F A and Rafferty C S 1996 *J. Appl. Phys.* **80** 9 p 4944
[8] Solmi S, Parisini A, Bersani M, Giubertoni D, Soncini V, Carnevale G, Benvenuti A and Marmiroli A 2002 *J. Appl. Phys.* **92** 3 pp 1361-1366
[9] Pavlov P V, Pashkov V I and Shitova E V 1967 *Works of the second Union conference on diffusion in semiconductors* pp 302 - 308

IOP Conf. Series: Materials Science and Engineering **498** (2019) 012036 doi:10.1088/1757-899X/498/1/012036

The influence of chemical structure of nitrogen-containing heterocyclic compounds on the character of ion mobility spectra

S S Grishin[1], K I Negru[1], E P Baberkina[1],A E Kovalenko[1], A A Gushchina[1], D A Aleksandrova[1], Y E Shutova[1], A P Zharikov[1], E V Dorskaya[1], Y R Shaltaeva[2], V V Belyakov[2], A V Golovin[2], E A Gromov[2], M A Matusko[2] and V F Khamraev[3]

[1]D. Mendeleev University of Chemical Technology of Russia, Miusskaya sq. 9, 125047 Moscow, Russian Federation
[2]National Research Nuclear University MEPhI (Moscow Engineering Physics Institute), Kashirskoe shosse 31, 115409 Moscow, Russian Federation
[3]North-Caucasus Federal University, Stavropol, ul. Pushkina, 1, 355009, Stavropol, Russian Federation

Corresponding author's e-mail address: shaltaeva@yandex.ru

Abstract. In this article, the trace amounts of nitrogen-containing compounds have been examined using ion-mobility spectrometry. Depending on the chemical structure of substances in the row of nitrogen-containing compounds ion-mobility spectra, different types of rules have been stated. The obtained ionograms have been analyzed to determine a probable ion-mobility value for certain functional groups.

1. Introduction

The problem of drug distribution in the modern world is still relevant. There is an urgent need for express analysis and preliminary samples detection due to an increase in synthetic drugs flow. One of the rapid drug detection methods is ion-mobility spectrometry. Ion-mobility spectrometers are mobile, compact; the analysis time takes split second; the well-known drugs such as morphine, amphetamine derivatives can be reliably detected by IMS. Recently, design drugs, e.g., spices, containing different bioactive N-heterocyclic compounds, have become widespread. In this article, the results of ion-mobility spectra analysis are represented for N-heterocyclic compounds. The values of ion-mobility have been found, and the analysis of spectra has been carried out.

2. Experimental equipment

Spectrograms were recorded by using KERBER ion-mobility spectrometer. Ionization was carried out by impulse corona discharge via atmosphere pressure chemical ionization (APCI) [1-4]. Detection of the examined substances was carried out in positive polarity due to the basic nature of many drugs.
The ions are formed in positive polarity according to the next scheme [5-7]:

$$M + H^+(H_2O)_n \rightarrow MH^+(H_2O)_n \rightarrow MH^+(H_2O)_{n-x} + \times H_2O$$

where M — sample, $H^+(H_2O)_n$ — reactant ion, $MH^+(H_2O)_n$ cluster ion, $MH^+(H_2O)_{n-x}$ product ion, $\times H_2O$ water.
In this equation, the product ion is called a protonated monomer. A protonated dimer $MH^+(H_2O)_n$ and other molecular ions are presumed to be formed.
The product ion in this equation is called a protonated monomer. The formation of a protonated dimer $MH^+(H_2O)_n$ and other molecular ions is possible [8].
Two mechanisms of formation of charged particles in positive polarity under the influence of corona discharge were proposed:

1. Charged particles formation via donor-acceptor interaction between positively charged reactant ion and aromatic system of the examined substances (π-complex).
2. Positively charged ions formation via protonation of basic functional groups and aromatic carbons and heteroatoms.

Atom charge calculations were carried out with Hypercube HyperChem 8.0 software using semi-empirical method PM3 to determine a probable protonation site. The electric charge distribution diagram has been constructed. Bond lengths and enthalpies of formations have been calculated for protonated particles. Total ion-dipole moment trend has been demonstrated. A conclusion concerning the molecular ion thermodynamic stability and its orientation in space under the constant electric field gradient in the device drift tube has been made according to this data. PM3 method demonstrates well-reproduced properties of a molecule (ion) such as enthalpy of formation, molecular geometry, which have been used to assume the possibility of ion formation.

3. Experimental results and discussion

The examined substances were divided into three groups.

The first group contained indoles with a carbonyl substituent in the position **3** as a privileged scaffold for synthetic cannabinoids (spices). The table 1 represents PM3 method calculation results.

Table 1. PM3 method calculation results in synthetic cannabinoids (spices)

Compound	Ion-mobility value K_0, $cm^2V^{-1}s$	Proposed molecular ion	The direction of the molecular ion-dipole moment vector	Molecular ion enthalpy of formation kcal/mol
Indole-3-carboxaldehyde 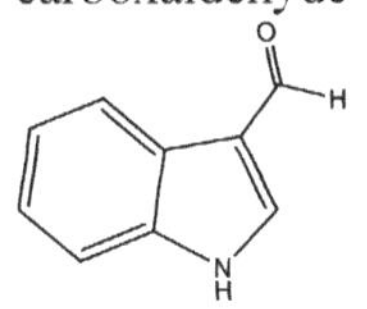	1.627	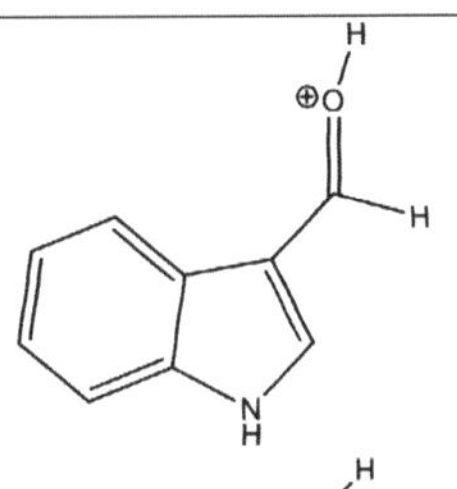	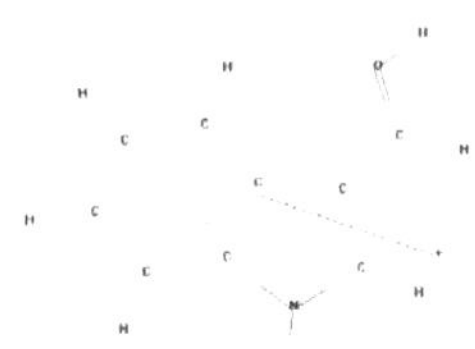	164,8
3-acetylindole 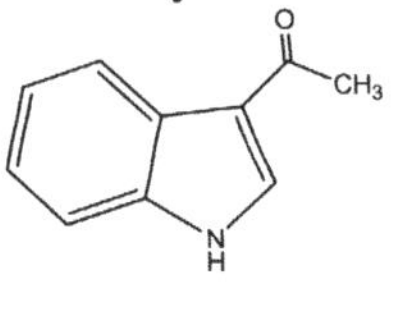	1,590	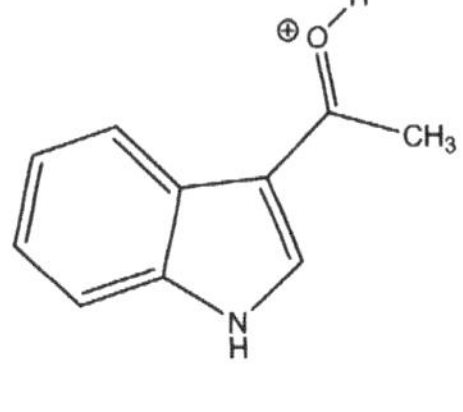	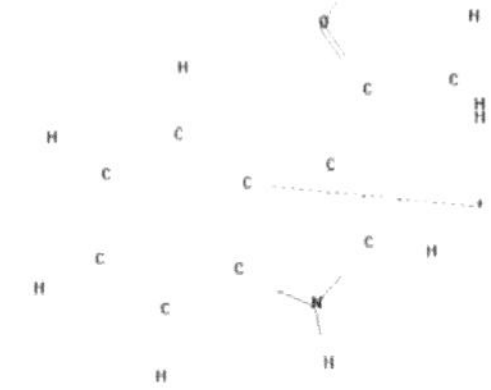	153,38

cyclopentyl(1H-indole-3-yl)methanone 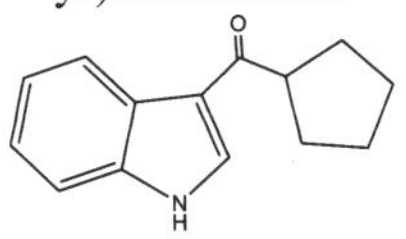	1,377	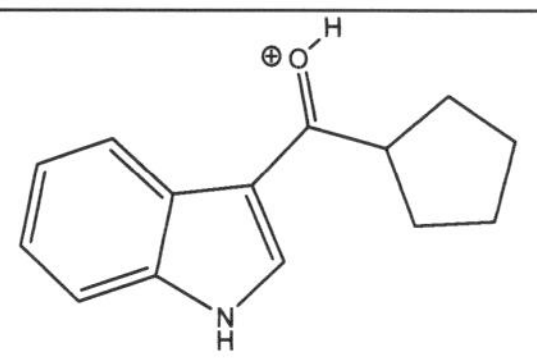	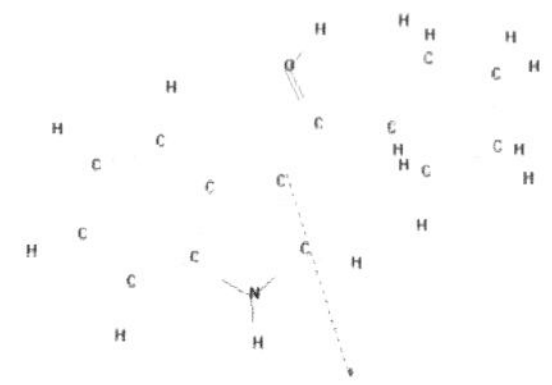	142,00
cyclohexyl(1H-indole-3-yl)methanone 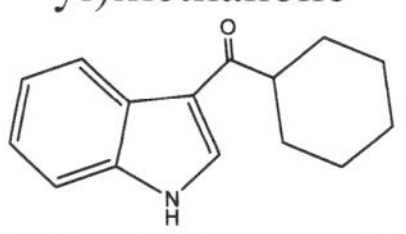	1,322	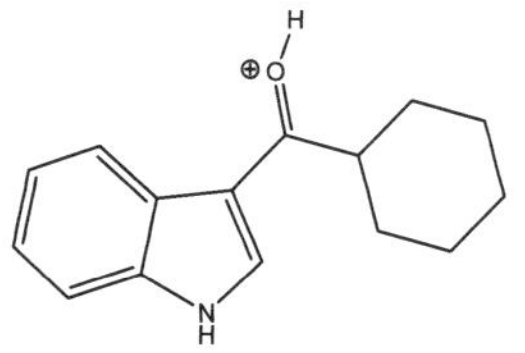	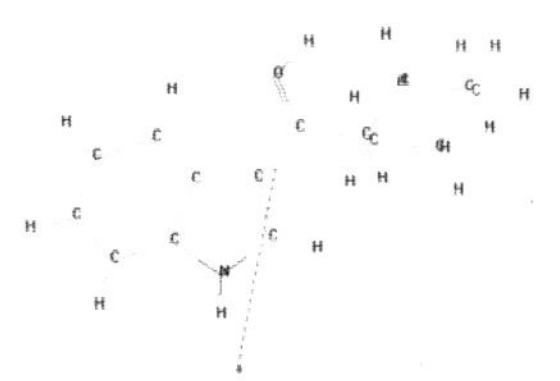	136,60

The figure **1** represents the general spectrogram of compounds from the first group.

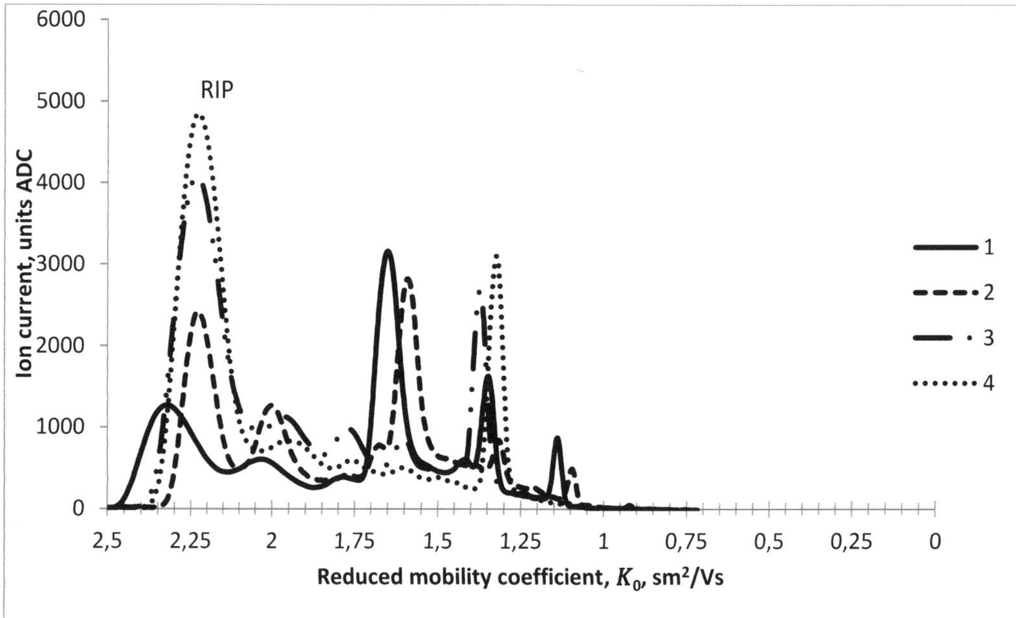

Figure 1. General spectrogram of the first group compounds.
1)Indole-3-carboxaldehyde; 2) 3-acetylindole; 3) cyclopentyl(1H-indole-3-yl)methanone; 4) cyclohexyl(1H-indole-3-yl)methanone

The comparison of spectrogram shapes of the first group substances demonstrates a decrease of the ion-mobility values of similar compounds with an increase of carbon atoms number in substituents at the position **3**. The most likely protonation site is oxygen according to PM3 semi-empirical calculation data. Besides identification peak, there are two more peaks related to protonated form in the positions **2** and **3** of indole-3-carboxaldehyde and 3-acetylindole. On the contrary, there are no such peaks on the cyclopentyl(1H-indole-3-yl)methanone and cyclohexyl(1H-indole-3-yl)methanone spectra due to steric hindrance in the positions **2** and **3** of the indole fragment.

The reduced ion-mobility dependence on the molecular weight of the compounds has been constructed to corroborate that all of three identification peaks from this group have similar molecular ion structure (Figure 2).

To confirm the thesis that all three identification peaks of this group of substances have the same nature of a molecular ion protonated by the oxygen atom, the dependence of the reduced ion mobility on the molecular weight of the substances was plotted (Figure 2).

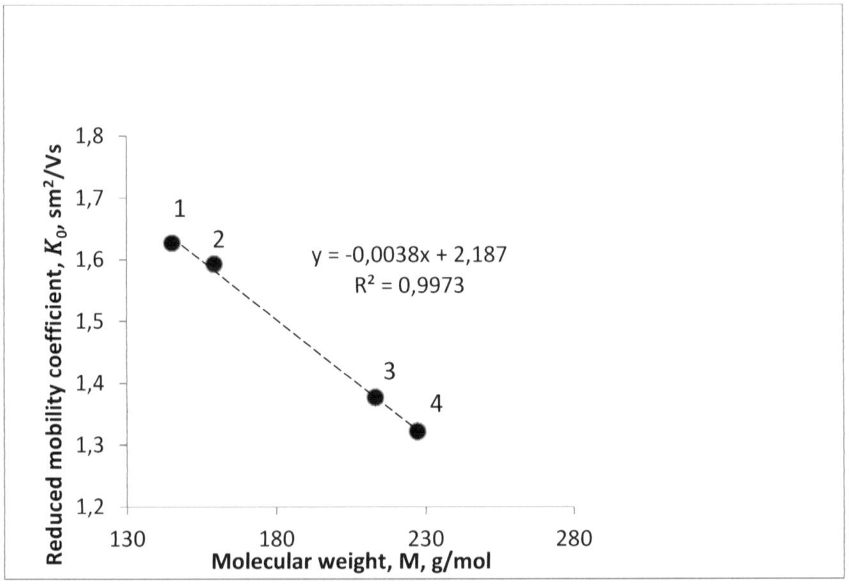

Figure 2. The reduced ion mobility dependence on the molecular weights of compounds 1)Indole-3-carboxaldehyde; 2)3-acetylindole; 3)cyclopentyl(1H-indole-3-yl) methanone; 4)cyclohexyl(1H-indole-3-yl)methanone.

The linear dependence states that all three molecular ions form in a similar way.

We have suggested that the motion of the ion in the tube depends not only on its charge value but also on the spatial orientation along the electric field. The directions of the total dipole moment vector of protonated ions from this group have been determined to estimate the correlation factor by PM3 method (table 1).

The dipole moment vector spatial orientation allows suggesting that indole-3-carboxaldehyde and 3-acetylindole ions will move from drift tube to collector faster than cyclopentyl(1H-indole-3-yl)methanone and cyclohexyl(1H-indole-3-yl)methanone. Alkylindoles represent the second group of the examined substances. The PM3 method calculation results for this group are shown in table 2.

Table 2. PM3 method calculation results for Alkylindoles

Compound	Ion-mobility value $K_0\ , cm^2V^{-1}s$	Proposed molecular ion	Direction of the molecular ion-dipole moment vector	Molecular ion enthalpy of formation kcal/mol
indole 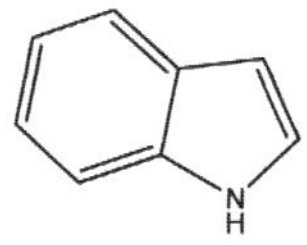	1,776	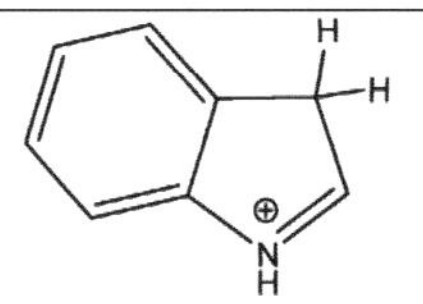	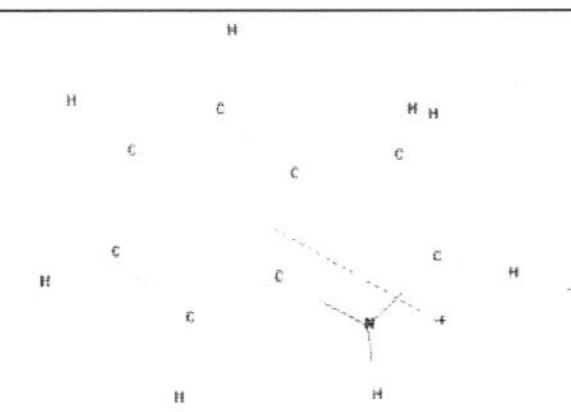	268,1
2-methylindole 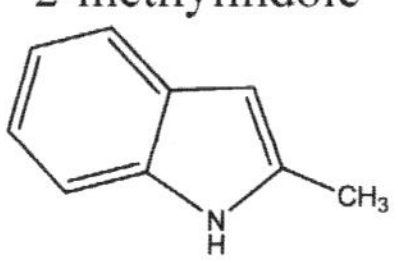	1,640	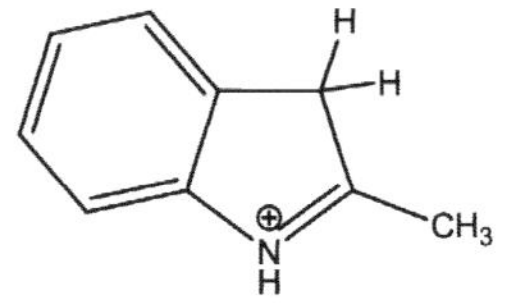	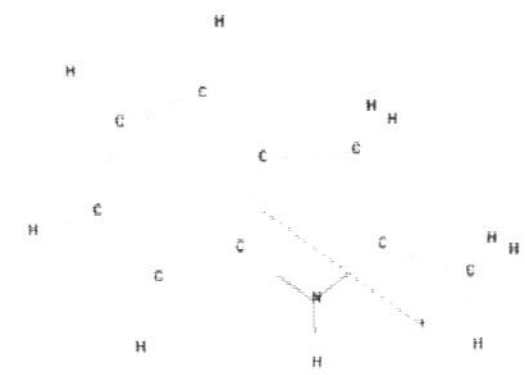	184,74
3-methylindole 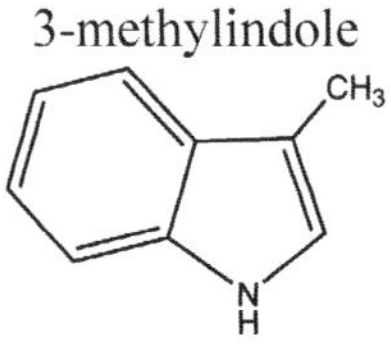	1,685	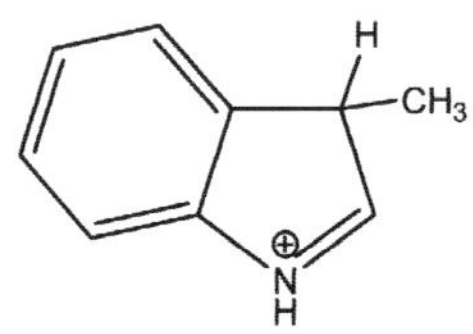	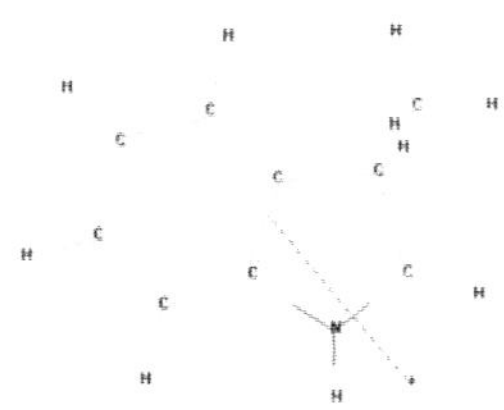	193,30
2,3,4,9,-tetrahydro-1H-carbazole 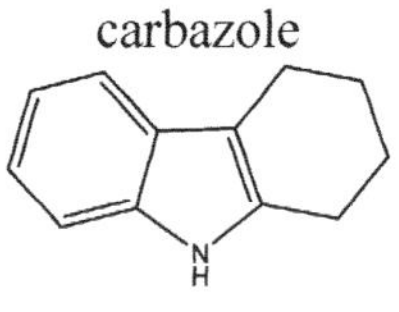	1.534	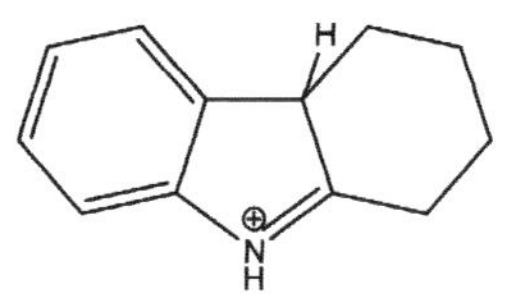	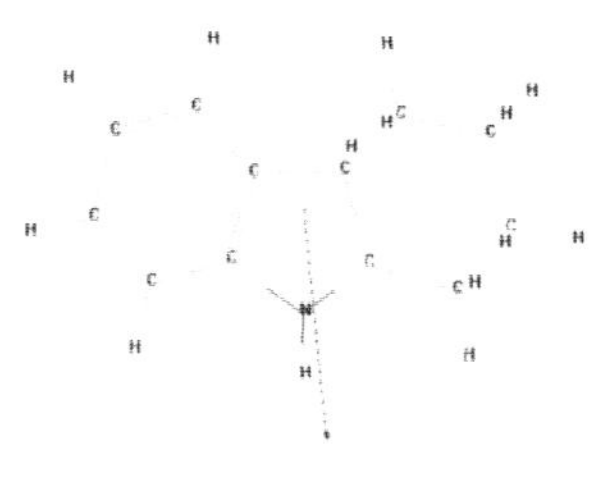	177.2

The Figure **3** represents general spectrogram of compounds from the second group.

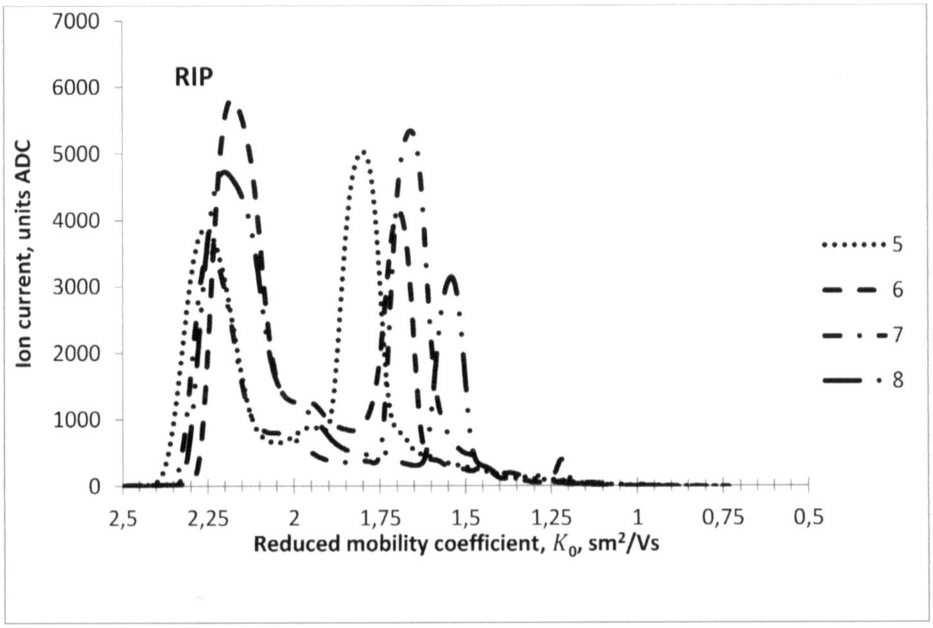

Figure 3. General spectrogram of the second group compounds.
5)Indole; 6)3-methylindole; 7)2-methylindole; 8)2,3,4,9-tetrahydro-1H-carbazole

The spectrogram analysis demonstrates that structural isomers 2-methylindole and 3-methylindole have different ion-mobility values in spite of spectra curves similarity. Furthermore, the indole ion-mobility is slightly different from 2-methylindole and 3-methylindole. This agrees with the fact that the dipole moment vectors of these compounds are almost the same. 2,3,4,9-tetrahydro-1H-carbazole molecular ion-dipole moment vector is different from the rest of the group hence it moves through the drift tube as a "flat" particle showing less ion-mobility value (table 2).

The third group is represented by unsubstituted heterocycles.

The table **3** shows PM3 method calculation results for the third group.

Table 3. PM3 method calculation results for heterocycles

Compound	Ion-mobility value K_0 , $cm^2V^{-1}s$	Proposed molecular ion	Direction of the molecular ion dipole moment vector	Molecular ion enthalpy of formation kcal/mol
indole 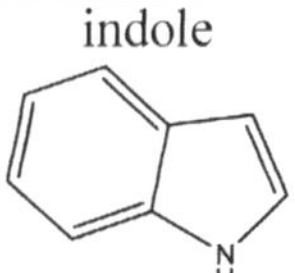	1,776	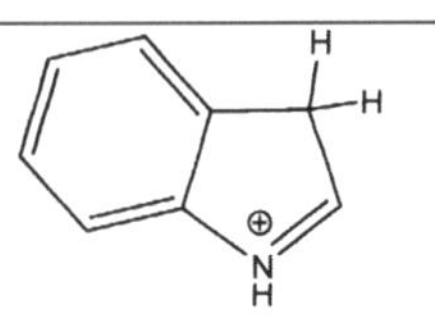	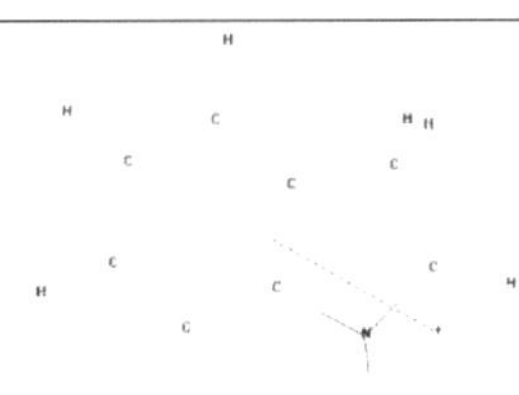	268,1

indazole 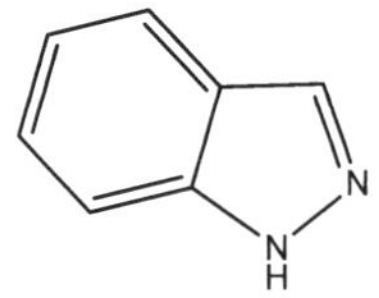	1,719	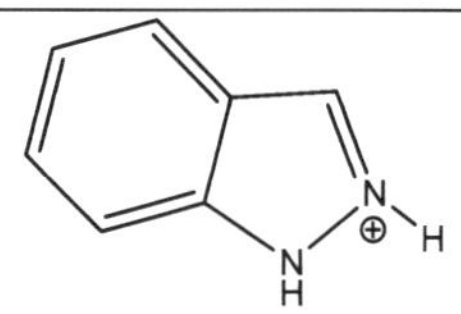	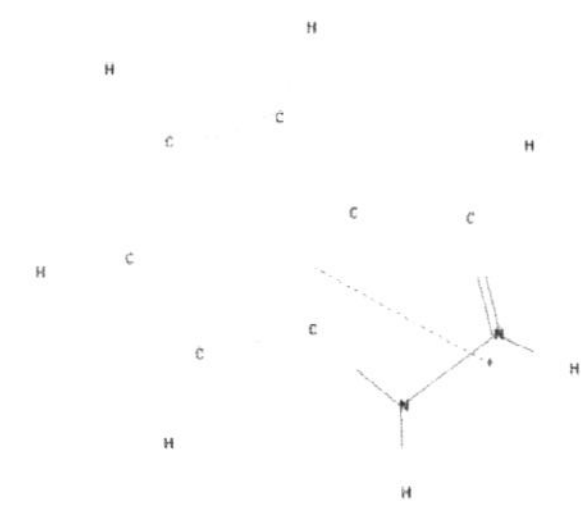	228.38
benzimidazole 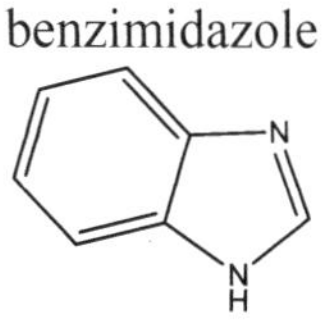	1,783	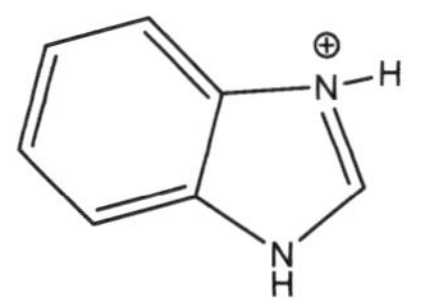	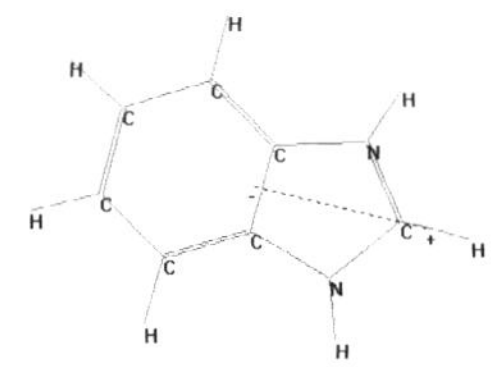	190.98

The Figure **4** represents general spectrograms of compounds from the third group.

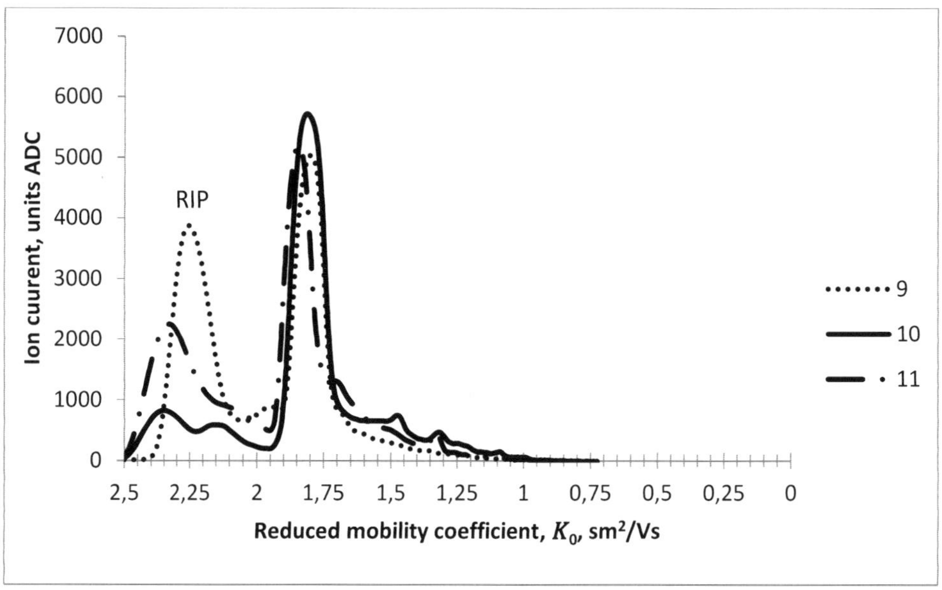

Figure 4. General spectrogram of the third group compounds 9) indole; 10) indazole; 11) benzimidazole.

All benzannelated heterocycles show protonated ion signals in a limited range of ion mobility.

Protonated heterocycles calculation results demonstrate that dipole moment vectors have almost the same values. This agrees with the ion-mobility values on the spectrograms of these substances (table **3**).
]

4. Conclusion

Systematic detection of ion-mobility signals of N-heterocyclic compounds of similar in structure and chemical properties has been carried out. A PM3 semi-empirical method was employed to account for spectra character. The probable direction and calculated dipole moment vector values for protonated molecules have been obtained by using this method. A correlation between the ion-mobility and dipole moment vector has been made. The dependence of reduced ion-mobility on dipole moment vector direction and on the spatial orientation of cation generated in the drift tube is observed in the row of examined substances. The dependence of ion mobility values on the molecular weight of ion is noted among heterocyclic ketones **1-4** (Figure 2).

5. References

[1] Pertwee R G 2008 *British journal of Pharmacology* **153** 2 pp. 199-215
[2] Bossong M G, Van Dijk G P and Niesink R J 2005 *Addict. Biol.* **10** pp. 321-323
[3] Maheux C R, Copeland C R, Pollard M M 2010 *Microgram J.* **7** pp. 42-49
[4] Kamata H T, Shima N, Zaitsu K, Kamata T, Nishikawa M, Katagi M, Miki A and Tsuchihashi H 2007 *Jpn J. Forensic Sci. Technol.* **12** pp. 97-106
[5] Reitzel L A, Dalsgaard P W, Müller I B and Cornett C 2012 *Drug. Test Anal.* **4** pp. 342-354
[6] Jankovics P, Váradi A, Tölgyesi L, Lohner S, Németh-Palotás J and Kőszegi-Szalai H 2011 *Forensic Sci. Int.* **210** pp. 213-220
[7] Nakajima J, Takahashi M, Seto T, Kanai C, Suzuki J, Yoshida M and Hamano T 2011 *Forensic Toxicol.* **29** pp. 95-110
[8] Sysoev A, Poteshin S, Chernyshev D, Karpov A, Tuzkov Y, Kyzmin V and Sysoev A 2014 *Eur. J. Mass Spectrom.* **20** 2 pp. 185-192

Temperature distribution on dielectric membrane structures for sensitive elements of semiconductor gas sensors

D.S. Veselov* and Y.A. Voronov

Micro- and Nanoelectronics Department, National Research Nuclear University MEPhI (Moscow Engineering Physics Institute), Kashirskoe shosse 31, 115409, Moscow, Russian Federation

*Corresponding author's e-mail address: *dsveselov@mephi.ru*

Abstract. The aim of this article is to show a simulation of temperature distribution on dielectric membrane structures with membranes of different dielectric materials in the measuring mode of a semiconductor gas sensor. The simulation results were obtained using the software package Synopsys TCAD. In various moments of the measuring cycle were obtained two-dimensional temperature distributions on membrane structures. Dependencies of temperature versus time for the variety of the coordinates on membranes, temperature versus coordinates at the sensitive layer area's highest heat value and temperature versus time in the center of the membranes are presented. Recommendations are given for the selection of membrane material for the heat-insulating membrane structure of sensitive elements for semiconductor gas sensors.

1. Introduction

In modern trends of MEMS technology development, due to the constant miniaturization of MEMS elements, the process of their design becomes increasingly complex and at this time cannot be efficiently implemented using existing analytical methods. Therefore, recently, in order to obtain high-quality results and reduce the time of design, computer-aided design is increasingly used. Modern simulations capabilities allow to visually demonstrate to a high degree of accuracy the final characteristics of the device under development and the processes occurring in them. It contributes to the optimization of material selection for device realization and technological modes of its production [1-3].

In recent times, for the analysis of gaseous environments are increasingly used semiconductor gas sensors [4, 5]. The measurement of gas concentration for such sensors requires the pre-heating of sensitive layers. It selectively accelerates the processes occurring on the surface and in the volume of the sensitive layers. Thereby, the selectivity of semiconductor gas sensors is provided for [6-10]. The possibility of achieving a high temperature of the sensitive layers is mainly influenced by heat exchange in the surrounding environment through the construction of sensitive elements. A distinctive design feature of such sensitive elements is the presence of a heat-insulating structure [11, 12]. The most promising heat-insulating structures for sensitive elements of semiconductor gas sensors are dielectric membrane structures. (Such a structure is shown in Fig. 1.) This is explained by that dielectric films have the lowest coefficient of thermal conductivity among all known materials [13, 14]. However, the formation of sensitive layers on such membrane structures is complicated by the fragility of dielectric membranes [15-17]. It requires the formation of sensitive layers exclusively before the formation of the membranes. The authors of this article have developed a new method to create sensitive elements based on dielectric membrane structures [18-21]. The main distinction of this method is that it involves the formation of membranes after the formation of the sensitive layers [22, 23].

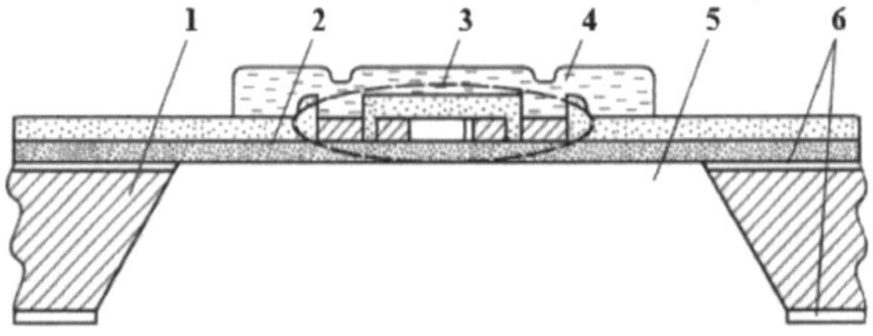

Figure 1. Structure of the sensitive element on the dielectric membrane: 1 – silicon substrate; 2 – dielectric membrane; 3 – structural elements (resistive heater, resistive temperature sensor, sensitive layer contact area); 4 – sensitive layer; 5 – cavity under the membrane; 6 – thermal silicon oxide layer.

The selection of the optimal dielectric membrane material is necessary to provide the required temperature conditions of gas concentration measurement and for the development of this new technology of sensitive elements for semiconductor gas sensors based on dielectric membrane structures. For this purpose, knowledge of temperature distribution on membranes of different dielectric membrane structures in measuring mode of the semiconductor gas sensor is required. In this article, the results of the simulation of the temperature distributions on dielectric membrane structures in the semiconductor gas sensor's measuring mode are shown, and were obtained using the software package Synopsys TCAD.

2. Simulation

The dielectric membrane structures are to be implemented on substrates of monocrystalline silicon of orientation (100), with a diameter of 76 mm and thickness of (380 ± 20) μm. The thickness of the membranes is to be about 1-3 μm.

For the simulation was built a model of the membrane structure with a platinum resistive heater in the center of the membrane, presented in Fig. 2. The thickness of the substrate was adopted to 10 μm. This was done because the actual substrate thickness (380 μm) sharply increases the simulation time, while not significantly affecting the results of the final simulation. The membrane size taken was equal to 1 × 1 mm^2. The thickness of the membrane was chosen to be 2 μm. The strip width of the resistive heater was made as 20 μm, the strip length was taken at 2.95 mm and the strip thickness adopted was 0.5 μm. The area of the heating region was calculated at about 0.3 mm^2. Calculated resistance of the heater at room temperature was about 31.5 Ω. The simulation of temperature distribution was carried out by the application to the resistive heater of a rectangular pulse with voltage level equal to 1 V and duration 100 ms. Rise time and fall time of the pulse amounted to about 10 ms each. The total time of simulation for the cycle of heating and cooling (measuring cycle) was equal to 210 ms.

The simulation of temperature distribution was carried out on membrane structures with membranes of silicon oxide, silicon nitride and silicon oxynitride. Since the simulation of composite compounds in the software package Synopsys TCAD is not provided for, silicon oxynitride membrane was modeled as a four-layer membrane. It was modeled with alternating layers of silicon oxide and silicon nitride with a thickness of 0.5 μm each. The properties of this membrane are close to that of silicon oxynitride membrane.

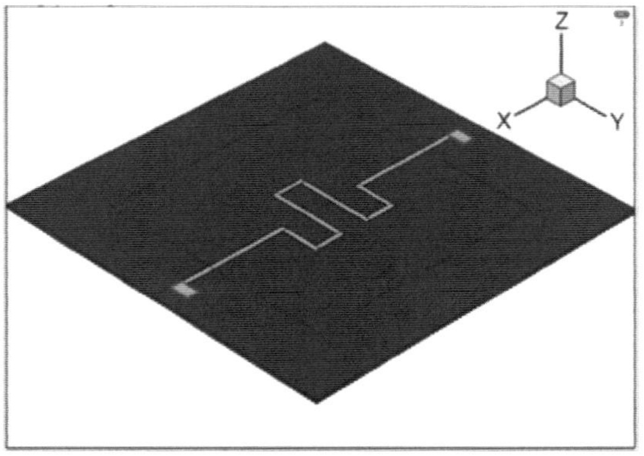

Figure 2. Model of the membrane structure (with a coordinate system that has zero in the center of the membrane).

For all membrane structures, at the application of the rectangular pulse on the resistive heater at various moments of the measuring cycle were obtained two-dimensional temperature distributions, presented in Fig. 3-5. Also, for all the membrane structures were obtained the dependencies of temperature versus time for the measuring cycle of a variety of coordinates on membranes, presented in Fig. 6-11. Besides, for all the membrane structures were obtained dependencies of temperature versus coordinates (with zero in the center of the membranes) at the moment of the highest heat for the sensitive layer area and with temperature versus time for the center of membranes, presented in Fig. 12, 13.

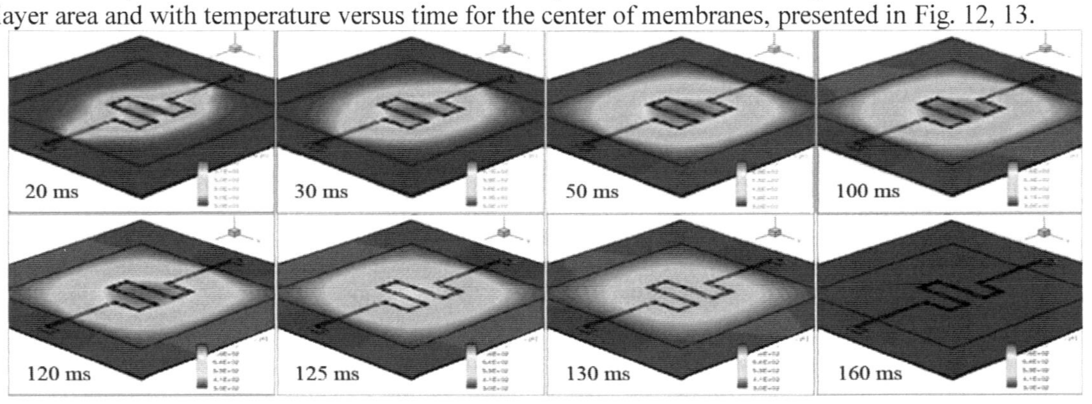

Figure 3. Two-dimensional temperature distribution on the silicon nitride membrane.

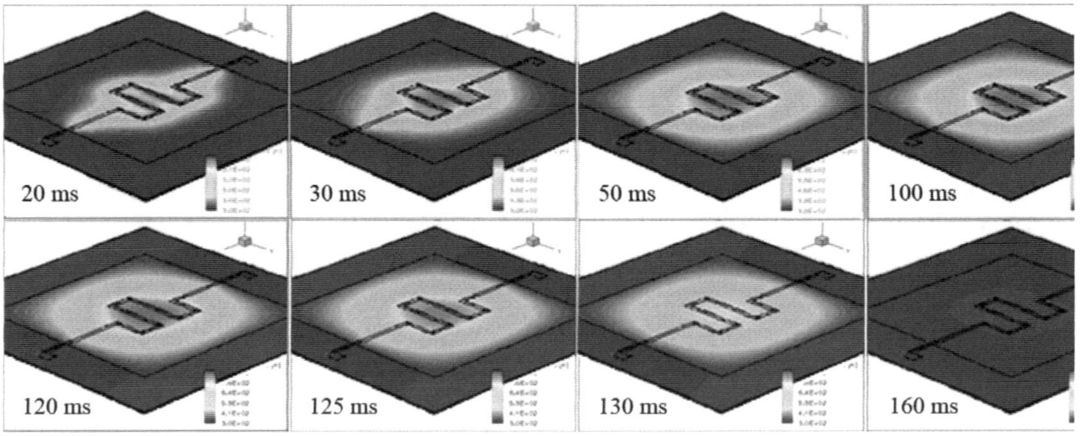

Figure 4. Two-dimensional temperature distribution on the four-layer membrane (close to the silicon oxynitride membrane).

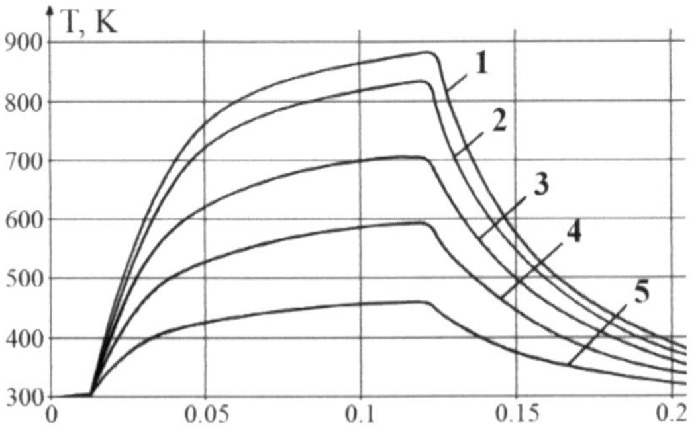

Figure 5. Temperature versus time for the measuring cycle for various coordinates X from the center of the silicon oxide membrane: 1 - 0; 2 - 100 μm; 3 - 200 μm; 4 - 300 μm; 5 - 400 μm.

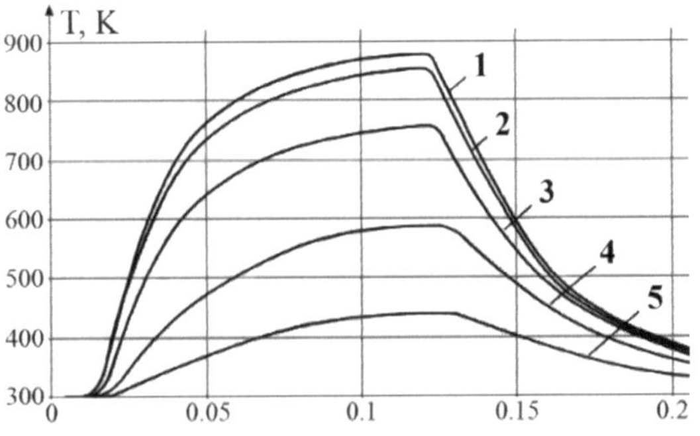

Figure 6. Temperature versus time for the measuring cycle for various coordinates Y from the center of the silicon oxide membrane: 1 - 0; 2 - 100 μm; 3 - 200 μm; 4 - 300 μm; 5 - 400 μm.

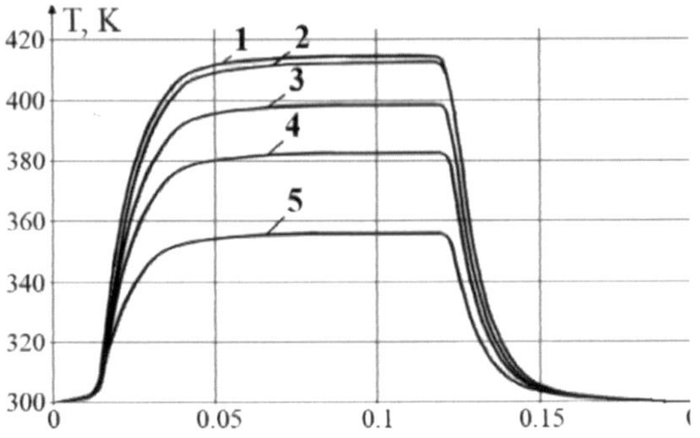

Figure 7. Temperature versus time for the measuring cycle for various coordinates X from the center of the silicon nitride membrane: 1 - 0; 2 - 100 μm; 3 - 200 μm; 4 - 300 μm; 5 - 400 μm.

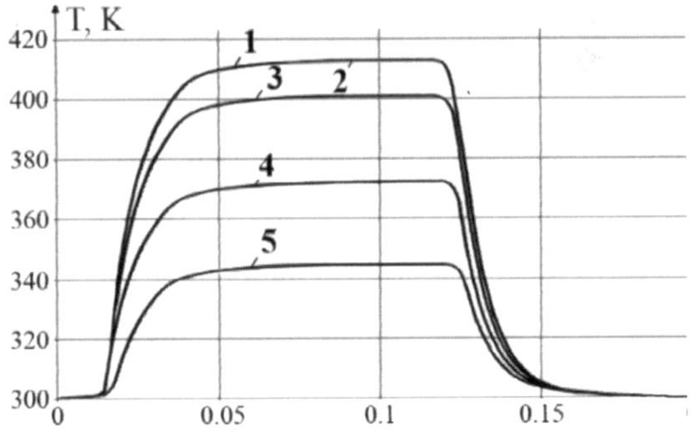

Figure 8. Temperature versus time of the measuring cycle for various coordinates Y from the center of the silicon nitride membrane: 1 - 0; 2 - 100 μm; 3 - 200 μm; 4 - 300 μm; 5 - 400 μm.

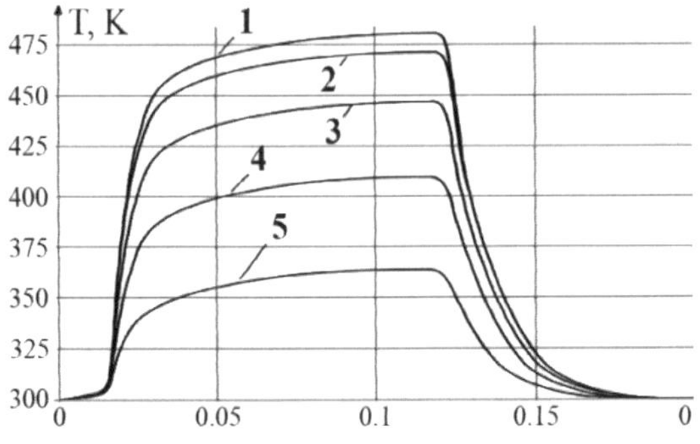

Figure 9. Temperature versus time for the measuring cycle for various coordinates X from the center of the four-layer membrane: 1 - 0; 2 - 100 μm; 3 - 200 μm; 4 - 300 μm; 5 - 400 μm.

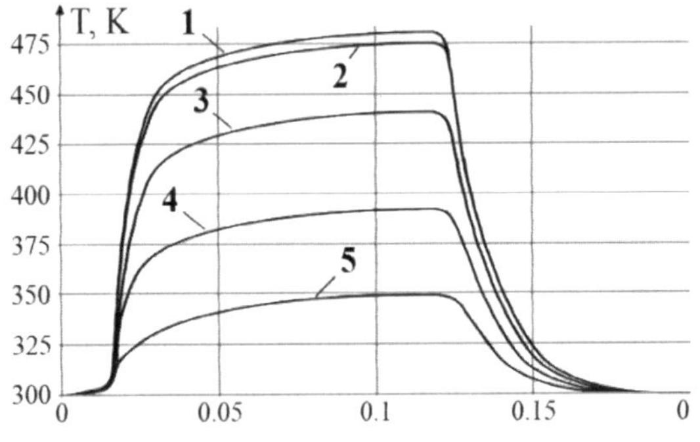

Figure 10. Temperature versus time for the measuring cycle for various coordinates Y from the center of the four-layer membrane: 1 - 0; 2 - 100 μm; 3 - 200 μm; 4 - 300 μm; 5 - 400 μm.

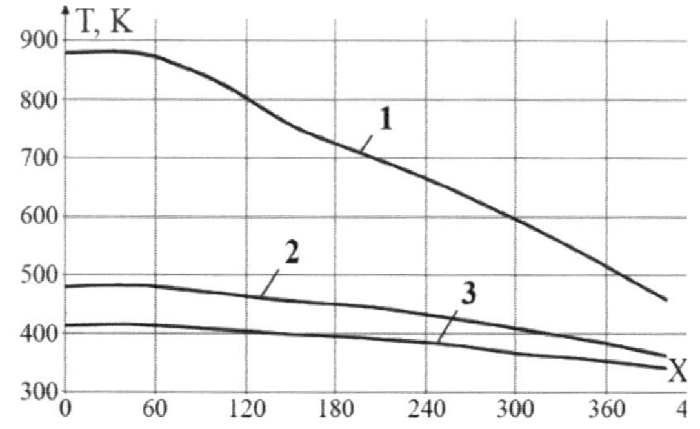

Figure 11. Temperature versus heat for coordinates X at the moment of the highest heat of the sensitive layer area: 1 - silicon oxide membrane; 2 - four-layer membrane; 3 - silicon nitride membrane.

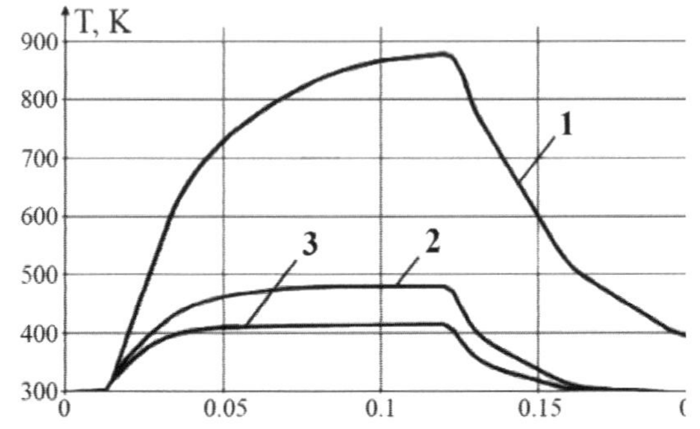

Figure 12. Temperature versus time of the measuring cycle in the center of the membrane: 1 - silicon oxide membrane; 2 - four-layer membrane; 3 - silicon nitride membrane.

3. Results and discussion

Temperature distributions for membranes of different dielectric materials in the measuring mode of a semiconductor gas sensor were obtained. The character of these distributions is uniform and similar for various membranes. This indicates the suitability of the simulated configuration of resistive heaters for semiconductor gas sensors.

Maximum temperature on all dielectric membranes was reached on the plot of 100 - 120 ms from the start of the measuring cycle. For the silicon oxide membrane, the maximal temperature was about 610 oC, for the silicon nitride membrane, it was about 140 oC and for the silicon oxynitride membrane (four-layer membrane) it was about 210 oC. It visually demonstrates the best heat-insulating properties of the silicon oxide membranes.

The simulated membrane structure with the area of the silicon oxide membrane equals 1 mm2 and the heating region area of the resistive heater about 0.3 mm2 provides the required heat-insulating properties to measure the concentration of a large number of gases. Further decreasing of the heating region area, in accordance with the simulated, surely, will make heat-insulating properties of this membrane structure suitable for use as the basis of sensitive elements for semiconductor sensors of any gases.

On the other hand, the simulation of temperature distribution on the silicon nitride membrane structure has demonstrated the significant heat dissipation through the substrate during the measuring cycle. It significantly decreases the temperature of the sensitive layer area and does not allow obtaining the required conditions for measurement of the concentration of most gases. Therefore, the usage of the silicon nitride membrane structure as the basis for sensitive elements for semiconductor gas sensors is advisable only in the case of improving its heat-insulating properties. For this purpose, it is necessary to reduce the heating region area of the resistive heater or, with the presence of technological opportunities, to increase the area of the membrane. At that, both of these solutions will significantly complicate the process of the production of sensitive elements. Thus, the selection of a silicon nitride film as the heat-insulating membrane material of sensitive elements for semiconductor gas sensors cannot be considered as optimal.

Along with the heat-insulating properties, not less important a selection criterion for the membrane material is the accuracy of achieving the temperature at measurement of the gas concentration and its stability during the measuring period.

As can be seen from the obtained and presented dependencies, silicon nitride membrane can be characterized by a lesser time to achieve and better stability of gas concentration measurement temperature, than can silicon oxide membrane. In the simulation, the silicon nitride membrane was characterized by a relatively short period of heating during the temperature measuring process, about 40 ms, and a long period of relative temperature stability for measurement, about 80 ms. The obtained dependencies for the silicon oxide membrane differ by the absence of temperature stability period in gas concentration measurement. Throughout the entire period of the pulse application, there was observed a significant increase in temperature. This would adversely impact on the selectivity of gas sensors. As such was the case, silicon nitride was more preferable as the membrane material for sensitive elements of semiconductor gas sensors than was silicon oxide.

This material allows the measurement temperature of gas concentration to be stabilized, however, it reduces heat-insulating properties of the membranes to a greater degree than silicon oxide.

Simulated dependencies of temperature versus time in the center of the membranes, made from different dielectric materials, show the greatest suitability of silicon oxynitride as the membrane material. With the same parameters of the pulse on the silicon oxynitride membrane, there was achieved a larger temperature of the sensitive layer area than on the silicon nitride membrane. In addition, during the heating, the silicon oxynitride membrane was characterized by the presence of a temperature stability period for gas concentration measurement with almost the same duration as would be for a silicon nitride membrane. The results of the simulation allow the following conclusions about nitrogen and oxygen contents in dielectric membrane films for sensitive elements of semiconductor gas sensors to be made:

- the increase of oxygen content in membrane film composition improves the heat-insulating properties of the membrane, which contributes to the improvement of gas sensor selectivity;
- the increase of nitrogen content in membrane film composition lengthens the duration of the temperature stability period for the measurement of the gas concentration, which contributes to the improvement of gas sensor sensitivity.

The optimal element composition of dielectric membrane films for sensitive elements of different semiconductor gas sensors should be selected individually, in accordance with the following. The membrane film should contain the minimum amount of oxygen, yet which is sufficient to guarantee the achievement of the required temperature of the sensitive layer during the measurement. Conjointly, in order to increase the period of temperature stability for measurement and improve the accuracy of achieving the required temperature, membrane film should contain the maximum possible amount of nitrogen. Moreover, that the amount of silicon in membrane film also has significant influence on heat-insulating properties of the membrane, time of achieving the measurement temperature of gas concentration and temperature stability during the measurement should be taken into account.

4. Conclusion

The simulation of the temperature distribution on membrane structures with membranes of different dielectric materials with the application to the resistive heater of a rectangular pulse with voltage equal to 1 V and duration equal to 100 ms was carried out, and the results were visualized and presented. In various moments of the measuring cycle of the semiconductor gas sensors were obtained two-dimensional temperature distributions on silicon oxide, silicon nitride and silicon oxynitride (four-layer membrane) membrane structures. Dependencies of temperature versus time for variety coordinates on membranes, temperature versus coordinates at the moment of the highest heating of the sensitive layer area and temperature versus time in the center of membranes are presented. The simulation results were obtained using the software package Synopsys TCAD. On the basis of obtained results were made recommendations in the selection of membrane film material for heat-insulating membrane structures of sensitive elements for semiconductor gas sensors.

The usage of pure silicon oxide or pure silicon nitride as the membrane film material of heat-insulating membrane structures for sensitive elements of semiconductor gas sensors is inexpedient. Silicon oxynitride is more suitable as membrane film material for such membrane structures. Moreover, it is expedient to individually find the optimal element ratio of membrane film for each semiconductor gas sensor depending on its destination. For example, if it is necessary to ensure the high temperature of gas concentration measurement, in the composition of the membrane film oxygen content should prevail over nitrogen content. If it is necessary to ensure the high accuracy of achieving relatively low temperature of gas concentration measurement, in the composition of membrane film, nitrogen content should prevail over oxygen content. Therefore, it is important to experimentally find the optimal element ratio of membrane film material for the sensitive elements of semiconductor gas sensors for various purposes.

Obtained and presented simulation results and made recommendations allow to reduce the field of required experimental study of heat-insulating materials at dielectric membrane structures development. This simplifies the selection of optimal membrane film composition for dielectric membrane structures of sensitive elements for semiconductor gas sensors.

Acknowledgement

This work was supported by the Competitiveness Program of NRNU "MEPhI" (M.H.U.).

References

[1] Hansen S E, Deal M D and Law M E 1993 *SUPREM-IV.GS Two Dimensional Proc. Simulation for Silicon and Gallium Arsenide*
[2] 2005 *User Guide X-2005.10* (Synopsys: U.S.A)
[3] 2013 *Synopsys TCAD News Apr.*
[4] Zaretskiy N P, Menshikov L I and Vasiliev A A 2012 *Sensors and Actuators B: Chem.* **175** pp 234-245
[5] Leghrib R, Pavelko R and Felten A 2010 *Sensors and Actuators B: Chem.* **145** pp 411-416
[6] Vasiliev A A, Pisliakov A V and Sokolov A V 2016 *Sensors and Actuators B: Chem.* **224** pp 700-713
[7] Samotaev N N, Vasiliev A A and Podlepetsky B I 2007 *Sensors and Actuators B: Chem.* **127** 1 pp 242-247
[8] Vasiliev A A, Pavelko R G and Gogish-Klushin S Yu 2008 *Sensors and Actuators B: Chem.* **132** 1 pp 216-223
[9] Samotaev N N, Podlepetsky B I, Vasiliev A A and Sokolov A V 2013 *Automation and Remote Control* **74** 2 pp 308-312
[10] Vasiliev A A and Malyshev V V 2013 *Sensors and Actuators B: Chem.* **189** pp 260-267
[11] Sakai G, Baik N S, Miura N and Yamazoe N 2001 *Sensors and Actuators B: Chem.* **77** pp 116-121
[12] Heule M and Gauckler L J 2003 *Sensors and Actuators B: Chem.* **93** pp 100-106
[13] Veselov D S and Voronov Yu A 2016 *J. of Phys.: Conf. Series* **747** 012022
[14] Akbari S, Thang P C and Veselov D S 2016 *Conf. Series: Materials Sci. and Engineering* **151** 012032
[15] Veselov D S, Voronov Yu A and Vanyukhin K D 2015 *Phys. Proc.* **72** pp 495-499
[16] Veselov D S and Voronov Yu A 2015 *Phys. Proc.* **71** pp 423-427
[17] Veselov D S 2015 *Appl. Mechanics and Materials* pp 799-800
[18] Veselov D S 2015 *Phys. Proc.* **72** pp 500-503.
[19] Veselov D S and Voronov Yu A 2015 *Appl. Mechanics and Materials* pp 919-922.
[20] Veselov D S, Bakun A D and Voronov Yu A 2016 *J. of Phys.: Conf. Series* **748** 012015
[21] Veselov D S, Bakun A D and Voronov Yu A 2016 *J. of Phys.: Conf. Series* **748** 012017

IOP Conf. Series: Materials Science and Engineering **498** (2019) 012038 doi:10.1088/1757-899X/498/1/012038

Modelling of saturation current of an organic field-effect transistor with accounting for contact resistances

V O Turin[1], B A Rakhmatov[1], G I Zebrev[2], C H Kim[3], B Iñiguez[4] and M S Shur[5]

[1]Orel State University after Ivan Turgenev, Orel, Russia
[2]National Research Nuclear University MEPhI, Moscow, Russia
[3]Gwangju Institute of Science and Technology, Gwangju, South Korea
[4]Rovira i Virgili University, Tarragona, Spain
[5]Rensselaer Polytechnic Institute, Troy, NY, USA

Corresponding author's e-mail address: voturin@ostu.ru

Abstract. The saturation current of an organic field-effect transistor was calculated numerically by bisection method taking into account the source and drain resistances. Dependences of the saturation current as a function of the gate voltage, centred on the threshold voltage, and as a function of the "extrinsic" (taking into account the source and drain resistances) saturation voltage are presented. The calculations were carried out using an iterative scheme assuming the saturation current to be zero for iteration zero and using different numbers of iterations. In addition, we carried out a compact modelling of the saturation current in the framework of the approach we proposed earlier. It is shown that when in the equation for the compact modelling initial value of saturation current of zero value (corresponding to zero iteration) is used, a good agreement with the bisection method of over a wide range of gate voltages is obtained.

1. Introduction

The approach to organic field-effect transistor (OFET) above-threshold drains current compact modeling (CM) that provides monotonic decrease of the output conductance with drain bias increasing [2-4] and with an analytical account of the source and drain resistances was proposed in [1]. An equation determining the dependence of the saturation current on "extrinsic" centered gate voltage was obtained in an implicit manner. Based on that equation, an iterative procedure was proposed for modeling the OFET saturation current. Linearization yielded an equation for the compact modeling of the OFET saturation current using the initial value of the saturation current obtained after several iterations.

In our paper, we simulate the saturation current of the OFET taking into account the source and drain resistances by the bisection method.

We also investigate the simulation of the saturation current of the OFET by the iterative method performing calculations for a different number of iterations. The value of the saturation current at zero iteration is assumed to be zero.

In addition, we carried out a compact modeling of the saturation current in the framework of the proposed in [1] approach. We investigate solutions of the equation for the saturation current of the OFET compact modeling when using the initial saturation current value obtained by iterative method with different number of iterations.

As a basis, in our work we use a compact model of OFET, proposed in the [5] and developed on the basis of a compact model MOSFET Level 1 for a long-channel MOSFET.

The saturation voltage for an OFET is given by

$$V_{SAT} = \alpha_S V_{GT}\ , \tag{1}$$

where α_S is a dimensionless parameter related to the influence of the substrate, $V_{GT} = V_{GS} - V_T$ is the gate-to-source voltage V_{GS} centered on threshold voltage with threshold voltage V_T.

The saturation current of the OFET is assumed to be equal to:

$$I_{SAT} = g_{CH} V_{SAT} = \alpha_S K \frac{\mu_0}{V_{aa}^{\gamma}} V_{GT}^{\gamma+2}, \tag{2}$$

where g_{CH} is the OFET channel differential conductance in linear regime:

$$g_{CH} = K \mu_{FET} V_{GT}, \tag{3}$$

where $K = W / L\ C_i$, W is the channel width, L is the channel length, $C_i = \epsilon_0 \epsilon_{ox}/d_{ox}$ is the insulator capacitance per unit area, $\mu_{FET} = \mu_0 (V_{GT}/V_{aa})^{\gamma}$ field-effect mobility, μ_0 is the conversion mobility set to 1 cm^2/V·s. In our work, we mainly use the parameters of the OFET model with a channel length L = 40 um and the gate width W = 1000 um from [6]: γ = 0.91, V_{th} = -12 V, μ_{FET} = 0.13 cm^2/Vs at the gate-to-source bias V_{GS} = -50 V, α_S =0.46, C_i = 3.3 nF/cm^2 and V_{aa} = 358 V. The value of the total resistance of the contacts R_T = 2.4 MΩ (R_S = R_D = 1.2 MΩ) were chosen sufficiently large to more clearly demonstrate the influence of these parameters on the results of calculations.

The OFET transconductance in the saturation regime, is determined by the following equation:

$$g_{mSAT} = \frac{\partial I_{SAT}}{\partial V_{GS}} = \alpha_S K \frac{\mu_0}{V_{aa}^{\gamma}} (\gamma + 2) V_{GT}^{\gamma+1} \tag{4}$$

when neglecting differential conductivity in the saturation regime. "Extrinsic" and "intrinsic" voltages on the drain, source, and gate are related as:

$$V_{ds} = V_{DS} + I R_T, \tag{5}$$

$$V_{gt} = V_{GT} + I R_S. \tag{6}$$

2. Implicit equation for saturation current in "extrinsic" case and its solutions by bisection method and by iterations

For An equation obtained in [1] determined the dependence of the saturation current on the "extrinsic" centered voltage on the gate V_{gt} implicitly:

$$I_{sat} = I_{SAT}\left(V_{gt} - I_{sat} R_S\right) = \alpha_S\ K\ \frac{\mu_0}{V_{aa}^{\gamma}} \left(V_{gt} - I_{sat} R_S\right)^{\gamma+2}, \tag{7}$$

where function I_{SAT} is given by equation (2) with argument V_{GT} given by equation (6) in case when current I is equal to saturation current I_{sat}.

We solve equation (7) by the method of bisection. But this equation is also suitable for calculating the saturation current I_{sat} by the iterative method:

$$I_{sat\ i+1} = I_{SAT}\left(V_{gt} - I_{sat\ i}\ R_S\right) = \alpha_S\ K\ \frac{\mu_0}{V_{aa}^{\gamma}} \left(V_{gt} - I_{sat\ i}\ R_S\right)^{\gamma+2}. \tag{8}$$

Here i is the iteration number. In this case, for the value of the saturation current at zero iteration $i = 0$ we take the zero value of the saturation current:

$$I_{sat\ 0} = 0. \tag{9}$$

For the first iteration i = 1, we have:

$$I_{sat\ 1} = I_{SAT}\left(V_{gt}\right) = \alpha_S\ K\ \frac{\mu_0}{V_{aa}^{\gamma}} \left(V_{gt}\right)^{\gamma+2}. \tag{10}$$

Note, that on this iteration we calculate saturation current considering V_{gt} as "intrinsic" voltage (without taking into account the source and drain resistances).

For the second iteration i = 2, we have:

$$I_{sat\ 2} = I_{SAT}\left(V_{gt} - I_{SAT}(V_{gt})\ R_S\right). \tag{11}$$

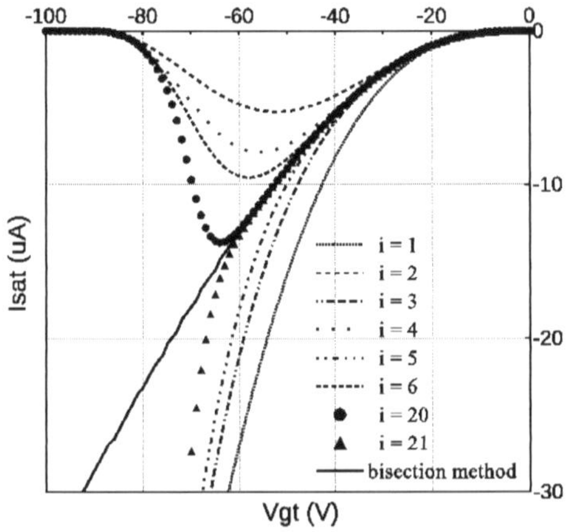

Figure 1. The saturation current as a function of the "extrinsic" centered gate voltage V_{gt}, obtained by the modelling with iterative method for a different number of iterations i = 1 .. 6, 20, 21 in comparison with the modelling with the method of bisection (solid line).

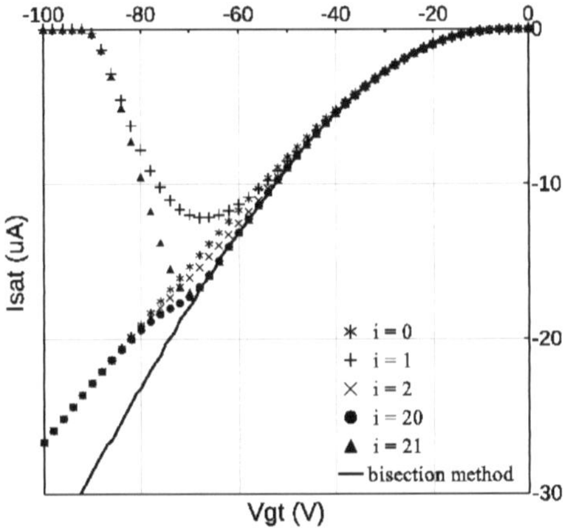

Figure 2. The saturation current as a function of the "extrinsic" centered gate voltage V_{gt}, obtained by compact modelling with the initial saturation current obtained from iterations with different i in comparison with the method of bisection (solid line).

3. Compact modelling of saturation current in "extrinsic" case

But for compact modeling, equations are needed that can be solved by quadrature. To obtain such an equation (7) has to be linearized [1]. In this case, the value of the current $I_{sat\ i}$, that is obtained after i iterations, is taken as the initial value of the saturation current for the linearization. The equation (7) for the saturation current is transformed as follows:

$$\begin{gathered}
I_{sat} = I_{SAT}(V_{gt} - I_{sat} R_S) = I_{SAT}(V_{gt} - (I_{sat\ i} + dI_{sat})\, R_S) = \\
I_{SAT}\left((V_{gt} - I_{sat\ i}\, R_S) - dI_{sat}\, R_S\right) = \\
I_{SAT}(V_{GT\ i} + dV_{GT}) \approx I_{SAT}(V_{GT\ i}) + \left.\frac{\partial I_{SAT}}{\partial V_{GS}}\right|_{V_{GT\ i}} dV_{GT} = \\
I_{SAT}(V_{GT\ i}) + g_{mSAT}(V_{GT\ i}) dV_{GT} = I_{SAT}(V_{GT\ i}) - g_{mSAT}(V_{GT\ i})\, dI_{sat}\, R_S = \\
I_{SAT}(V_{GT\ i}) - g_{mSAT}(V_{GT\ i}) \cdot (I_{sat} - I_{sat\ i})\, R_S = \\
I_{SAT}(V_{gt} - I_{sat\ i}\, R_S) - g_{mSAT}(V_{gt} - I_{sat\ i}\, R_S) \cdot (I_{sat} - I_{sat\ i})\, R_S.
\end{gathered} \tag{12}$$

Leaving only the last line, replacing the approximate equality by the exact equality and replacing the notation I_{sat} with $I_{sat\ CM}$, an equation is obtained that implicitly determines the saturation current for compact modeling:

$$I_{sat\ CM} = I_{SAT}(V_{gt} - I_{sat\ i}\, R_S) - g_{mSAT}(V_{gt} - I_{sat\ i}\, R_S) \cdot (I_{sat\ CM} - I_{sat\ i})\, R_S\,. \tag{13}$$

From the resulting equation, the saturation current of the OFET is easily expressed in the form suitable for compact modeling:

$$I_{sat\ CM} = \frac{I_{SAT}(V_{gt} - I_{sat\ i}\, R_S) + g_{mSAT}(V_{gt} - I_{sat\ i}\, R_S)\, I_{sat\ i}\, R_S}{1 + g_{mSAT}(V_{gt} - I_{sat\ i}\, R_S)\, R_S}. \tag{14}$$

Taking as the initial value $I_{sat\ 0}$, we have:

$$I_{sat\ CM\ 0} = \frac{I_{SAT}(V_{gt})}{1 + g_{mSAT}(V_{gt})\, R_S}. \tag{15}$$

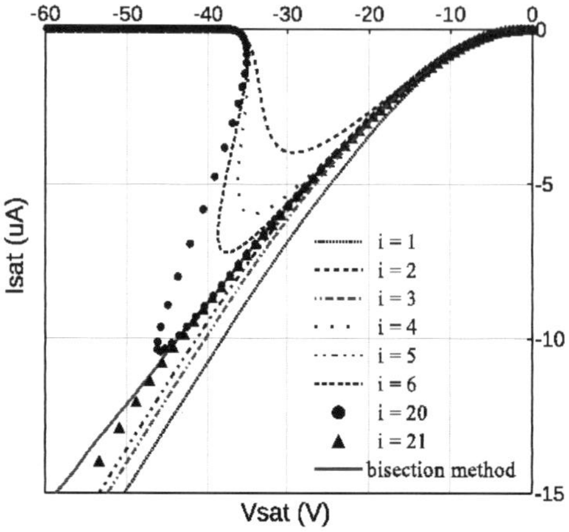

Figure 3. The saturation current as a function of the saturation voltage, obtained by the modelling with iterative method for a different number of iterations i = 1 .. 6, 20, 21 in comparison with the modelling with the method of bisection (solid line).

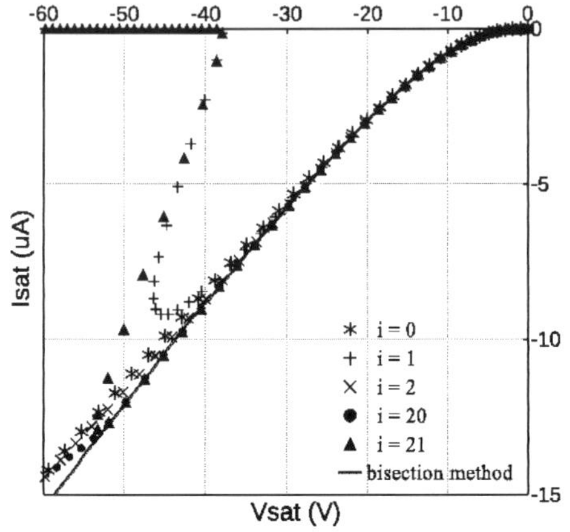

Figure 4. The saturation current as a function of the saturation voltage, obtained by compact modelling with the different initial value of the saturation current obtained from iterations with different i in comparison with the method of bisection (solid line).

If for the initial value we will take $I_{sat\,1} = I_{SAT}\left(V_{gt}\right) = \alpha_S\, K\, \frac{\mu_0}{V_{aa}^{\gamma}}\, V_{gt}^{\gamma+2}$, we will get:

$$I_{sat\,CM\,1} = \frac{I_{SAT}\left(V_{gt}-I_{sat\,1}\,R_S\right)+g_{mSAT}\left(V_{gt}-I_{sat\,1}\,R_S\right)I_{sat\,1}\,R_S}{1+g_{mSAT}\left(V_{gt}-I_{sat\,1}\,R_S\right)R_S}. \quad (16)$$

It is this equation that was used in the [1].

If we will take $I_{sat\,2} = I_{SAT}\left(V_{gt} - I_{SAT}\left(V_{gt}\right) R_S\right)$ as the initial value, we will get:

$$I_{sat\,CM\,2} = \frac{I_{SAT}\left(V_{gt}-I_{sat\,2}\,R_S\right)+g_{mSAT}\left(V_{gt}-I_{sat\,2}\,R_S\right)I_{sat\,2}\,R_S}{1+g_{mSAT}\left(V_{gt}-I_{sat\,2}\,R_S\right)R_S}. \quad (17)$$

The "extrinsic" saturation voltage is easy to calculate by use of equation (5) with V_{sat} instead of V_{ds} and I_{sat} instead of I:

$$V_{sat} = V_{SAT}\left(V_{gt} - I_{sat}\,R_S\right) + I_{sat}\,R_T = \alpha_S\left(V_{gt} - I_{sat}\,R_S\right) + I_{sat}\,R_T\ , \quad (18)$$

where function V_{SAT} is given by equation (1) with argument V_{GT} given by equation (6) in case when current I is equal to saturation current I_{sat}. We can use equation (18) with the value for the "extrinsic" saturation current I_{sat}, obtained by any method previously discussed: bisection method, iterative method or using the equation for compact modeling.

4. Results and discussions

Figures 1 and 3 show that an even number of iterations, starting with a certain "extrinsic" centered voltage on the gate V_{gt} or saturation voltage V_{sat}, correspondingly, leads to a quick conversion of the calculated saturation current in comparison with the value calculated by the bisection method. With an odd number of iterations some increase of the calculated value of the saturation current is observed in comparison with that calculated by the bisection method. Note that the curve for i = 1 in Figure 1 corresponding to the first iteration, can be considered as the saturation current dependence on the "intrinsic" centered gate voltage.

The dependence in Figures 2 and 4 represented by asterisks * was obtained by the compact modeling using equation (15) (i = 0); the dependence represented by the pluses + is obtained using equation (16) (i = 1); the dependence represented by a cross-sign of multiplication × is obtained using equation (17) (i = 2); the circles correspond to the calculation using equation (14), where the initial value of the saturation current is taken as the value obtained by the iterative method for i = 20; triangles correspond

IOP Conf. Series: Materials Science and Engineering **498** (2019) 012038 doi:10.1088/1757-899X/498/1/012038

to the calculation using equation (14), where the initial value of the saturation current is taken as the value obtained by the iterative method for $i = 21$.

It can be seen from the Figures 2 and 4 that taking the starting point for compact modeling of saturation current using the results of calculation of the saturation current by the iterative method for odd values of the number of iterations ($i = 1$ and $i = 21$ are presented on Figures) results in a rapid convergence of the saturation current to zero from a certain value of V_{gs} or V_{sat}, correspondingly.

But for even values of the number of iterations ($i = 0$, $i = 2$ and $i = 20$ are presented on Figures), the form of the saturation current dependence on an "extrinsic" centered voltage on the gate V_{gt} or saturation voltage V_{sat}, correspondingly, obtained from compact modeling, is not much different from the results of modeling by the method of bisection.

Comparison of the saturation current calculation results by the iterative method, for a given number of iterations i, (Figures 1 and 3) and by the compact modeling (Figures 2 and 4), using the initial value for the saturation current, obtained by the iterative method with the same number of iterations i, allows us to conclude that compact modeling gives a good accuracy in a much wider range than the calculation using the iterative method that using the same number of iterations.

Finally, we conclude that even compact modeling of the saturation current using equation (15), taking as the starting point the saturation current equal to zero gives an acceptable accuracy for a sufficiently large range of an "extrinsic" centered voltage on the gate V_{gt}.

5. References

[1] Turin V, Rakhmatov B, Zebrev G, Kim C and Shur M 2016 *Proc. II Int. Conf.: Modern Problems of Physical and Mathematical Sciences (Orel)* (Orel: OSU) pp 135-151 (in Russian)

[2] Turin V, Rakhmatov B, Kim C and Iñiguez B 2016 *IOP Conf. Ser.: Mater. Sci. Eng.* **151** 012044

[3] Turin V O, Sedov A V, Zebrev G I, Iñiguez B and Shur M S 2009 *Proc. SPIE* **7521**, International Conference on Micro- and Nano-Electronics

[4] Turin V, Zebrev G, Makarov S, Iñiguez B and Shur M 2014 *Int. J. Numer. Model.: Electronic Networks, Devices and Fields* **27** 863

[5] Estrada M, Cerdeira A, Puigdollers J, Reséndiz L, Pallares J, Marsal L F, Voz C and Iñiguez B 2005 *Solid-State Electronics* **49** 6 p 1009

[6] Kim C, Castro-Carranza A, Estrada M, Cerdeira A, Bonnassieux Y, Horowitz G and Iñiguez B 2013 *IEEE Trans. Electron Devices* **60** 6 p 1136

IOP Conf. Series: Materials Science and Engineering **498** (2019) 012039 doi:10.1088/1757-899X/498/1/012039

Optimization of the 3D-IC model structure

M N Aralov[1], V F Barabanov[1], O Ja Kravets[1], A M Nuzhny[1] and N V Varlamov[2*]

[1]Voronezh State Technical University, Moskovsky Prospekt 14, 394026 Voronezh, Russian Federation
[2]National Research Nuclear University (MEPhI), Kashirskoe shosse 31, 115409 Moscow, Russian Federation

[*]Corresponding author's e-mail address: nvvarlamov@mephi.ru

Abstract. This article describes an algorithm of optimizing the placement of the main constituent elements of 3-D integrated VLSI circuits. To solve the formulated problem of optimization, it is proposed to use a modified genetic algorithm. The proposed algorithms, models and methods have been implemented in the software. The article presents the structural scheme and the results of the developed software.

1. Introduction

The 3-D integrated circuit (3D-IC) is a multi-layer object within which non-stationary physical processes occur [1, 11]. Given the high degree of integration, the problem of heat distribution within 3D-IC is very important [2, 5].

In designing 3-D integrated VLSI, the developer can choose arrangement of the crystals and arrangement of IP blocks on the chip area. Obviously, close arrangement of highly heated elements will result in appearance of local overheat areas, which, in turn, adversely affect operation of the entire device [8]. Numerous components make the task of finding the optimal arrangement highly non-trivial, therefore developers of 3D-ICs have the need for special software tools that can automate the arrangement process.

In changing the geometric structure of the object, it is important to consider possible restrictions on arranging layers or elements in the layer. For example, an element or layer may be fixed, or should take a certain coordinates range.

The presence of various, hard to formalize and conflicting limitations, availability of various optimization criteria allow attributing the problem of 3D-IC components arrangement to complex problems of optimization [4, 5].

2. Formulation of the optimization task

The optimization task is divided into two subtasks:

1) Obtaining a set of variants of heated elements arrangement inside the selected layers (layer structure optimization);

2) Obtaining a set of variants of arrangements of the selected layers themselves with regard to the obtained results of layer structure optimization.

Let us reduce the problem of arranging the elements in the layer to the well-known problem of arranging rectangles in a plane. The initial data of the problem are the number and the size of elements, as well as their thermal characteristics. The criterion for assessing quality of arrangement is aimed at the uniformity of distribution of heat sources.

Let us denote the area of possible elements arrangement as D. The area of this zone $S_D = w_D h_D$, where w_D and h_D are the layer dimensions. Let's take n for the number of arranged elements and denote the set of arranged elements (rectangles) as $E = \{e_1, ..., e_n\}$. Each arranged element $e_i \in E$ is characterized by three parameters: its length (w_i), width (h_i) and the maximum heating power (p_i). The solution of the problem is vector $ke=(ke_1, ke_2, ..., ke_n)$. Here $ke_i=(x_i, y_i, \alpha_i)$ specifies location of the e_i item, i.e. the coordinates of its center (x_i and y_i) and the rotation angle (α_i). The internal structure of element ei is considered uniform, so α may be determined by a one-bit variable that specifies (or does not specify) pi rotation by 90°.

In solving the problem of arrangement, the following limitations are to be observed (1) – (5).

The area of possible arrangement cannot be less than the total area of the atrranged elements:

$$\sum_{i=1}^{n} S(e_i) \le S(D) \quad (1)$$

The area of possible arrangement should include all arranged elements:

$$e_i \in D, \; i = \overline{1,n} \quad (2)$$

The elements should not overlap:

$$e_i \cap e_j = \varnothing, \; i \ne j; \; i, j = \overline{1,n} \quad (3)$$

Arrangement of the element may be anchored:

$$k_{ei} = \text{const} \quad (4)$$

The coordinates of the anchor point may be limited to the specified range:

$$x_i \in [x_i^{min}; x_i^{max}], \; y_j \in [y_i^{min}; y_i^{max}] \quad (5)$$

Thus, the arranged elements may be divided into three groups:

1. anchored (the coordinates of the anchor points are determined and cannot be changed);
2. freely positioned (the area of arrangement is limited only to area D);
3. conditionally freely positioned (indication of intervals of possible changing anchor points coordinates in the coordinate axes).

The function of assessing mutual arrangement of two elements e_i and e_j (with p_i>0 and p_j>0) looks as follows:

$$f(k_{ei}, k_{ej}) = l_{i,j}^2 / p_i p_j = ((x_i - x_j)^2 + (y_i - y_j)^2) / p_i p_j \quad (6)$$

Where $l_{i,j}$ is the distance between the centers of the elements e_i and e_j in vector k_e.

The task of optimizing arrangement may be described as follows: in view of the constraints (1-5), one has to find such a vector $\overline{k}_e$, with which

$$F(\overline{k}_e) = \max\left\{\sum_{i=1}^{n-1} \sum_{j=i+1}^{n} f(k_{ei}, k_{ej})\right\} \quad (7)$$

After obtaining one or more variants of the optimum arrangement of elements inside the layers, arrangement of the layers is to be optimized.

The solution to optimization of layers arrangement problem will be vector $\overline{k}^L=(k^L_1, k^L_2,.., k^L_N)$. Here $k^L_i=(n^L_i, \alpha^L_i, q_i)$, n^L_i is the number of layer l_i, α^L_i is rotation of layer l_i, and q_i is the number of the variant of elements arrangement within layer l_i. At the same time, α^L_i can be one of the four values indicating the rotation of the layer by 0, 90, 180 and 270 degrees.

For any vector $\overline{k}^L$, the following constraints are introduced (8)-(10).

The sequence number of layers should not coincide:

$$n_i^L \neq n_j^L, i = \overline{1,N}, j = \overline{1,N}, i \neq j \tag{8}$$

The sequence numbers of layers should not exceed the total number of layers:

$$n_i^L \leq L, i = \overline{1,N} \tag{9}$$

Layer location may be anchored:

$$k_i^L = \text{const} \tag{10}$$

The number of layers is taken as N. A set of layers is denoted as $L=\{l_1,l_2,..,l_N\}$.

To estimate thermal parameters of the object, a three-dimensional matrix P with dimensions N×m×n is introduced. Each element p_{kij} of matrix P characterizes the power of the heat source located on layer l_k with coordinates i, j. If there is no heat source with these coordinates, then $p_{kij}= 0$.

The goal of optimizing layers arrangement is ensuring uniform distribution of heat sources inside a multilayer object for reducing the probability of occurrence of local zones of overheating. In other words, the more powerful two heat sources are, the farther from each other they should be located. Based on this, the function quality assessment for some arrangement k^L is defined as

$$F(\bar{k}^L) = \min\left\{\sum_{k_1=1}^{N}\sum_{i_1=1}^{n}\sum_{j_1=1}^{m}\sum_{k_2=k_1}^{N}\sum_{i_2=i_1}^{n}\sum_{j_2=j_1}^{m}(p_{k_1i_1j_1}\cdot p_{k_1i_1j_1}/r^2)\right\} \tag{11}$$

Where $p_{k_1i_1j_1}$ and $p_{k_1i_1j_1}$ are powers of heat sources with coordinates $k_1i_1j_1$ and $k_2i_2j_2$, and $r=\sqrt{(k_1-k_2)^2+(i_1-i_2)^2+(j_1-j_2)^2}$ is the distance between them.

Equation (11) assesses the impact of all elements in matrix P on each other. This requires large resources of the computing system, in case of using (11) as the target function when the genetic algorithm is used, since the target function is calculated with each iteration of the cycle (see Fig. 2). For a flat multilayer object, the influence of the elements in matrix P is assessed separately for each column P. The quality of heat sources arrangement inside a column of matrix P with coordinates x, y is assessed with equation

$$f_{column}(x,y) = \sum_{i=1}^{N-1}\sum_{j=i+1}^{N}(p_{ixy}\cdot p_{jxy}/(i-j)^2) \tag{12}$$

The modified target function for genetic algorithm of optimizing arrangement of layers is as follows:

$$F_{column}(\bar{k}^L) = \min\left\{\sum_{i=1}^{n}\sum_{j=1}^{m} f_{column}(i,j)\right\} \tag{13}$$

3. The use of the modified genetic algorithm

Adaptation of the genetic algorithm (GA) to the problem of layers arrangement is in defining the following parameters:

1. the method of population coding;
2. the method of forming a population;
3. implementation of genetic operators of selection, recombination and mutation;
4. presentation of the target function for assessing the proposed variant of arrangement;
5. forming the shutdown criteria; and
6. the optimum values of changed parameters of the genetic algorithm.

The genetic algorithm of optimizing the arrangement of elements on the layer is shown as a diagram in Figure 1. The project information is represented by binary coding. The coding chromosome (CCH) is a string, in which information about each arranged element is represented in the binary code: the coordinates of the anchor point and the element rotation angle. In this method of presentation, the length of CCH is determined by formula:

$$L = N(\log_2 w_d + \log_2 h_d + 2), \tag{14}$$

where N is the number of elements; w_d h_d is the length and width of arrangement area D.

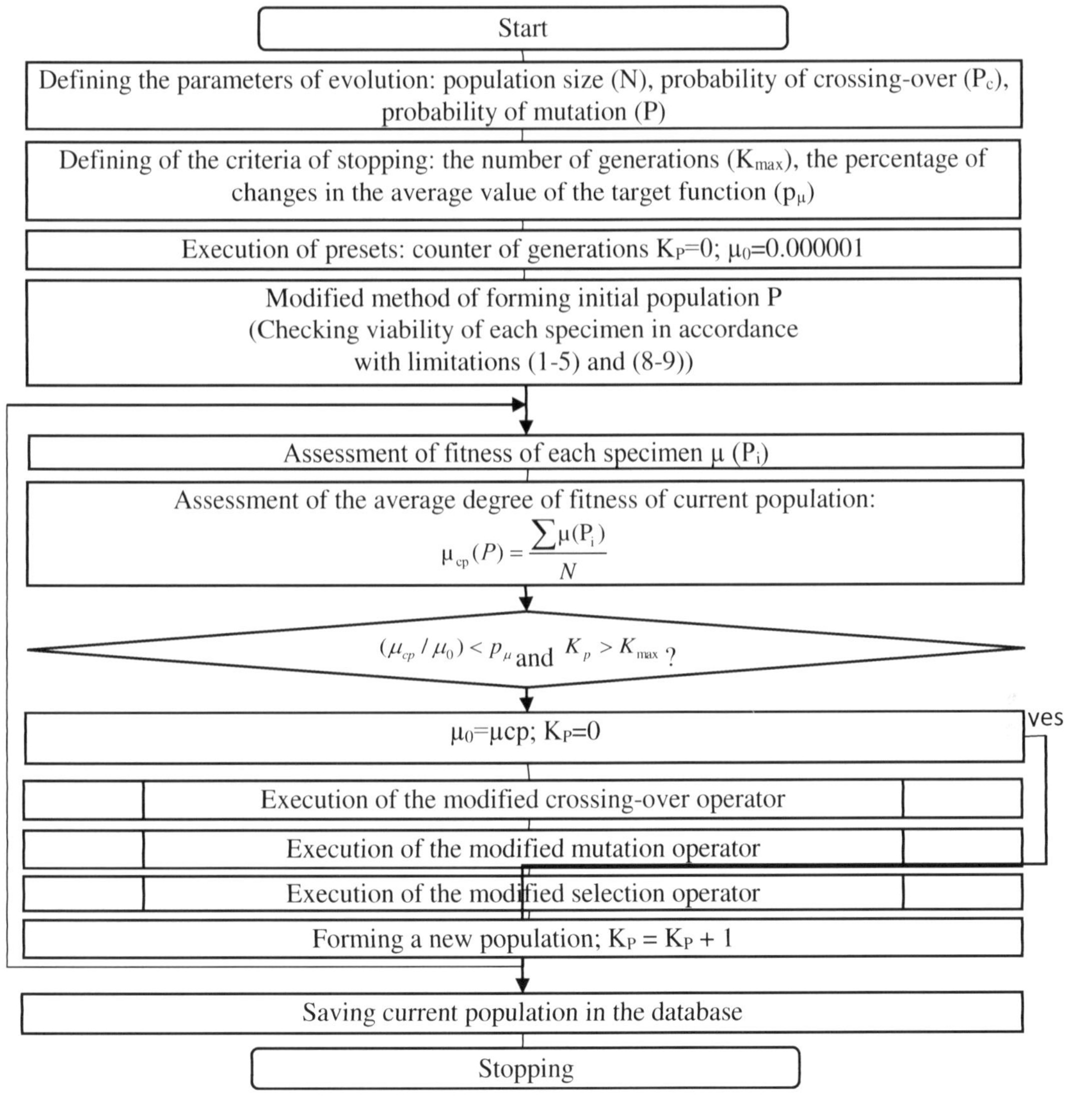

Figure 1. Diagram of GA for optimization of arrangement of 3D-IC elements

Modification of genetic operators is aimed at minimizing the risk of looping the algorithm in the area of the local extremum of the target function.

4. The software implementation of the proposed optimization algorithm

The software modules for optimizing arrangement of elements within the layer and optimizing arrangement of layers have a similar structure shown in Figure 2.

The graphical user interface of the system for optimizing arrangement of heating elements contains information about the simulated object, selected layer, and arranged elements. Here GA parameters are adjusted, and the limitations for arrangement of the selected element are set. The status of operation, errors, and other information are displayed in the console. The results of optimization can be displayed as a diagram of changes in the TF, and in text form.

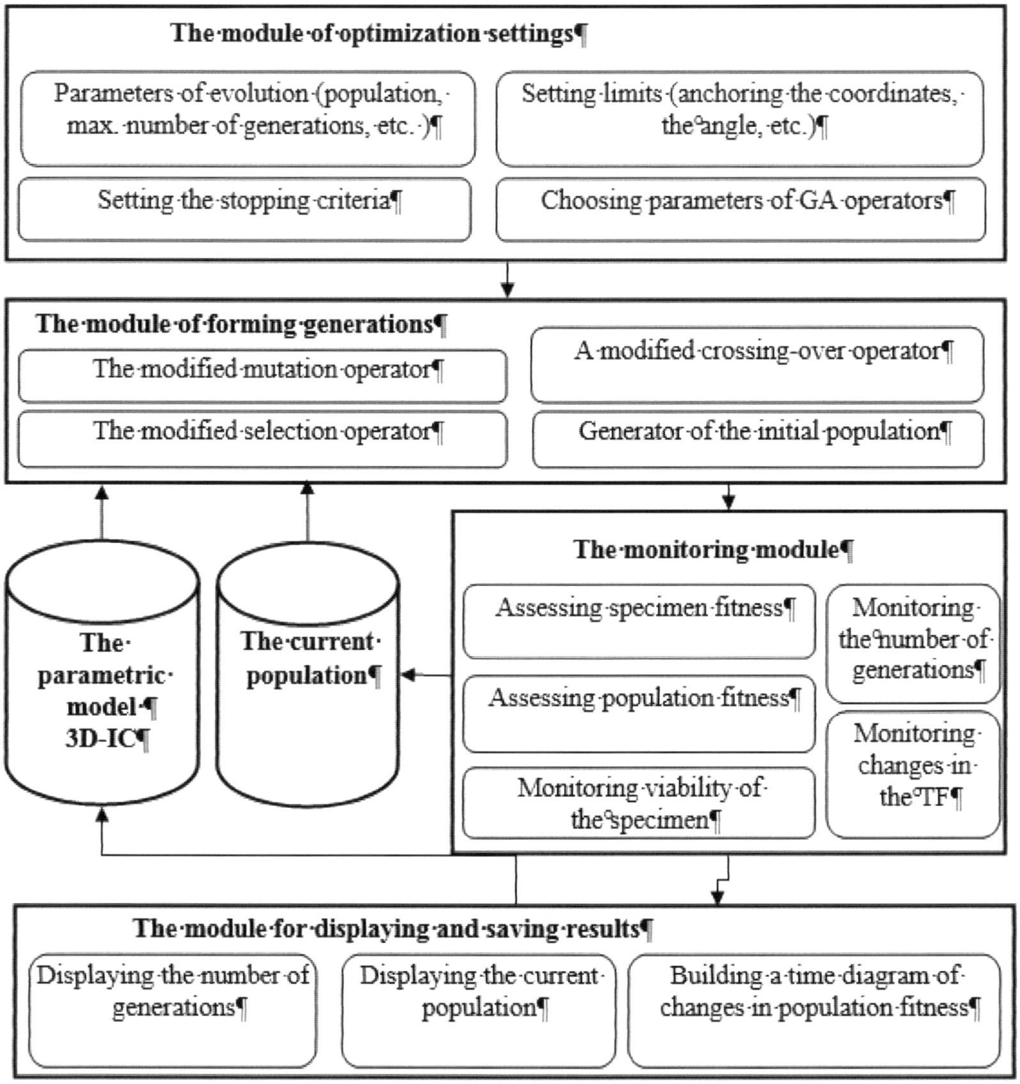

Figure 2. Block diagram of the GA-based optimization module

The graphical user interface of the system for optimizing arrangement of the layers contains information about the simulated object: the object name, the quantity and size of layers, etc. Here GA parameters are adjusted, and the limitations for arrangement of the selected layer are set. The status of operation, errors, and other information are displayed in the console. The results of optimization can be displayed as a diagram of changes in the TF, and in text form.

5. The results of the proposed software

To check correct operation of the application, an existing model of an 8-bit microcontroller was used. The freely arranged and conditionally free elements are schematically presented in Figure 3. The dimensions of the elements are indicated in the brackets.

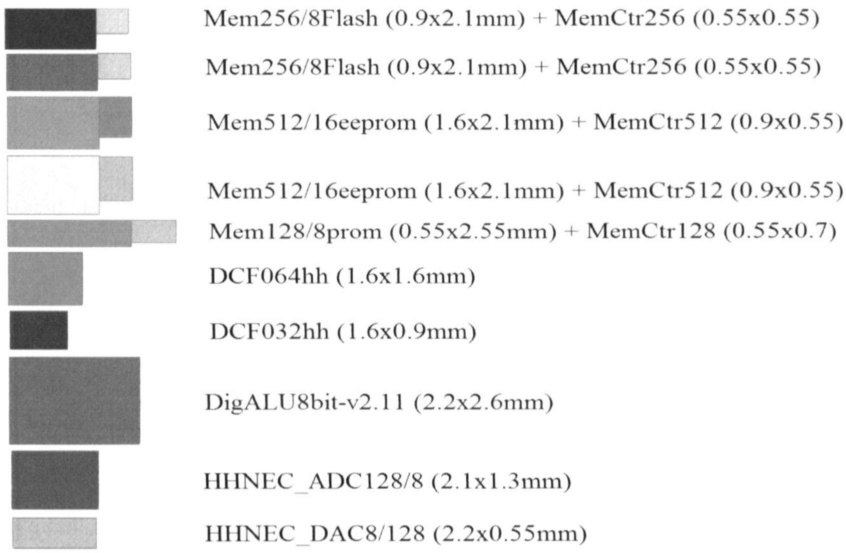

Figure 3. The freely arranged elements of the modeled object

Memory elements (Mem256/8Flash, Mem512/16eeprom, Mem128/8prom) and elements of control (MemCtr128, MemCtr256, MemCtr512) are rigidly attached to each other, i.e. they are conditionally free. Elements DCF064hh, DCF032hh, DigALU8bit-v2.11, HHNEC_ADC128/8 and HHNEC_DAC8/128 are freely arranged.

The area of arrangement, i.e. the space of the crystal with anchored elements is shown in Figure 4.

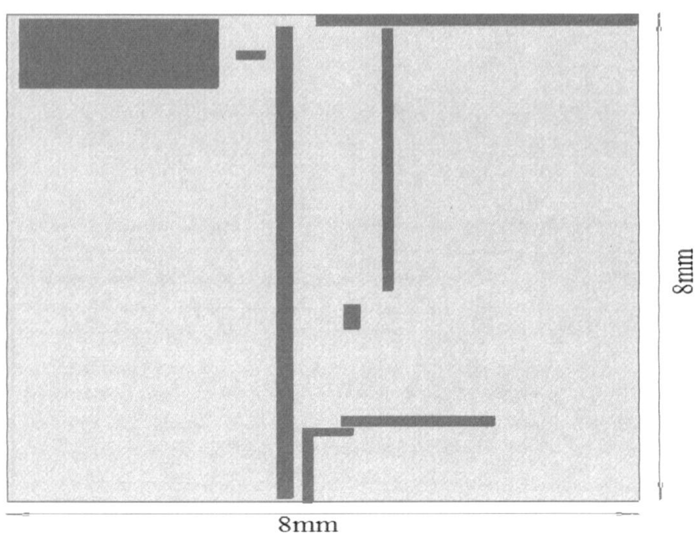

Figure 4. The area of arrangement with anchored elements.

As a result of studying the model with the use of the approximate method based on the modified genetic algorithm, 10 possible options of arrangement were identified, 3 of which have been rejected for technical reasons. After building and analyzing thermal models, the final variant of arrangement has been chosen, which is schematically shown in Figure 5. The selected variant coincided with the arrangement obtained with the use of the Cadence Design Systems software and implemented on the chip (see Figure 6).

IOP Conf. Series: Materials Science and Engineering **498** (2019) 012039 doi:10.1088/1757-899X/498/1/012039

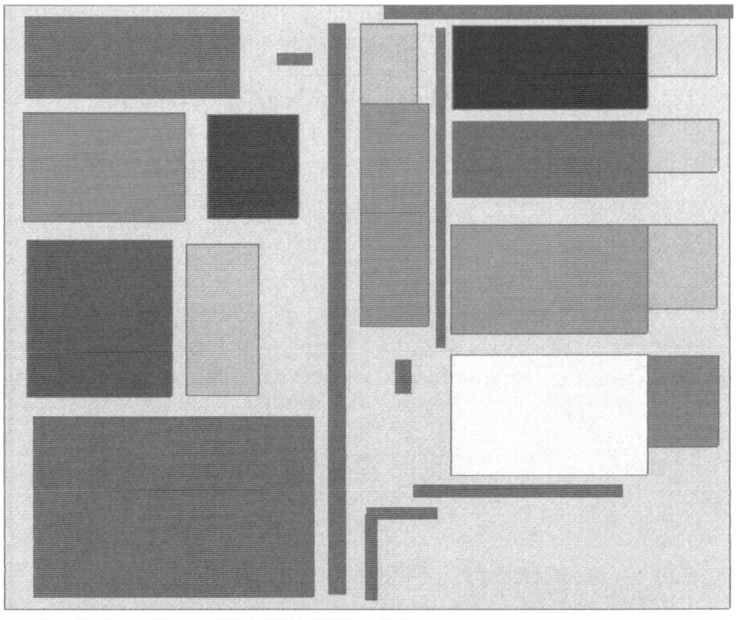

Figure 5. The final way of elements arrangement

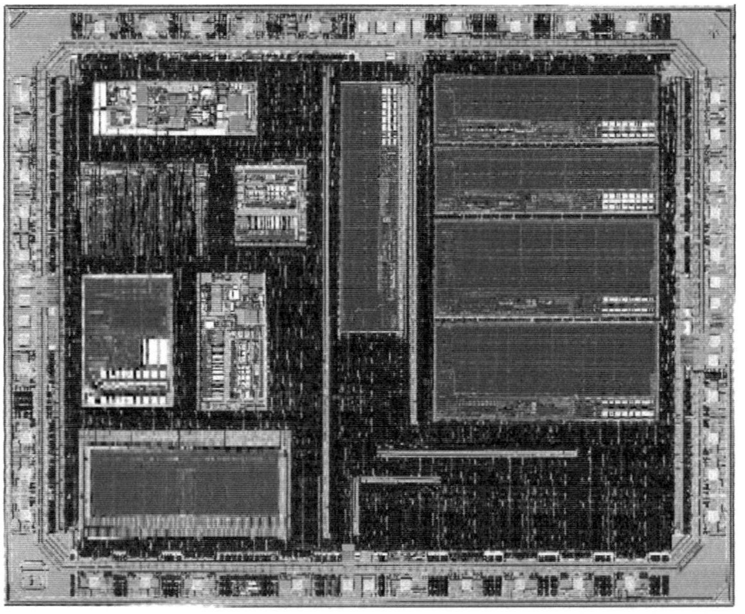

Figure 6. The variant of arrangement obtained in the Cadence software

6. Conclusion

The proposed software module is part of the software suite for modeling thermal distribution inside multilayer objects [6, 7]. Simulation of thermal distribution with the use of the developed software suite, and comparison of the results with the data obtained by a thermal imager were also performed for a number of other semiconductor devices. Analysis of the results obtained with with the use of software suite showed good adequacy of generated models [3, 9]. The performed experiments confirm the working efficiency of the modules included into the developed software suite, and, as a consequence, correctness of theoretical calculations.

IOP Conf. Series: Materials Science and Engineering **498** (2019) 012039 doi:10.1088/1757-899X/498/1/012039

References

[1] Woychik Ch G, Lee S, McGrath S, Tosaya E and Arkalgud S 2016 *Circuit Insight* http://www.smtnet.com/library/files/upload/scalable-3d-ic-assembly.pdf
[2] Peter Ramm et al. 2010 *Proc. ESSCIRC/ESSDERC 2010 Sevilla, Spain, p. 9-16 (IEEE Xplore 978-1-4244-6664-1/10).* doi:10.1109/ESSCIRC.2010.5619857
[3] Håkan Nilsson 2014 *Eur. J. Appl. Physiology* **92(6)**:714-6
[4] Baker R J *2010 CMOS: Circuit Design, Layout, and Simulation* Third Edition (Wiley-IEEE)
[5] Lin M, Gamal A El, Lu Y-Ch and Wong S 2006 *IEEE Trans. Comput.-aided Des. Integrated Circuits Syst.* doi:10.1145/1117201.1117219
[6] Aralov M N Barabanov V F 2016 *Monitoring Module and 3D Visualization of Thermal Fields in Multilayer Objects* (Moscow: Federal Institute for Industrial Property, Certificate of State Registration of Computer Program No. 2016618405 dated 28.07.2016) (in Russian)
[7] Aralov M N Barabanov V F 2016 *A System of Modeling of Thermal Fields in Multilayer Objects* (Moscow: Federal Institute for Industrial Property, Certificate of State Registration of Computer Program No. 2016618435 dated 29.07.2016) (in Russian)
[8] Aralov M N 2015 *Modeling Inform. Technologies: Selected Pap. Intern. Sci. School "Paradigma"* (Yelm, WA, USA: Science Book Publishing House) pp 24-28
[9] Bahar Y N, Landrieu J, Pere Ch, Nicolle Ch 2014 *Intern. J. Technol.* **1:** 3-13
[10] Podlepetsky B I 2016 *IOP Conf. Ser.: Mater. Sci. Eng.* **151(1)** doi:10.1088/1757-899X/151/1/012007

Simulation of characteristics of microwave transistors with AlGaN heterostructures

E Ph Pevtsov[1*], T A Demenkova[1], V I Indrishenok[1] and N V Varlamov[2]

[1]Moscow Technological University (MIREA), Prospekt Vernadskogo 78, 119454 Moscow, Russian Federation
[2]National Research Nuclear University MEPhI (Moscow Engineering Physics Institute), Kashirskoe shosse 31, 115409 Moscow, Russian Federation

Corresponding author's e-mail address: *pevtsov@mirea.ru

Abstract. In work results of researches in the area of simulation of electrical characteristics of the AlGaN/GaN field-effect transistor are provided. The considerable difference of transport processes in case of the strong and feeble electrical polarization is shown. The role of capture of centers in volume of a buffer layer GaN is analyzed and it is shown that deep interruptions can influence considerably on distribution of electrostatic potential and density of electrons in a buffer layer. In case of high concentration of interruptions there is a sharp lowering of density of the free electrons with increase of distance from the channel, the current density in the depth of a buffer layer decreases.

1. Introduction

GaN field-effect heterojunction transistors became the main type of high-power semiconductor microwave devices. Large band-gap width of gallium nitride (3.4 eV) and low collisional ionization coefficient enable operation of devices with high bias voltages. AlGaN/GaN heterojunction transistors provide quite high electron density in the channel (about 10^{13} cm^{-2}), which taking into account large value of saturated electron velocity ($2.7 \cdot 10^{7}$ cm/s) may provide drain current of several amperes per millimeter of channel width.

At present, a number of publications are dedicated to aspects of instrumental-technological modeling of GaN-based transistor structures (see, for example, [1-5]), but the results obtained in these publications are not sufficiently complete. Within the framework of this paper, we propose methods for modeling effects associated with existence of deep centers and polarization and their influence on performance of microwave transistors and present corresponding numerical calculations.

2. Formulation of the problem

Mechanisms of high electron density appearance in the channel were considered in [6 and 7]. These mechanisms are associated with the effects of spontaneous and piezoelectric polarization in nitride semiconductor materials with wurtzite type crystal lattice. Hexagonal wurtzite lattice has no center of symmetry, and interatomic bond contains a considerable proportion of ionic component. Therefore, spontaneous polarization P_{SP} occurs in such crystals. P_{SP} vector is parallel to axis *c* of the lattice and is directed opposite to the direction of normal to crystal face, on the surface of which cations (Ga^{+}, Al^{+}) are located. Polarization value depends on chemical composition of the material. In $Al_xGa_{1-x}N$ P_{SP} value is approximately linearly dependent on Al content, and at x=1 it is three times higher than in

GaN. Piezoelectric polarization P_{PE} depends on deformation of crystal lattice under mechanical stresses. Relationship between P_{PE} value and direction with tensor of mechanical stresses is determined by tensor of piezoelectric coefficients.

Polarization pattern in heterostructure typical of gallium nitride transistors is shown in Fig. 1.

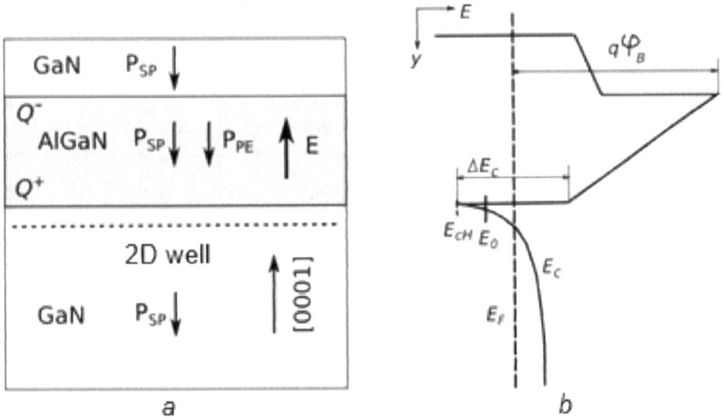

Figure 1. Location of layers (*a*) and dependence of energy of *Ec* conductivity band bottom on distance from the surface (*b*) in the structure of field-effect transistor. Energy diagram corresponds with thermodynamic equilibrium condition and presence of metal electrode on the surface.

Structure oriented along the crystal axis [0001] consists of thick lower GaN layer, AlGaN layer and thin upper GaN layer. Spontaneous polarization P_{SP} occurs in each of these layers. Due to difference in lattice parameter *a* in AlGaN and GaN, middle layer is exposed to extension, so piezoelectric polarization P_{PE} coincident in direction with spontaneous polarization occurs in it. In upper rather thin GaN layer piezopolarization is insignificant due to relaxation. Condition for continuity of electrical induction in the presence of difference in polarization values in different sides of heterojunction is taken into account by introducing Q^+ and Q^- charges on corresponding heterointerfaces. Surface density of Q^+ charges may exceed surface density of qN_Dd donors in $Al_xGa_{1-x}N$ layer with thickness d. As calculated in [4], $Q^+/q=1.4x10^{13}cm^{-2}$ at x=0.25; where q – value of elementary charge.

Positive charge on heteroboundary pulls free electrons and, as illustrated in Fig. 1b, potential well with surface electron density n_S is formed on the surface of lower buffer layer. Value of this electron density may be determined using formula given in [3], which defines the change of *Ec* on the thickness of middle barrier layer assumed that there are no charges in it,

$$\frac{dq^2}{\varepsilon\varepsilon_0}\left(\frac{Q^+}{q}-n_s\right)=q\varphi_B-\Delta E_c+\left(E_F-E_{cH}\right) \quad , \tag{1}$$

where ε_0 – electrical permittivity of vacuum, ε – material relative permeability, $q\varphi_B$ – potential barrier height, ΔE_c – discontinuity of E_c on the boundary, $E_F–E_{cH}$ – distance between Fermi level and conduction band bottom of buffer layer on the boundary. The latter value may be represented in the following form $E_F-E_{cH}=(E_F-E_0)+(E_0-E_{cH})$, where E_0 – absolute value of energy of the bottom of lower subband in two-dimensional quantum well of triangular profile. Energy intervals included in the latter equation may be expressed in terms of n_S:

$$\left(E_F-E_0\right)=\frac{\pi\breve{h}}{m}n_S \text{ and } \left(E_0-E_{cH}\right)=\left(\frac{9\pi\breve{h}q^2}{g\varepsilon\varepsilon_0\sqrt{gm}}n_S\right)^{2/3} , \tag{2}$$

where *m* – effective mass of electrons. It should be noted that in the above formulas variables and material parameters depend on composition of $Al_xGa_{1-x}N$. According to numerical calculation performed in [3] $n_S=1\cdot10^{13}$ cm^{-2} at *d*=20 nm and *x*=0.25.

3. Modeling of GaN field-effect transistor

Modeling of characteristics of AlGaN/GaN field-effect transistors was performed in Sentaurus TCAD modeling system. Structural components of the considered transistor are shown in Fig.2. Source/drain contacts are buried, so that they overlap the thickness of channel area of buffer layer. Distance between source and drain is 5 μm, gate length is 0.8 μm, drain is 3 μm away from the gate.

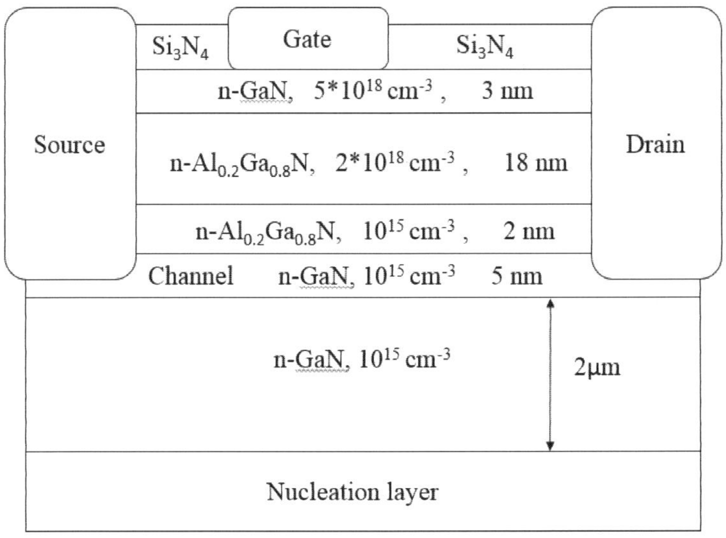

Figure 2. Field-effect transistor structure.

Modeling of structures with heterojunctions and corresponding calculations in particular are based on generalization of Shockley diode theory (Anderson model) and special models of charge transfer process in heterojunctions, which take into account recombination of carriers through surface states, tunneling and both effects simultaneously. At initial stage we investigated dependence of states of spontaneous and piezoelectric polarization of $Al_XGa_{1-X}N$ barrier layer on aluminum content in its composition and the way it affects characteristics of such devices. According to model proposed in [4] at pseudomorphic growth of AlGaN on GaN with AlGaN barrier layer thickness of more than 15 nm, carriers concentration is in quadratic dependence on Al molar content in AlGaN. However, spontaneous polarization is determined not by the absolute value of dipole moment averaged by volume, but by difference of polarization in two different system states, at that the value of macroscopic polarization is also determined by piezoelectric polarization arising due to mismatch of the lattice parameters on heterointerfaces. Polarization effects considered in the program – batch file of characteristics modeling, may be seen in band diagrams and space-charge density distributions shown in Fig. 3.

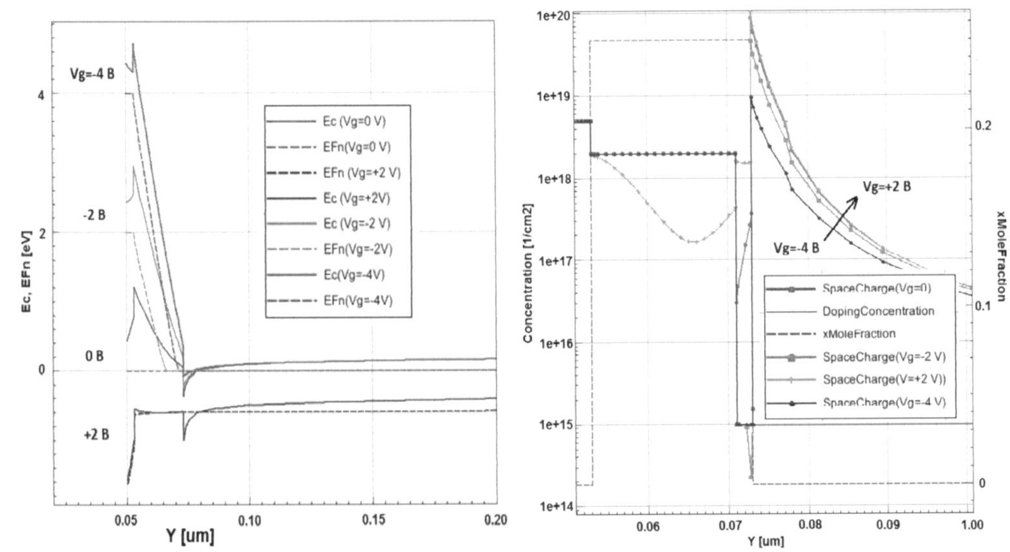

Figure 3. Band diagrams (on the left) and space-charge density distributions (on the right) in section passing through the middle of the gate length at different voltages V_{gs} on it and at voltage on drain V_{ds}=0.

$I_d(V_{gs})$ output current-voltage characteristics are shown in Fig. 4.

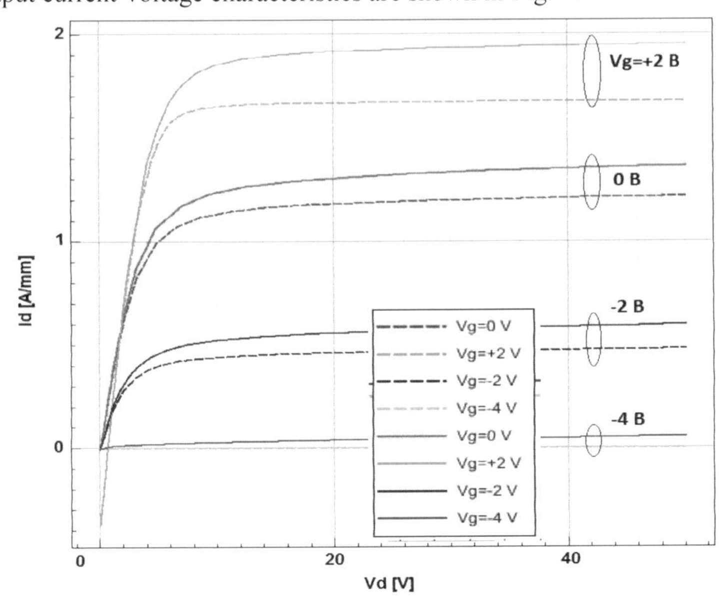

Figure 4. Families of output current-voltage characteristics of the model without (solid lines) and with (dashed lines) traps in GaN layer.

Increase of negative voltage on the gate reduces electron density n_S in the channel; however, permanence of Q^+ polarization charge causes the intensity of electric field in the barrier to increase, and almost all V_{gs} voltage falls on the barrier layer. At positive V_{gs}, n_S increases and voltage on barrier layer decreases. With further growth of V_{gs}, accumulation of n_S is limited due to electrons pass to the gate through ΔE_c permanent barrier. Fermi quasi-level shift in buffer layer is shown in Fig. 4 and is probably related to the use of ohmic contacts to barrier type source/drain in initial program. This figure shows the decrease of current increment and, consequently, the decrease of transmission steepness at positive gate bias, which is connected with the above-mentioned dependence of n_S on V_{gs}. Maximum value of steepness in transistor without traps is 3.7 S/mm. In order to clarify the role of traps in GaN they were included in batch file for modeling of instrument characteristics. Specified concentration is 10^{16} cm^{-2}, energy position is in the middle of band-gap, electrons and holes capture cross section is 10^{-18} cm^2. The results obtained are presented in Fig. 5.

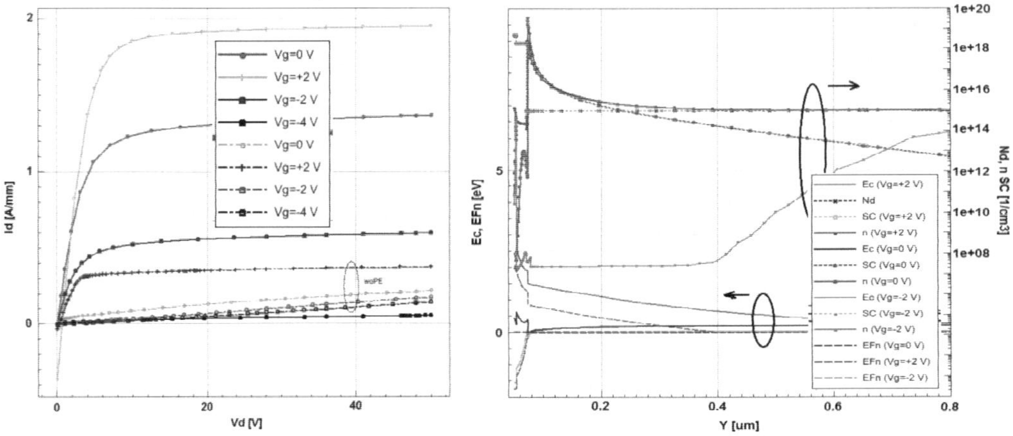

Figure 5. Left field shows output characteristics without (dashed lines) and with (solid lines) taking into account polarization. Right field shows energy diagrams (axis of ordinates is on the left), distribution of electron density and space charge density (axis of ordinates is on the right) in section passing through the middle of the gate length at various V_{gs} and at V_{ds}=0.

As can be seen from Fig. 5, such traps reduce drain current in saturation region by about 10%. Of particular interest is the comparison of modeling results with and without taking into account polarization effects, for which purpose polarization models call commands were excluded from the program. Design and material parameters remained the same. Absence of polarization results in sharp decrease of drain current and change of characteristics of charge carrier transfer in the channel. Insufficient concentration of donors in barrier layer cannot provide high surface electron density in the channel quantum well even at V_{gs}=0 (see the right-hand side of Fig. 5). At negative gate voltage the well is emptied and electric field penetrates into the depth of buffer layer. At V_{gs}=-2 V, thickness of depletion region takes significant part of this layer. Drain current is determined by transfer of electrons through non-depleted part of the buffer.

4. Conclusions

The authors believe that the following provisions and results are new in this paper. Modeling of electrical characteristics of AlGaN/GaN field-effect transistor was performed. Significant difference in transfer of charge carriers in case of strong and weak electric polarization in structural elements is shown. In the absence of polarization effects, localization of current in the region of quantum well is noticeably reduced. Consideration of the role of trapping centers in the volume of GaN buffer layer showed that deep traps may significantly affect distribution of electrostatic potential and electron density in the volume of buffer layer. In case of high concentration of traps, their capture of electrons results in sharper decrease of free electron density with increasing distance from the channel, current density in the depth of buffer layer becomes smaller.

Acknowledgments

This work was supported by Ministry of Education and Science of Russian Federation, (project 112/8.5098.2017).

References

[1] Ramnish, Arya S K and Ahlawat A 2014 *Int. J. of Advanced Science and Technology* **66** p 33
[2] Vitanov S, Palankovski V, Maroldt S, Quay, Murad S, Rödle T and Selberherr S 2012 *IEEE Transactions on Electron Devices* **59** 3 pp 685-693
[3] Kuzmik J *et al* 2012 *Japanese J. of Appl. Phys*. **51** 054102
[4] Radchenko D, Sbitnev K and Maleev N 2009 *Components and technologies* **7** 8 pp 57-61 (in Russian)
[5] Vinichenko A N and Vasil'evskii I S 2016 *AMNST2015. IOP Conf. Series: Materials Science and Engineering* **151** 012037
[6] Ashok A, Vasileska D, Goodnick S M and Hartin O L 2009 *IEEE transaction on electron devices* **56** 5 pp 998-1007
[7] Ambacher O, Smart J and Shealy J R 1999 *Appl. Phys.* **85** 6 pp 3222 – 3233

Algorithmization and optimization of processes for integrated electronics engineering

O Ja Kravets[1], O N Choporov[1], A P Preobrazhenskiy[2] and N V Varlamov[3*]

[1]Voronezh State Technical University, Russia, Voronezh, Moscow district, 14
[2]Voronezh Institute of High Technologies, Russia, Voronezh, Lenina street, 73a
[3]National Research Nuclear University (MEPhI), Kashirskoe shosse 31, 115409 Moscow, Russian Federation

Corresponding author's e-mail address: * nvvarlamov@mephi.ru

Abstract. This paper discusses the challenges of taking a technological solution based on certain requirements, with the existence of several alternative options (alternatives) with regard to the technical requirements in the manufacture of integrated circuits, as listed in the following. The algorithm of the choice of process equipment during the production of integrated electronics is developed. The functional block diagram of an adaptive system with an adjustable model is given. The structure of the information schema implementation of the adjustment model of technological production of integrated electronics is given. The developed model of the technological process of production of integrated electronics is presented. The stages of optimization of production of integrated electronics are given. The algorithm of control of technological process of production of integrated electronics is developed.

1. Introduction

Engineering of a new series or type of chips gives an expert a number of uncertainties [1-3]. For example, it is necessary to evaluate the possibility to engineer this series or party of products while considering the specifications of the existing equipment and to evaluate the expected yield percentage.

In fact, a process solution needs to be made, based on certain requirements with several versions (alternatives), considering the specifications.

A process plant selection algorithm of engineering with several alternatives, considering the uncertainties that an expert has, at evaluation of criteria quality, is proposed for engineering application in the parameters of units on the basis of optimization methods. The algorithm allows for the selection of an alternative, which brings us nearest to the "ideal value" (i.e. "an ideal point") in the case of uncertainty [4, 5].

2. Algorithm for the selection of the process plant at integrated electronics engineering

Let us assume some number of alternatives n for selection of a processing method or selection of specific equipment (plant), out of which a decision-maker (an engineer) shall select, considering the influencing criteria (quality criteria of the products under the specifications) in an amount of m:

IOP Conf. Series: Materials Science and Engineering **498** (2019) 012041 doi:10.1088/1757-899X/498/1/012041

$$A_1 = \{a_{11}, a_{12} \ldots a_{1n}\}$$
$$A_2 = \{a_{21}, a_{22} \ldots a_{2n}\}$$
$$\ldots$$
$$A_m = \{a_{m1}, a_{m2} \ldots a_{mn}\}$$

Let us introduce the concept of an ideal alternative:

$$A_{id} = \{a_{id1}, a_{id2} \ldots a_{idn}\},$$

such that any value of a_{idi} is better than the a_{ji},, i=1…n, j=1…m for any i and j. Thus, it is necessary to determine such an A_j,, which would be most closely approximate to A_{id}. However, we must also consider the importance of each criterion (in the quality criteria of engineered products). With this purpose, let us introduce an importance option for each criterion U_i. Let us set the probability distribution function as p_j [6, 7]. Let us consider the effect of evaluation of the relative importance of each j criterion, by expressing it through the probability distribution:

$$p_{ji} = \frac{F_{E_j}(a_{ji})}{\sum_{k=1}^{m} F_{E_j}(a_{ji})},$$

where E_j is a fuzzy set, p_{ij} – membership degree evaluation, F_{E_j} - function of preferability on each criterion. To estimate the uncertainty Δ, let us use the Shannon measure for the uncertainty [1, 8]:

$$\Delta = -K \sum_{j=1}^{m} p_j \log_2 p_j$$

where K is a norming quantity to bring the uncertainty to the range from 0 to 100% (0 to 1). The overall uncertainty will be:

$$\Delta_{\text{î áù}} = -K \sum_{j=1}^{m} \sum_{i=1}^{n} p_{ji} \log_2 p_{ji}.$$

The uncertainty of a certain quality criterion at chipping will be:

$$C = \frac{\omega_j \Delta}{\Delta_{all}}$$

where ω_j - is a factor of quality importance, given by an expert.

As a result, the actual deviation $\sigma[A_j]$ of the alternative A_j from A_{id} shall be determined.

The mean square deviation may be written as:

$$\sigma[A_j] = \sqrt{\sum_{j=1}^{m} C\left[1 - F_{E_j}(a_{ji})\right]}.$$

Having determined the deviation for all j=1…m, the alternative (the best engineering version) with the minimum deviation from A_{id} should be selected. This selection algorithm is shown in Figure 1.

In step 1, the process (processing procedure) is selected. Then, the list of equipment and the operation modes available are loaded from the database. Then, the expert is provided with all preset alternatives of A_j.

In step 1, the process (processing procedure) is selected. Then, the list of equipment and the operation modes available are loaded from the database. Then, the expert is provided with all preset alternatives of A_j.

IOP Conf. Series: Materials Science and Engineering **498** (2019) 012041 doi:10.1088/1757-899X/498/1/012041

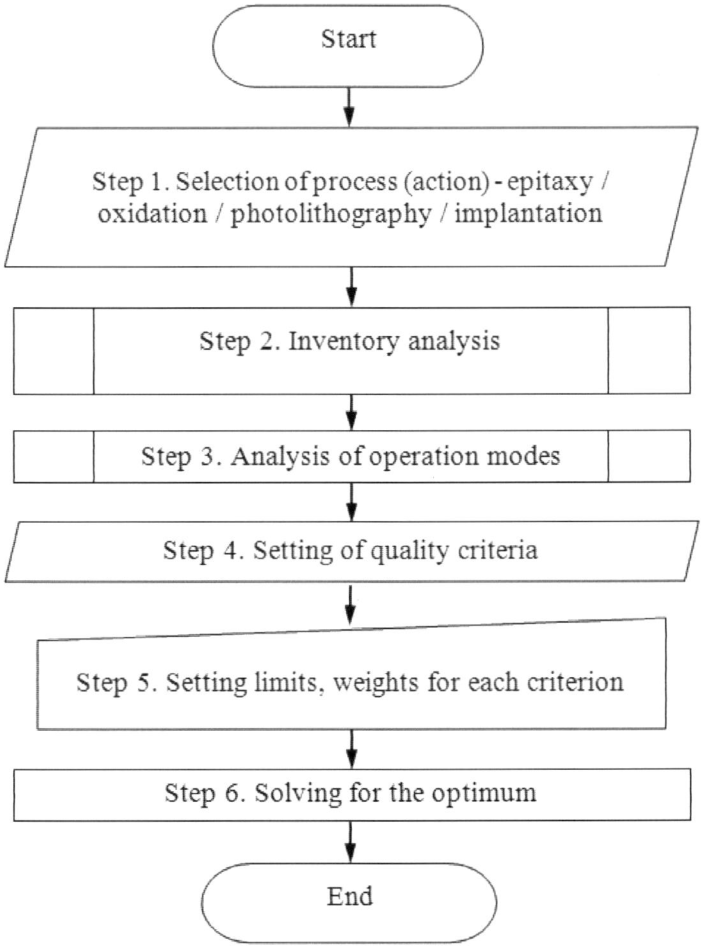

Figure 1. Algorithm for the selection of process plant via integrated electronics engineering

Then an ideal selection is made. The next step includes the entering of $F_{E_j}(a_{ji})$, which characterizes a preference of alternative i as of j quality parameter and weight coefficients for all i=1..n and j=1…m.

The deviations are calculated as described above and the expert is provided with the computed solution.

This algorithm is used in the unit for the selections of equipment and operation mode of the engineering optimization system for the chips, allowing for (if required) adjustment of the road sheet, to determine the most appropriate process plant for this series of chips, considering the specifications. That is, there is a feedback of the algorithm and the process for the selection of an operation mode and applied plant composition.

The past experience of the algorithm provides for three options:

1. The products cannot be engineered further to the specifications, using this equipment; either specifications need to be adjusted, or the equipment needs to be upgraded.
2. There is the only one implementation option.
3. There are several implementation options. In this case, the best solution is to be found according to the above algorithm for the selection of the best alternative.

Thus, the above algorithm allows for making a rational selection of operation modes and the composition of the process plant for engineering of a new series of products, considering the requirements, as well as for determining the manufacturability at the process engineering stage.

IOP Conf. Series: Materials Science and Engineering **498** (2019) 012041 doi:10.1088/1757-899X/498/1/012041

3. The forming of an adaptive system

Given the fact that in the period of production of chips, for the accounting of uncontrolled variables, the existing process model or operations thereof need to be adjusted to solve this problem. Further, for the results of the above analysis, it is proposed to use the adaptive systems with adjustable models.

The adaptive system, shown in Figure 2, can be operated in two modes - information gathering and control. At information gathering, values $y(t)$, $u(t)$ are recorded to the database for k time intervals (at a lack of process information). The proposed control algorithm is as follows:

Processing of the controlled object information and entry of the current process settings to the database.

Adjustment of the existing (adjustable) model parameters, considering the new data about the current process state (if required). If the system parameters are stable, the model is not adjusted.

Determining the best values of a controlling action as of the adjusted model. Implementation of the controlling action so that the system behavior meets the specified criterion.

The information diagram of this system is shown in Figure 3.

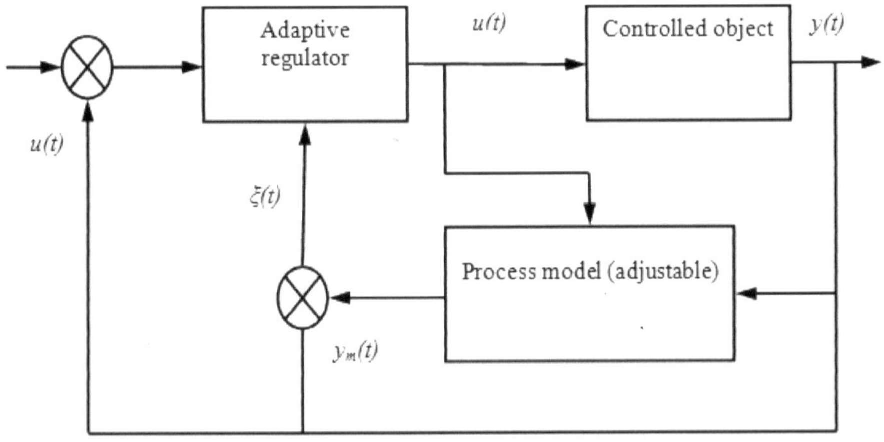

Figure 2. Functional diagram of the adaptive system with adjustable model

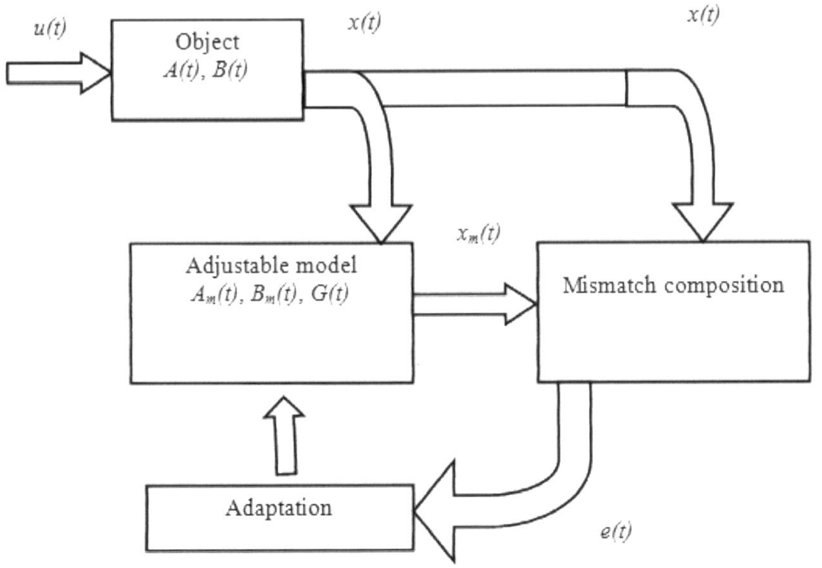

Figure 3. The information diagram for adjustment of the process model for integrated electronics engineering

4. The algorithm of optimization the system

As described above, the non-ideality and aging of processing plants and effects of random or uncontrolled parameters for integrated electronics engineering can cause alteration of input and output variables for each operation and the process in general in time, so the process optimization should consider the effect of uncontrolled parameters.

The process model is shown in Figure 4. In general, the process optimization steps are shown in Figure 5.

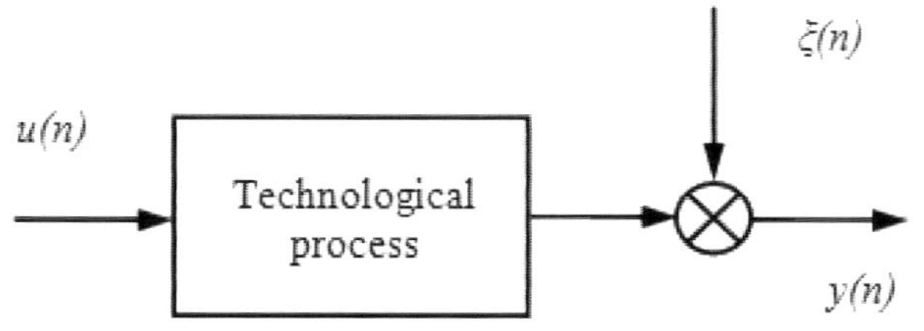

Figure 4. Process model for integrated electronics engineering

Let us develop an algorithm for the calculation of adaptation coefficients.

Let us put the object model with adaptation, which is well-known from literature [9]:

$$B \cdot u(t) = \sum_{m=0}^{M} K_n \cdot \frac{d^m y}{dt^n} + \sum_{m=0}^{M} H_n \cdot \frac{d^m y}{dt^n} \tag{1}$$

where K_n is the object characterizing matrix and H_n is the correction unit matrix (parameter adjustment) [9].

In this case, the number of object parameters and adjustment parameters is the same, equal to m, which does not always happen in practice.

The process control task is to find proper coefficients for minimal deviation of the actual state of the process.

The adjustment model [9] can be similarly written as:

$$B \cdot u(t) = \sum_{m=0}^{M} A_n \cdot \frac{d^m y}{dt^n} \tag{2}$$

where A_n is the adjustable model matrix.

Expressions (1) and (2) show that the ideal adaptation with a zero error requires:

$$\begin{gathered} K_m + H_m = A_m \\ H_m = H_m(t) \end{gathered} \tag{3}$$

If no restrictions are imposed on H_m, and if (3) is fulfilled, then the coefficients in the two previous expressions will be identical.

Considering (2):

$$\frac{dH_m}{dt} = \frac{\partial J(\xi)}{\partial A_m} \tag{4}$$

Using the expression (2), we get:

$$\frac{dH_m}{dt} = \sum_{m=0}^{M} \left(-C \bullet \frac{d\xi(t)}{dt} \bullet \frac{du_m(t)}{dt} \right) \tag{5}$$

Now, let us develop a control algorithm for the process of integrated electronics engineering.

In general, (5) is a system of nonlinear equations. The search for solutions for this system may take a long time and is often impossible [10-12].

Since control and output variables have a limited range of possible values, one can use polynomial approximations, such as the linear (for n parameters (n=1,2...)):

$$Y_{out} = \sum_{i=1}^{n} k_i X_i + \sum_{i=1}^{n} k_i X_i Y_i + \tilde{n}_0$$

where k_i, c_0 are some coefficients. To find the best solution, the extremum of the optimization function F needs to be found.

In practice and in [14], the authors indicate the need to bring the equations to a linear form [14, 15] for process control tasks, for faster calculations at solution finding.

To validate the mathematical model, we propose to calculate the deviation of the estimated value from the actual one and to compare it with a permitted deviation value Δ:

$$\left| F_{fact\_i} - F_{calc\_i} \right| \geq \Delta$$

The actual value is calculated by substituting the current values $Y_{тек}$ in the objective function with the values measured at the process outcome.

We propose to find the estimated value by substitution of values of the Y_{1i} and Y_{2i} output variables, calculated by a mathematical model.

At this, if the difference modulus does not exceed a permissible value, it is considered that the mathematical model adequately describes the real process and can be used for process control. In other cases, the mathematical model requires clarification.

The values of control and output variables for the i-th moment are denoted by U_{1i}, U_{2i}, U_{1i}, U_{2i}.

We propose to organize the interaction with the control system as follows.

Data transmission to the control system starts at time t=t_0=0.

Some time (from the time t=t_0 to t=t_i, i=1,2...n) after the start of the process is required to gather information about the current state of the system: hardware settings, given chipping algorithms and others.

At time t=t_i the control algorithm starts.

Switching of the control system (that is, the process control) is started at time t_i, that is, at the moment the process plant is switched on.

The management process enters the transmitted data to the control system and records them to the history files of the process after a certain period of time.

At the optimization of the process, the transmission of now-used-data to the control system should be limited: it is necessary to use the data for the last period of time - Δt, as the previously recorded data is outdated (Fig.5).

IOP Conf. Series: Materials Science and Engineering **498** (2019) 012041 doi:10.1088/1757-899X/498/1/012041

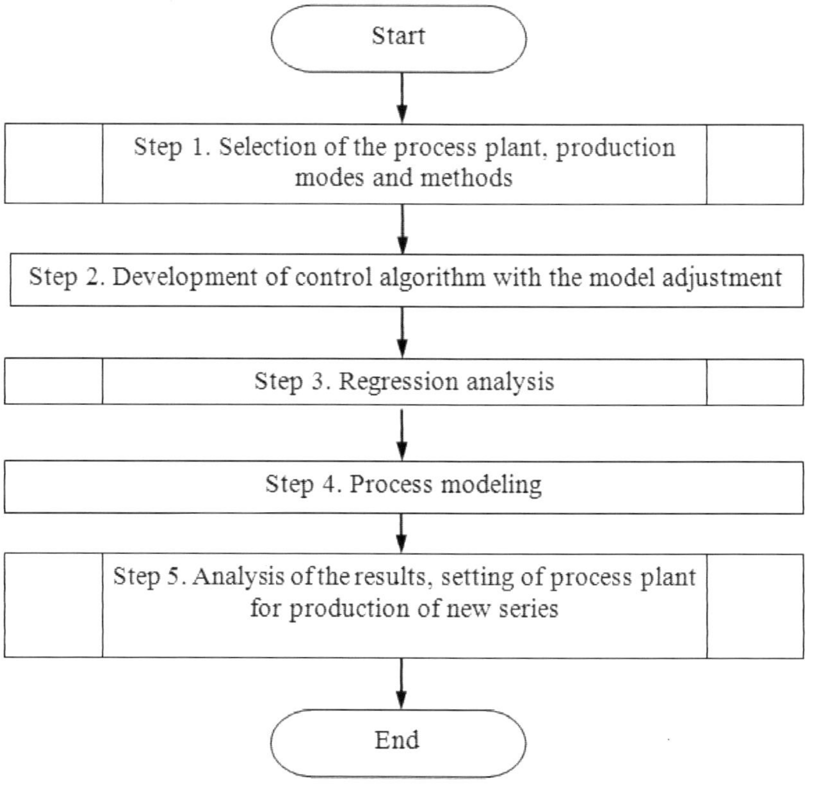

Figure 5. Steps to optimize the integrated electronics engineering

The moment of time t=0 is the time of the first measurement of process parameters.

The value for Δt is selected considering the time of each process step.

Thus, an automated process control system receives and transmits data, necessary for adaptation to the actual state of the process. To determine the influence of the control systems on the actual state of the process for the production of the subsequent series of products, there are used regression analysis and data, which are transmitted to the automated process control system.

The developed algorithm is shown in Figure 6.

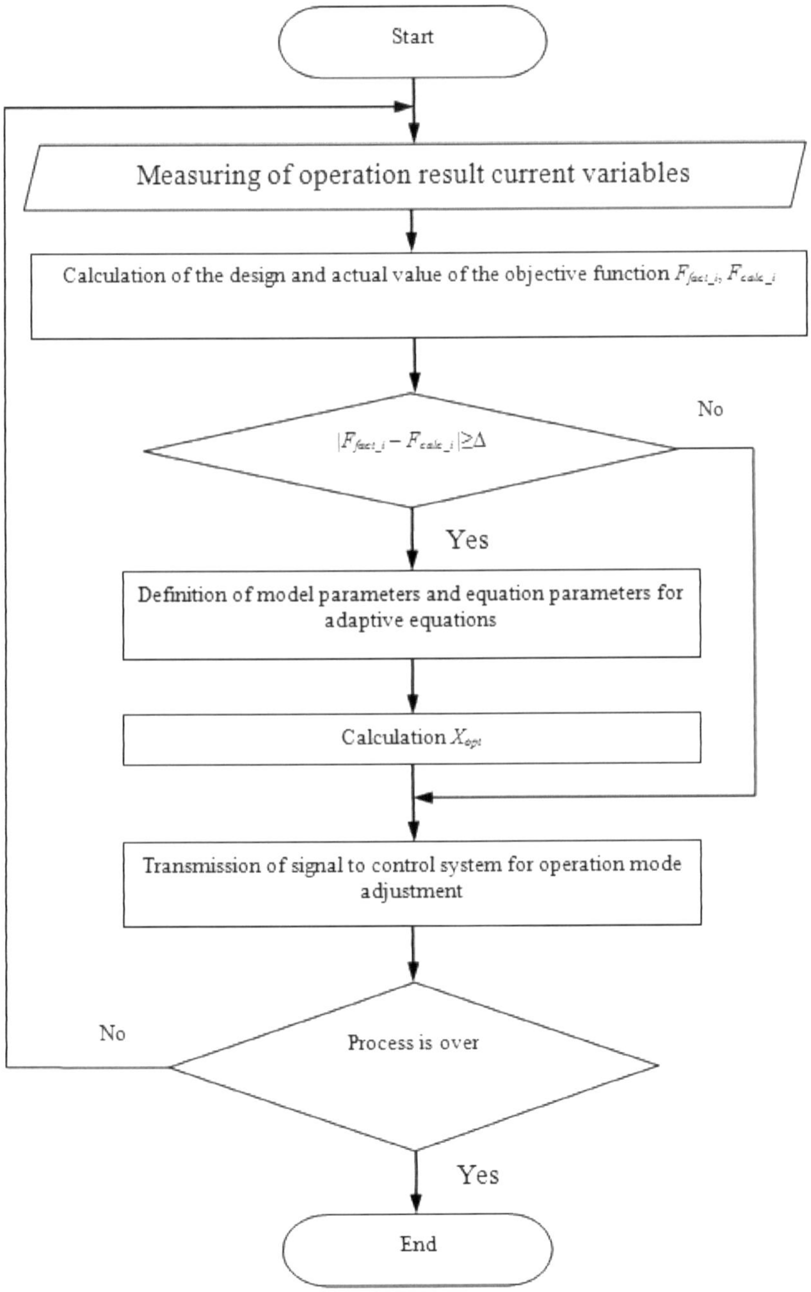

Figure 6. Control algorithm for the process of integrated electronics engineering

5. Conclusion

The paper presents developed algorithms to manage the production process in the manufacturing of integrated circuits. Within it, the algorithm of the choice of technological equipment, a model of the production process, and the algorithm of the control of the technological process of production are developed, and recommendations for the optimization of the production of integrated electronics are given.

IOP Conf. Series: Materials Science and Engineering **498** (2019) 012041 doi:10.1088/1757-899X/498/1/012041

References

[1] Borisov A N, *et al* 1989 *M. Radio i Sviaz* p 304 (in Russian)
[2] Lisiev G.A. Popova I.V. 2011. *Decision Support Technologies: Textbook* (2nd ed., tereotype. - Moscow: FLINTA,) p 133
[3] Podlepetsky B I, *et al* 2016 *IOP Conf. Ser.: Mater. Sci. Eng.* **151** 012007
[4] Ivashov E N, Korpachev M Yu and Kostomarov P S 2012 *Mater. Int. Sci. Tech. Conf.* **7** 7 pp 82-87
[5] Krasheninnikov V R 2003 (*Ulyanovsk: UlSTU)* p 150
[6] Kramer G 1976 *Mathematical methods of statistics* (Moscow: Mir, 648 p.) (in Russian)
[7] Yarushkina N G 2004 *Fundamentals of the theory of fuzzy and hybrid systems: textbook* (Moscow: Finance and Statistics) p 320 (in Russian)
[8] Shannon K 1963 *Works on the cybernetics and information theory* (Moscow: Publishing House of Foreign Literature) p 830 (in Russian)
[9] TsypkinYa Z 1984 *Fundamentals of the Information Theory of Identification* (Moscow: Science) p 320 (in Russian)
[10] 1969 *Adaptive systems. Reference book* ed Chinaeva P I (Kiev: Naukova Dumka) p 600
[11] Semenkin E S, Semenkina O E and Korobeinikov S P 1996 *Optimization of technical systems. Textbook* (Krasnoyarsk: SIBUP) p 284 (in Russian)
[12] Chulichkov A I 2003 *Mathematical models of nonlinear dynamics* (Moscow: FIZMATLIT) p 296
[13] 1973 *Adaptive systems for automatic control of complex technological objects* ed Alexandrovsky N M (Moscow: Energia) p 272
[14] Klyachkin V N 2007 *Automation Modern Technol* **7** pp 3-6
[15] Lotov A V and Pospelova I I 2008 *Multicriterion decision-making problems* (Moscow: MAX Press) p 197

IOP Conf. Series: Materials Science and Engineering **498** (2019) 012042 doi:10.1088/1757-899X/498/1/012042

Perspective of use of shungite cathode for microbial fuel cell design

A P Parhaeva, E S Bulycheva, D I Kadochnikov, M A Kulieva, E G Losevskaya, O E Sharapova, E A Shuvalova, E V Maslovskaya, D V Sosin, Y R Shaltaeva, A O Sultanov, N V Siglovaya and S A Shostachenko

National Research Nuclear University MEPhI (Moscow Engineering Physics Institute), Kashirskoe shosse 31, 115409 Moscow, Russian Federation

Corresponding author's e-mail address: yrshaltayeva@mephi.ru

Abstract. The article is devoted to one of the directions of alternative energy – the creation and use of microbial fuel cell (MFC). It is used anode made of plate of macro porous silicon doped with antimony. The cathode is a plate of shungite rock. Shungite is a natural stone containing fullerenes. The possibility of functioning and effective operation of this system is investigated

1. Introduction

The microbial fuel cell (MFC) is an environmentally friendly alternative source of electrical energy, in which microorganisms are used as biocatalysts to oxidize organic and inorganic matter and generate current [1]. Despite the fact that MFC strongly loses in efficiency to modern energy sources, which are widely used in industry there are able to produce a large amount of electricity. Unique feature of MFC is the ability to generate electricity from organic waste. Increase of MFC efficiency and spread of microbial fuel cell into various spheres of people's lives, will allow recycling into electricity toxic waste that pollute the environment.

In MFC, electrogenic microorganisms oxidize the organic substrate and transfer electrons to the anode. Further, the electrons flow on the electric circuit to the cathode and are connected to protons, oxygen, provided from air, and electrochemically activated catholyte [2].

It is known from the literature [3] that the maximum theoretically achievable potential difference between the anode and the cathode is about 1.1 V. In practice, it was achieved the voltage of 0.7-0.8V on the studied laboratory models of MFC [4]. This demonstrates a close to linear dependence of electrical power on the volume of the anode chamber and the area of the anode current-collecting surface.

2. Experimental equipment

The study of two-chamber MFC were carried out in the Lyceum 1511 with the sample taken from the river Chertanovka, Moscow. The design of the MFC cell is consisted of two interconnected chambers separated by an ion exchange system. A container (15 x 2,5 x 3 m*10^{-2}) consisting of 7 cells for the possibility of assembling several MFC at the same time was used as the body (Fig 1.I). One cell is not used.

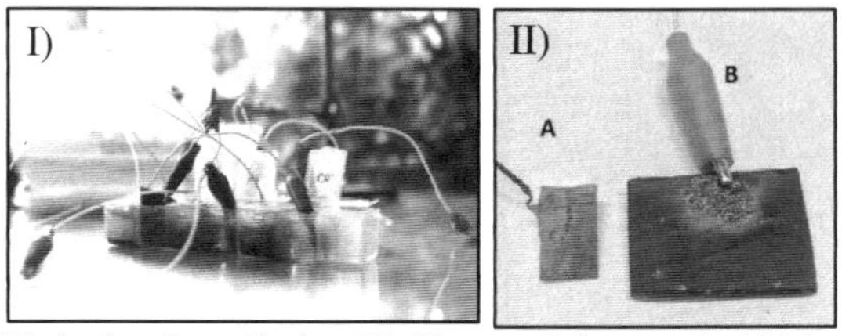

Figure 1. I) Photo of MFC during installation II) Photo of the anode (A) its surface of macro porous silicon doped with antimony and of cathode (B) of shungite C60, etc

The working volume of the anode and the cathode compartment $15.75*10^{-6}$ m^3. The chambers contain electrodes: the anode is made of a plate of macro porous silicon doped with antimony 1x1 $m*10^{-2}$, which for our experiment was produced in MEPhI. Of particular interest in the design of MFC is the use of high – tech materials as the anode and cathode-macroporous silicon and shungite (Fig 1.II).

Before using the silicon plate as an electrode, it was carried out microscopic study on the microscope OLIMPUS BX51M (Japan) (Fig.2).

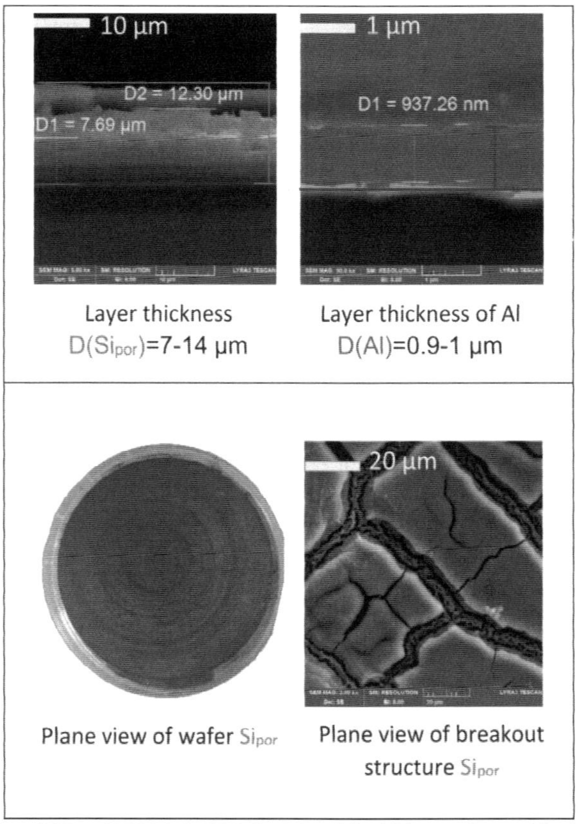

Figure 2. Photo of anode of Si_{por} (LYRA3 TESCAN) with Al layer for soldering and microphotograph of a plate of macro porous silicon doped with antimony with pore size measurements.

Electrodes of macro porous silicon was prepared by electrochemical etching method. Pore size varies in 120-150 nm range. It is observed cracking with crack size 10-20µm. Single-crystal electron antimony-doped silicon wafers Si_{por}, diameter d=76mm (silicon grade in Russian specification 76 KES) were used as a starting material. It was carried out etching in a solution of hydrofluoric acid and water at a current value of $120*10^{-4}$ mA/m^2. The current source is – AKTAKOM ATN – 2232. Diffusion doping of antimony samples was used to increase electrical conductivity. The shungite plate 2x3x0,2 $m*10^{-2}$ is used as a cathode. It is selected as an experimental material with a low cost. It was used ion exchange system based on the polymer monopolar heterogeneous ion-selective membrane MK-40, reinforced with nylon fabric (manufacturer OHK "Shchekinoazot"). The membrane was glued between anode and cathode chambers using glue kernel.

The cathode chamber is filled with a solution of 0.8% sodium chloride. The anode chamber is filled with mud (silt) from the Chertanovka river. It was used methylene blue $C_{16}H_{18}ClN_3S$ in concentration 0,99 mol/m^3 as a mediator in the anode chamber. To prevent oxygen from entering the chamber a float made of extruded polystyrene foam was placed above the liquid of the anode chamber. The top cover of the anode chamber is closed. The electrode was immersed completely in the liquid. The shungite plate was immersed in the cathode chamber by 50% to access oxygen for reducing reaction. The cathode chamber is open for better access of air.

The developed surface of macro porous silicon serves as an adhesive basis for microbial flora, which is part of the river silt. MFC design based on literature data [5]. The wires of the electrode parts of the installation have an output from the top of the structure.

The measurement is made by MASTECH MY-65 multimeter. The measurements were taken when the load was set in the form of a 10 kOhm resistor on the electric circuit.

Interest in the use of silicon is due to its high chemical resistance, manufacturability, the ability to obtain porous structures with a developed surface, to control the electrical conductivity of the volume and hydrophilic behaviour of the surface of such structures, as well as the ability to use basic technologies of microelectronics [6].

3. Experimental results and discussion

The experiment lasted for 9 days. The results are presented in Fig. 3 and 4.

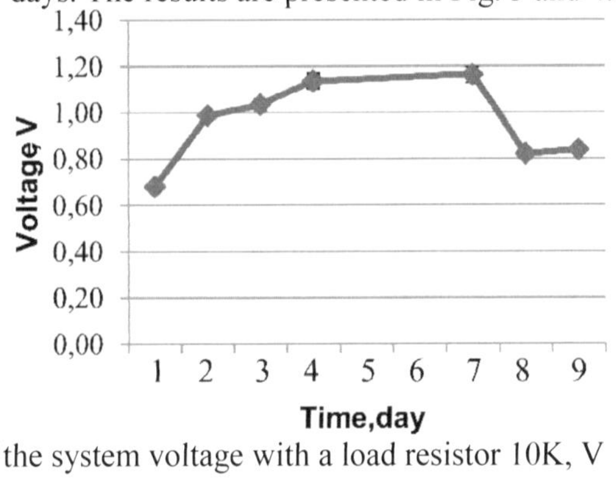

Figure 3. The dynamics of the system voltage with a load resistor 10K, V

Starting from the first day, the voltage in the system there was a tendency to increase and by the 7th day a was maximum of 1.16 V. Further, the voltage dropped from the maximum value more than to 30%.

The maximum power value was recorded from the 4th to the 7th day of the experiment, amounting to 0.16 mW.

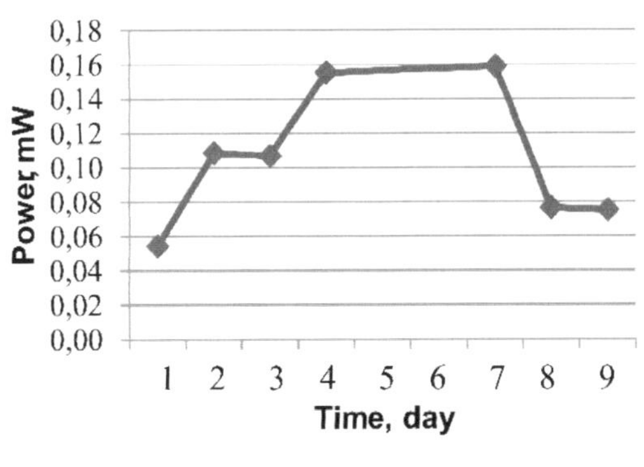

Figure 4. Dynamics of changes in the system capacity, mW

In our opinion, the decrease in the power of the MFC is due to a decrease in the concentration of nutrients and the achievement of the maximum density of bacteria in the anode chamber.

4. Conclusion

The Design of MFE using as electrodes of porous silicon and shungite not less efficient then counterparts found in the scientific literature [9]. During the experiment, the voltage value equal to 1.16 V was achieved, which corresponds to the maximum theoretically achievable value. It is planned to study the operating temperature range of this MFC.

Summing up we can say that the field of development of microbial fuel cells is actively developing and promising. Creation of new laboratory samples with experimental electrode materials is a promising area of research.

5. References

[1] Alferov S V, Alferov V A, etc 2014 Appl. Biochem. Microbiol. **570**
[2] Bukach O V, Myakinkova L L, 2014 Innovation and expertise **2** (13)
[3] Kalyuzhny, S V, Fedorovich, 2007 Chemistry and life. **36**
[4] Samkov A A, Volchenko N N, Hudokormov A A, Kalashnikov A A, Veselovskaya M.V, 2014 Scientific journal of Kubsau **101** 7
[5] Logan, Bruce E, 2008 Microbial fuel cells **200**
[6] Greenberg V. Shkundin and Niinikoski E, 2012 Modern electronics **10**
[7] Gorokhov M N, Lebedev K S, Platonov V V, 2014 Izvestiya Tulgu. Natural Sci. **236**
[8] Mirzayev R, Mustonen K, Monazam M R, Mittelberger A, Pennycook T J, Mangler C, Susi T, Kotakoski J and Meyer J C 2017 Sci. Adv. 3 (6)
[9] Sheudzhen T M, Volchenko N N, Samkov A A, 2017 Problems Sci. **15**

Rationale of the need to the development of semiconductor industry in Russia with the 28 nanometer semiconductor device fabrication node and below

A V Fomina*, V V Frantsuzova, Ya I Petrenko, D V Kornachev and A R Avanesyan

National Research Nuclear University MEPhI (Moscow Engineering Physics Institute), Kashirskoe shosse 31, 115409 Moscow, Russian Federation

*Corresponding author's e-mail address: AVFomina@mephi.ru

Abstract. This article describes the systemic problem of the development of the Russian semiconductor industry due to the lack of national production of integrated microcircuits with topological standards below 65 nm. Here are the results of the analysis of the feasibility of production lines of integrated circuits developing, depending on the forecasted volume of demand for various categories of semiconductor products, the technological backlog of Russian organizations and the estimated cost of the project. In this article, there are suggestions on possible sources of project financing and measures of non-financial support. Also, it describes the general technological parameters of the created production. It contains an assessment of key performance indicators of the project and the impact of its implementation on related technological areas.

1. Introduction

If we talk about the changes taking place during the last decade in the scientific, industrial and social spheres, it should be noted that most of them are related to advances in information technologies [1]. Moreover, the formation and distribution of such a cyber-physical system as the Internet of Things, predicted by leading analytical communities, will lead to further growth of sales markets and stimulation of the development of semiconductor technologies [2], [3]. The accumulated economic effect on a global scale for the period 2016-2025 is estimated at $12.8 trillion [4], for Russia this value for the same period can be from 70.1 to 153.5 billion dollars [1].

Digitalization of the economy, especially - industry, the administrative apparatus and defense industry involves the introduction of systems that allow information flows managing, through which direct influence on the functioning of physical systems is possible. This circumstance determines the strategic importance for the country's security of development of the national semiconductor industry, based on semiconductor production.

At the same time, it should be noted that the systemic problem of the modern national semiconductor industry is the closure of the scientific and production activities of industry organizations in the market's segment of defense industry. This circumstance determines a number of restrictions imposed on applicable models of the industry development in the medium and long term planning (up to 15 years).

2. Analysis of current situation

In the defense industry labor productivity is significantly influenced by a high level of requirements for reliability, the number and types of products tested. It should also be noted that military products have low seriality level, which imposes limitations on the possibility and rationality of the level of automation of the process of its creation. Due to the influence of these factors, much more man-hours are spent on the production of military products than in civil products, where production processes are much more unified and automated. As a result, the level of labor productivity, other things being equal, in the military industry will be significantly lower than in the civilian for the above reasons.

Support for the development of semiconductor product lines for advanced topological standards has an extremely high capital intensity (see Figure 1). Taking into account that labor productivity in the largest Russian semiconductor industries in 2016 did not exceed 4 million rubles per person per year, it can be concluded that those semiconductor industries, which focused on the production of military products are unable to ensure independent technological development

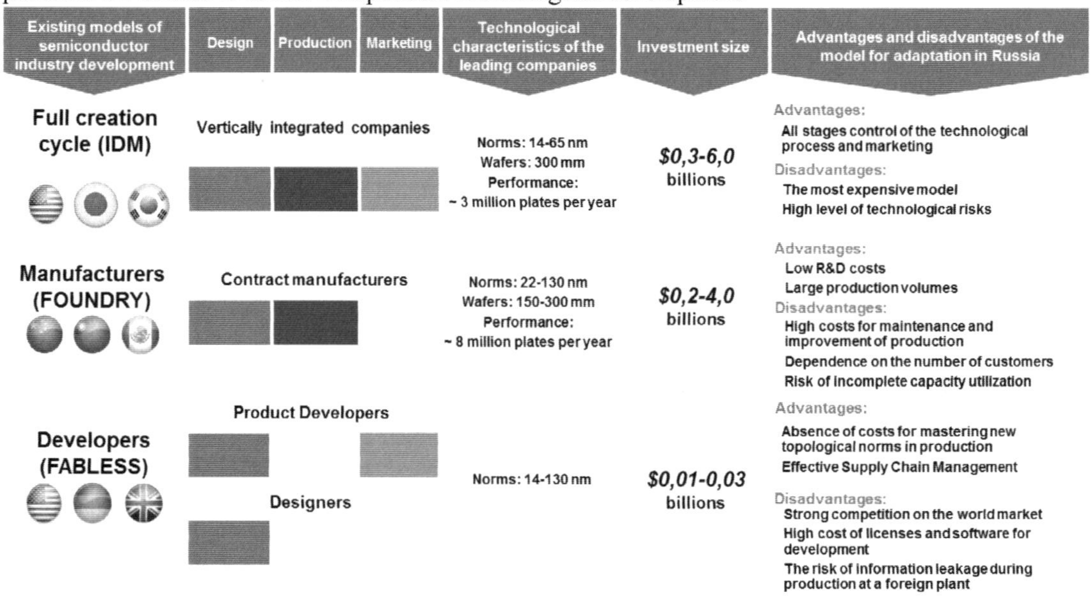

Figure 1. The cost of developing various models of semiconductor production

With the current world level of topological standards of 10-7nm on 300mm plates, the most advanced level available to Russian science is 65nm. Insufficient range of the microchips production technologies hinders the development of the industry.

At the moment, Russian design centers are already developing integrated circuits for 28nm and below technologies. They are forced to place their orders on the largest world factories (TSMC, GF SMIC, etc.). At the same time, the demand for semiconductor products with a technological level of 28 nm and below has already been formed and will only grow (see Table 1).

IOP Conf. Series: Materials Science and Engineering **498** (2019) 012043 doi:10.1088/1757-899X/498/1/012043

Table 1. Potential demand for Russian design centres the technology level of 28 nm and below.

Technology	Production Volume	Product Type
28 HPM	2 millions of IC per year	Processors
16 FFc	5 millions of IC per year	Processors
28 HT/ 22 FDx	0,1 millions of IC per year	Processors
28 HPP	2 500 pieces of wafers / 300 mm per year	PC, routers, network equipment
28 HPP	5 170 pieces of wafers / 300 mm per year	Processors, multiprocessor servers
22 FDx	8 000 pieces of IC per year	DSPC
22 FDx/ 14 нм	4 000 pieces of IC per year	Radiation-resistant SoC
28 нм	2 000 pieces of IC per year	Multimedia VLSI
22 FDx	500 pieces of wafers 300 mm per year	Processors
28 нм	200 000 pieces of wafers / 300 mm per year	IC for meters
28 HPM	800 pieces of wafers / 300 mm per year	SoC for telecommunication devices
28 HPM	5300 pieces of wafers / 300 mm per year	SoC for personal devices
16 FFc	600 pieces of wafers / 300 mm per year	SoC for server devices

According to world trends in 2017, the most of the silicon wafers were produced using technologies 28 nm and below (see Figure 2).

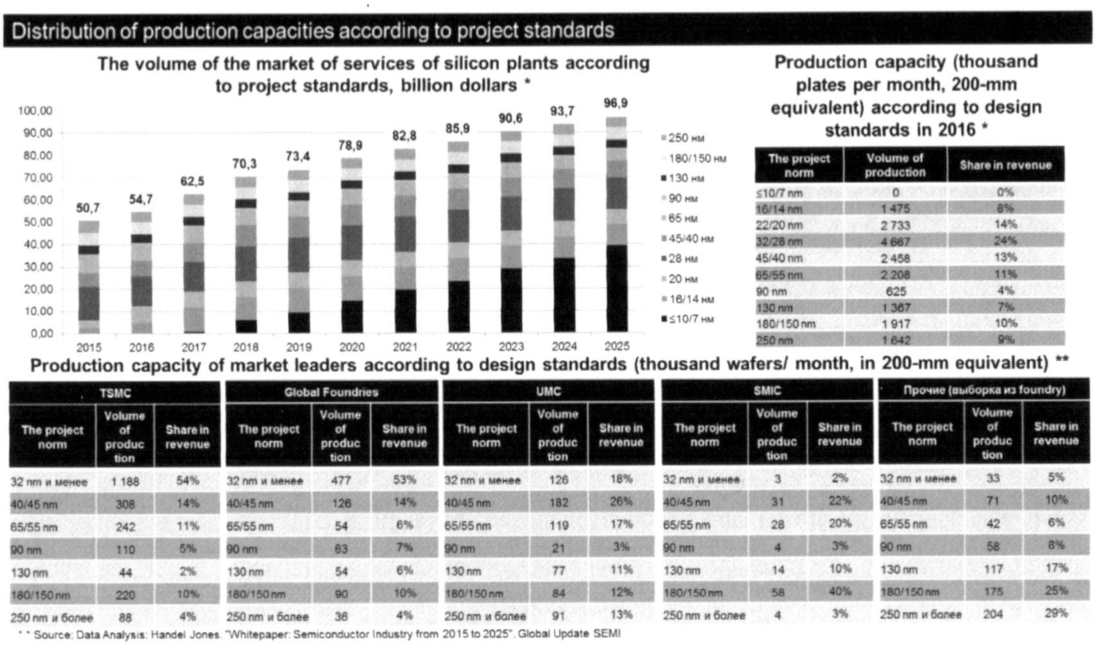

Figure 2. Distribution of silicon wafer production by technological level

Table 2 shows the estimated cost of the project, taking into account the existing technological reserve of leading Russian semiconductor companies

Table 2. Estimated cost of project cost.

Cost category	Cost of work
R&D	8%
Operating expenses	7%
CMP*	37%
Equipment	48%
Total:	from 55 to 80 billion rubbles.

3. Result and discussion

In order to ensure a sustainable long-term development of semiconductor production in Russia, it is necessary to form a set of regulatory and legal measures. A key condition for achieving profitability of production is the provision of direct state and quasi-public demand for the products of the factory (see Fig. 3).

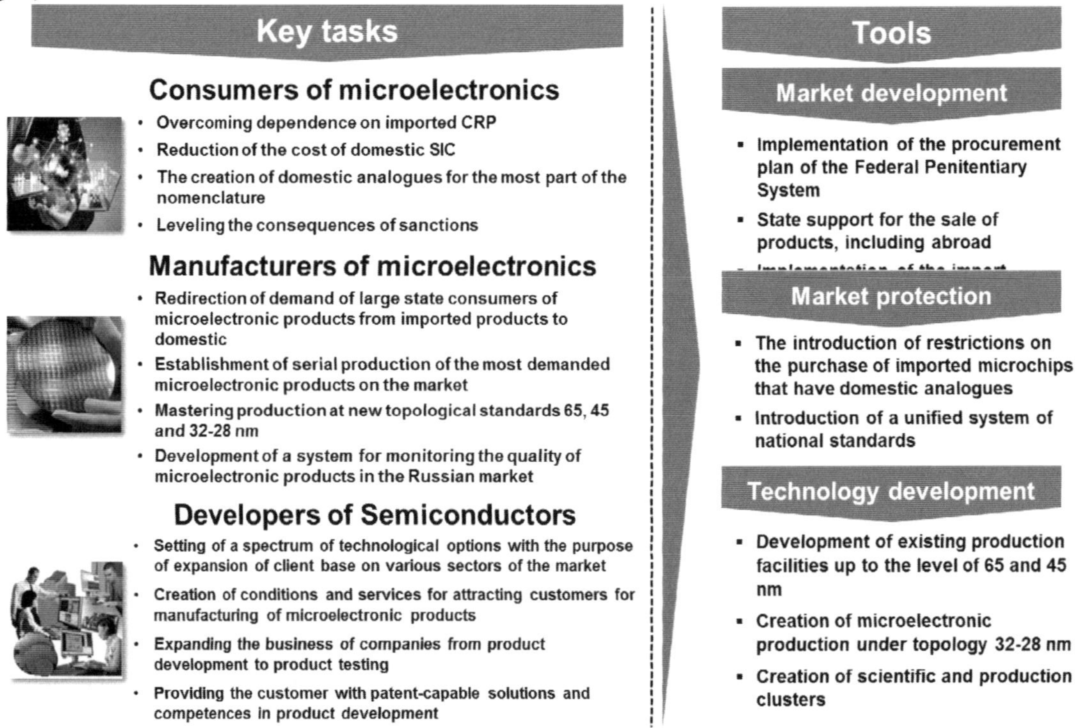

Figure 3. Solution of key tasks for the development of semiconductor industries in Russia

It should be noted that after the stage of construction and adaptation of technology, which is estimated to take 4 to 5 years, depending on the technology chosen, the first 1-2 years the basis of revenue will be military and dual-use products and only after that by 2023-2024 it is possible to exceed the share of orders of a civil character in the general order book of this factory due to the conversion and development of civil semiconductor production in Russia.

If we rely on the analysis of companies that currently have a technological process of 28 nm or below (see Figure 4), potential technological partners were chosen to create a semiconductor production with a 28 nm technological level on 300 mm diameter wafers, taking into account geopolitical and economic factors, companies represented in Table 3

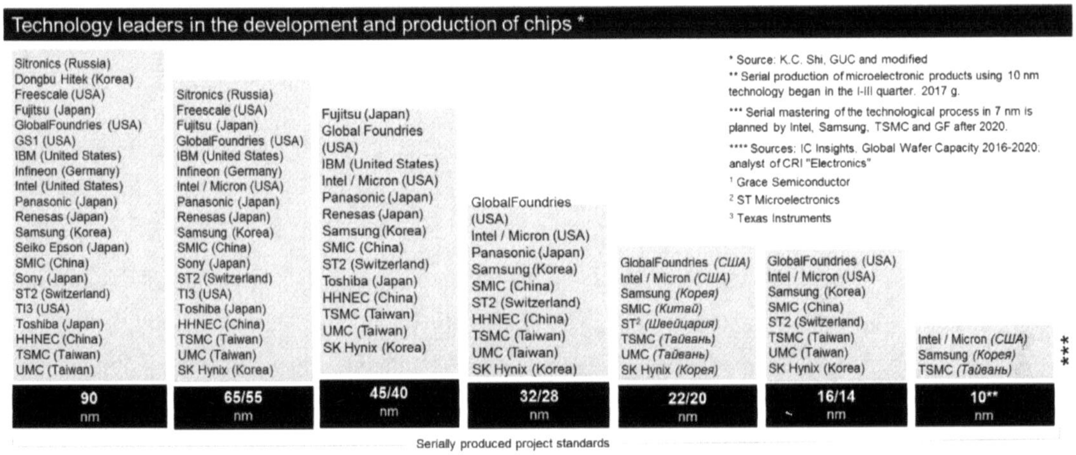

Figure 4. Technology leaders in the development and production of chips

Table 3. The list of potential technological partners.

Company	Country	Type of company
UMC	Taiwan	Foundry
GF	USA / UAE	Foundry
Imec	Belgium	Science Center
SMIC	China	Foundry
ST	France / Italy	IDM
Samsung	Korea	Foundry
TSMC	Taiwan	Foundry

4. Conclusions

According to the analysis of international experience, it can be argued that even with full government funding and the full required set of regulatory and legal support measures, it is highly likely that the best result of the factory's operations will be the achievement of a break-even point.

Nevertheless, this project will be strategically important, being the basis for carrying out the policy of import substitution in civil products, the creation of national cyberphysical systems, equipping state institutions and departments with trusted equipment.

Also, this project will be especially important in the field of preservation and development of the human resources potential of the Russian semiconductor industry, which is one of the most highly qualified and difficult to recover in the economy.

IOP Conf. Series: Materials Science and Engineering **498** (2019) 012043 doi:10.1088/1757-899X/498/1/012043

References

[1] 2017 *Digital Russia: New Reality* (Digital McKinsey) **133** p 8 (in Russian)
[2] Briggs B *et al* 2017 *Tech Trends 2017. The kinetic enterprise* (Deloitte Development LLC.) **140** pp 2-3
[3] 2017 *Tech Trends 2017: The kinetic enterprise* (Deloitte University Press) **140** p 25
[4] Makushin M, Martynov V, Suhoroslova Y 2016 *ELECTRONICS: Science. Technology. Business* **124** p. 61
[5] Martin Ch, Snabe J H, Nanterme P 2017 *World Economic Forum. - Executive Summary* p 71

IOP Conf. Series: Materials Science and Engineering **498** (2019) 012044 doi:10.1088/1757-899X/498/1/012044

Ultimate design and testing TPTS-based control systems with using full-scaled physical models of nuclear power plants

I M Zhukov and A O Tolokonsky

National Research Nuclear University MEPhI (Moscow Engineering Physics Institute), Kashirskoe shosse 31, 115409 Moscow, Russian Federation

Corresponding author's e-mail address: AOTolokonskij@mephi.ru

Abstract. Radiation degradation rate of base current in SiGe HBTs was experimentally investigated using X-ray irradiation source with Cu anode at room and low temperatures. The dependences of base and collector current on the emitter-base voltage of the transistors were measured during radiation impact and presented for different total dose levels and irradiation conditions.

1. Introduction

Development of I&C for Nuclear Power Plants is complicated by difficult of processes, count of signals and equipment. One else difficulty is requirements to equipment of I&C. Russian design power plants used TPTS equipment produced by VNIIA, which certificated by ISTEC (all classes of protection IEC 61226).

Equipment of VNIIA is 90% I&C of nuclear power plants. Design of all of these parts of I&C is a task of a few big project institutes (JSC Atomenergoproject, JSC Atomproject etc.). These institutes are participated to process of design for all lifecycle. It produces concept project which contains algorithms which will be programs for microcontrollers. One else concept project contains description of electricity connection and requirements to reliability. It's main task of project because it influents on count of equipment for I&C (count of microcontrollers) and its location on power plants. I&C for «Belarussian» NPP (is the most modern NPP of Russian design) involves over than 3500 TPTS controllers, which resolve different types of tasks of control systems. Such difficult distributed control systems need special methods for development and testing.

2. Experimental equipment

For programming of TPTS controllers used special program which name is GET-R1. Developer can create not trivial logic which can involve function of reliability and diversity and these algorithms can be translated into «machine» code (STEP code) and loaded into controller by this program. GET-R1 certificated by international institutes and can be used on NPPs.

Specialists of cathedra №2 of MEPhI developed special programming module (SimuNT) which was involved into GET-R1 for testing and modeling algorithms with using full-scaled model of power plants.

Full-scaled model of power plants is integrated information system which includes models of neutron kinetics, thermal hydraulics and model of power electricity. All of these models are integrated by special information system which name is ENICAD. This program lets create full-scale models with over than 3*106 variables. Full-scale model is used in simulators for NPP and testing algorithms of I&C.

IOP Conf. Series: Materials Science and Engineering **498** (2019) 012044 doi:10.1088/1757-899X/498/1/012044

Programming module SimuNT lets create special testing scripts on language Python to estimate reliability of control systems and algorithms. These scripts can imitate inputs (or even states of inner memory cells) of TPTS based algorithms or model inputs (for example, output value of pressure sensor) according to technology instructions. Script estimate state of equipment (get data from full-scaled model) and estimate state of controllers after imitation. An automatic rollback hardware or model changes happens in the end of script execution.

Scripts can be executed with full-scaled model (Model testing). And the same scripts can be executed with real hardware (Hardware testing). Exist opportunity to execute package of scripts for testing all controllers and algorithms. Diagram functionality of script engine is presented in Fig.1.

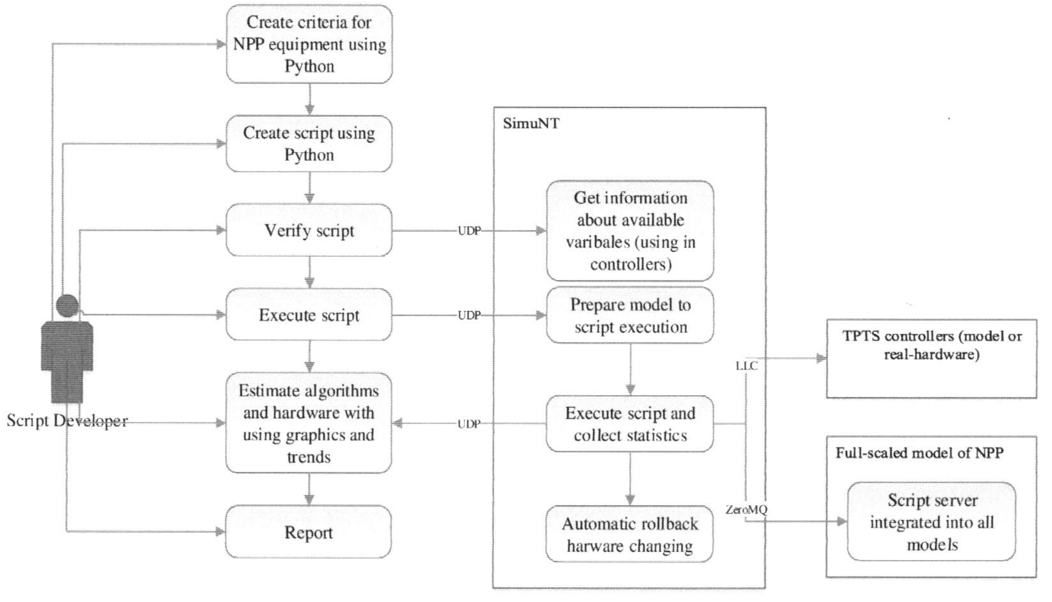

Figure 1. Diagram functionality of script engine

PyTest used for script engine. This framework allows create scripts with using special syntax. User can split tests on independent parts (it's the main part of unit module testing) with using this syntax, verify scripts in convenient IDE (PyCharm, Sublime, Visual Studio and others) and debug script. Python allow create template of scripts with using library Mako or create report in convenient format with using special python libraries.

Bellow example of the script is presented in Fig.2.

```
#!bin/python
# -*- coding: utf-8 -*-

from udp_client import *
import time

cli = Tclient()
##Allow test rewrite memory
cli.fs = 1
#Imitate 0.012 MPa of process function
cli["20KWB50CP001_HWIN"] = 0.012;
#Pause 1 sec
time.sleep(1);
#Check switch off equipment
if not cli["20KWB50AP001_XB01"]:
    raise Exception("Not switch off 20KWB50AP001");
##Rollback rewrite memory
cli.clear_force()
```

Figure 2. Example of testing script

This script allows check control function, which used in technology of NPP. It's technology protection of equipment (pump).

Script tests send messages with using UDP protocol to SimuNT module. SimuNT can be connected with full-scaled model of NPP. Library ZeroMQ is used for such connection. ZeroMQ is cross-platform, multithreaded library and can be used in supercomputers (full-scaled models is executed on such machines often).

SimuNT uses LLC protocol and hardware specific library for connection with TPTS controllers. Library uses certified crypto functions for protection. Crypto protection is the main part of script engine, because these scripts can be executed in any time of NPP lifecycle, even when power plants on 100% energy efficiency. This technology can be used for any NPP hardware of I&C. And it can be a standard of hardware testing for modern NPP. Example of using this testing system – is verification of algorithms control systems

BRU-A in the technological system of steam pipes of fresh steam. Verification was done by JSC Atomenergoproject with Modeling and Simulators Design Department of VNIIAES. Technology system with BRU-A equipment used in algorithms of safety system.

In the course of the work on the assessment of control loops, the work of the BRU-A was checked during emergency cooling of the reactor.

BRU-A is necessary to ensure the reliability of the steam pipe system and protects against bursting when the pressure is exceeded. When this event occurs, the steam from the steam pipeline is discharged into the atmosphere. The operation of the BRU-A is an emergency mode, because the BRU-K, PGI-PG units can cope with excess pressure, which allow to preserve the "working medium" and not lose the efficiency of the reactor installation.

For the NPP-2006 project, the emergency setpoint of the operation of the BRU-A is 6.8 MPa. After this pressure is reached, the surplus of steam is released into the atmosphere and the BRU-A enters the pressure maintenance level at 6.57 ± 0.1 MPa. To test the technological algorithms, a process model was developed that described the pressure dependence in the steam pipeline from the position of the BRU-A.

3. Experimental results and discussion

The verification program included simulating perturbations in the steam pipeline and checking the control algorithms. The mathematical model of the BRU-A regulator is presented below in Fig.3.

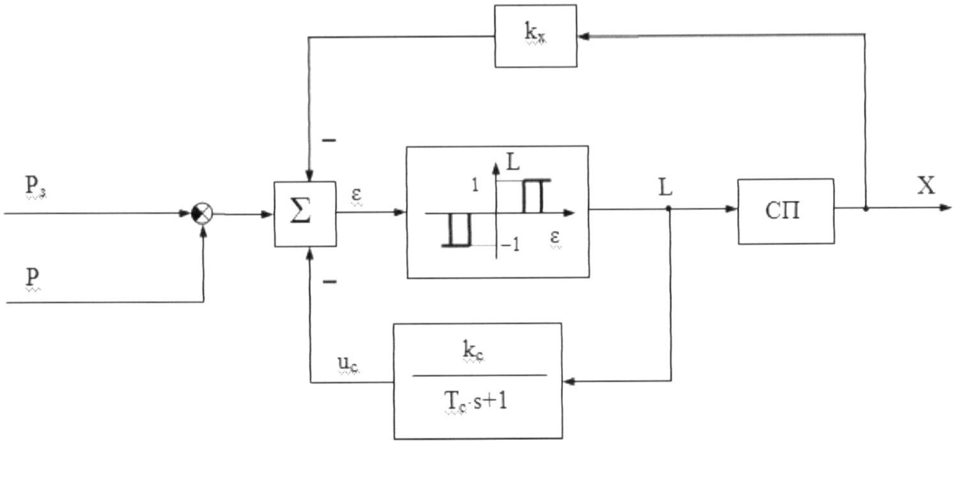

Figure 3. Structural diagram of the regulator BRU-A

The total signal is fed to the relay element: P – current pressure in the steam pipeline, MPa; Ps – setting pressure, MPa; kx – feedback gain by valve position, MPa/%; X – valve position, %; Uc – correction signal in the circuit of the three-position relay element, MPa.

The developed testing algorithms simulate an input disturbance of 6.81 MPa and show the process of maintaining pressure at the level of 6.57 MPa. The test results are presented below. The results allow to conclude that the main regulatory criteria are satisfied.

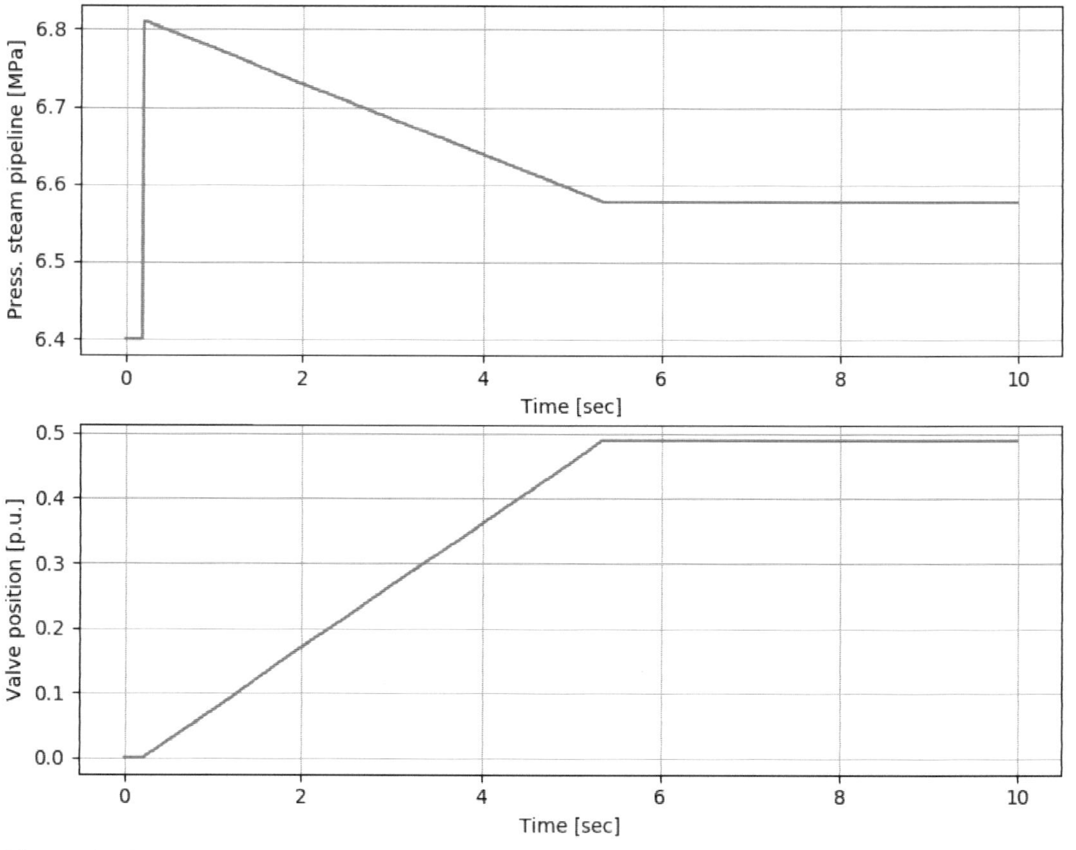

Figure 4. The process of maintaining the pressure in the GHG with the help of BRU-A at a pressure excess of more than 6.8 MPa

4. Conclusion

After the static testing of the task for automation, the controllers are programmed. The test scripts are connected to the equipment and simulate control actions, as well as on the model of the task for automation. The same execution of the results on the equipment and on the model allows to conclude that the validation procedure of the control system has been performed

References

[1] Bozhenkov O L and Kabachnikov A B 2009 *Nucl. Meas. Inform. Tech.* (Yadernyye izmeritel'no-informatsionnyye tekhnologii) **2** pp 27–30 (in Russian)

[2] Zverkov V V 2014 Automated control system for NPP technological processes (Moscow: MEPhI). **558** 150 (in Russian)

[3] Dunayev V G, Chernakov V A and Ljalyuk N G 2010 *Mater. Sci. Conf. VNIIAES.* (Materialy nauchnykh konferentsiy VNIIAES) p 20 (in Russian)

[4] Chernakov V A, Korolyov S.A., Zhukov S M 2009 *Abstracts Sci. Session MEPhI Nucl. Phys. Energy* (Tezisy dokladov nauchnoy sessii MIFI Yadernaya fizika i energetika) **1** p 173 (in Russian)

IOP Conf. Series: Materials Science and Engineering **498** (2019) 012045 doi:10.1088/1757-899X/498/1/012045

Multiple-trapping formalism for the description of charge transport in disordered organic semiconductors with correlated energy disorder

V R Nikitenko and A Yu Saunina

National Research Nuclear University MEPhI (Moscow Engineering Physics Institute), Moscow, 115409, Russia

*Corresponding author's e-mail address: vladronik@yandex.ru

Abstract. An applicability of transport level concept, as well as multiple trapping formalism for the description of charge transport in disordered organics with correlated disorder is supported by analytic modeling of temperature and field dependence of mobility. Results of this modeling are in good agreement with well-known results of Monte-Carlo simulations. The reasons for the description of transport by the multiple trapping model and the respective parameters are discussed.

1. Introduction

Transport properties of organic semiconductors are very important for the operation of organic electronic devices [1,2]. The modeling of transport in such materials is a specific problem, since, unlike inorganics, it occurs due to hopping of charge carriers between randomly distributed in energy localized states [3]. Numerical methods were widely used for the theoretical analysis of hopping transport – Monte-Carlo (MC) simulations [4] or numerical solution of the balance equation [5]. These methods are hardly applicable, however, to time-dependent and spatially non-uniform processes. On the other hand, the concept of transport level (TL) is known as a powerful tool for analytic modeling [6,7], because it allows describing of the hopping transport by a formalism of multiple trapping (MT) model [8-10] (the transport level is an analog of the mobility edge). The energy correlations in disordered organics can explain so-called Poole-Frenkel-type field dependence of mobility in a best way [4].

The TL concept was successfully used to modeling of charge transport in disordered organics in the framework of the uncorrelated Gaussian disorder model (GDM) [3], but its non-applicability for the case of correlated disorder was argued [11]. We will use below the term "Correlated disorder model" (CDM) for the case of long-range correlations, produced by the rather large randomly oriented molecular dipole moments, which is the case in polar organics [4]. Indeed, the first jump from initial state occurs usually to a neighbor state with proximal energy in the case of CDM. Consequently, the rate of this jump cannot be described as rate of a thermal activation to some final state of the energy, which is independent on initial state, as it is proposed in the TL concept. However, the rather deep initial state is surrounded by quasi-Coulomb potential well in the CDM, because the energy correlation function is in inverse proportion to the distance from the initial state, if this distance exceeds the typical inter-site distance [12]. Thus, the release of the carrier is a multi-step diffusion-type process, rather than one jump. In this work, the arguments in favor of applicability of TL concept for the organic semiconductors with long-range energy correlations are shown, and parameters of MT model for this case are discussed.

2. Field-dependent mobility by the use of transport level concept

It was shown [12,13] that in CDM the carrier release from a rather deep initial state could be treated as a release from Coulomb-type potential well:

$$U(r,\theta) = -\alpha |E_i| a/r - eFr\cos\theta\,, \tag{1}$$

where F – electric field strength, e – elementary charge, a – average intermolecular distance, θ is an angle between the direction of external electric field and radius vector of carrier, starting at the initial state, E_i – initial state energy, α – numerical constant. In this case the *effective* transport energy concept can be used, if the whole potential well is considered as a "collective trap". This approach was used to model the field dependence of mobility of charge carriers [12]. It was shown in the work [13], that it is the energy correlations which are responsible for the very form of Poole-Frenkel field dependence and the most known expression for this dependence in CDM was derived:

$$\mu = \mu_0 \exp\left(C_T + C_F\sqrt{\frac{eaF}{\sigma}}\right),$$
$$C_T = -\frac{9}{25}\left(\frac{\sigma}{kT}\right)^2, \quad C_F = 0.78\left[\left(\frac{\sigma}{kT}\right)^{3/2} - 2\right], \tag{2}$$

where F – electric field strength, e – elementary charge, a – average intermolecular distance, T – temperature, k – Boltzmann's constant and σ – width of Gaussian energy distribution of localized states, $g(E)$:

$$g(E) = \frac{1}{\sqrt{2\pi\sigma^2}}\exp\left[-\frac{E^2}{2\sigma^2}\right]. \tag{3}$$

The expression (2) is based on MC simulation results, thus the general applicability of the numerical constants is debatable. The aforementioned model [12] with the use of TL concept, on the other hand, is fully analytical and contains only one numeric constant, α. Field dependence of mobility is calculated according to equation by the use of Einstein relation:

$$\mu = \frac{1}{6}\frac{e}{kT}\frac{r_{ef}^2}{\langle t\rangle} \tag{4}$$

where $\langle t\rangle$ is the field-dependent average time of carrier release from the potential well and r_{ef} is the typical diffusion length of a carrier during the time $\langle t\rangle$ [12]. The time $\langle t\rangle$ could be expressed as the product of a reciprocal "attempts to escape rate" $\nu = \nu_0\exp(-2\gamma a)$, where ν_0 is a typical phonon frequency, and the reciprocal probability for a carrier to leave the well after the first jump $\eta(E_i)$, averaged over the initial erengy. In a weak field limit and neglecting energy disorder, $\eta(E_i) = \exp(E_i/kT)$ [14], one has to mention that $E_i < 0$. However, results of calculations show that an agreement with MC-results is sufficiently improved by the use of expression $\eta(E_i) = \exp\left[-(E_C - E_i)/kT\right]$, which represents the reduction of the potential barrier due to the disorder. In the course of random walk in a quasi-Coulomb potential well, a charge carrier gains a certain energy that is approximately equal to $E_C - E_i$ in a weak electric field and decreases with increasing field. Thus, average carrier energy after the release from a deep state does not depend on the initial state energy E_i. Averaging over initial energies yields [12] the analytic expression for the low-field mobility,

$\mu \sim \nu \exp\left[-0.5(\sigma/kT)^2 - E_C/kT\right]$, the same as for the case of uncorrelated disorder [10].

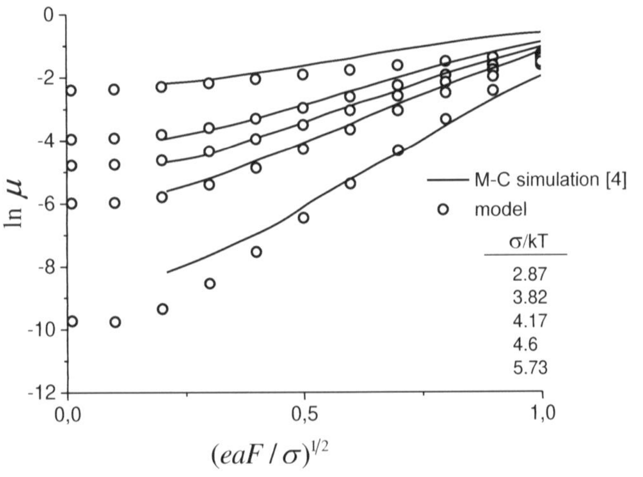

Figure 1. Field-dependent mobility in comparison with the results of Monte-Carlo simulation [4], using formula from ref. [12] and $\alpha = 0.68$.

Results of calculations of field-dependent mobility according to model [12] with $\alpha = 0.68$ are shown in Fig.1. This value of α is proximal to the constant from the estimated correlation function in the CDM, $C(r) \approx \alpha\sigma^2 a/r$, $r > a$: $\alpha = 0.74$ [13]; $\alpha \approx 0.7$, $r = a,\ 2a$ [5]; $\alpha \approx 0.65$, $a < r < 5a$ from our previous MC-simulations [15]. The value $\alpha = 0.68$ yields better agreement with MC data at strong disorder, than $\alpha = 0.74$, which was used in the work [12]. The temperature-dependent values of E_C were calculated in the same way, as in GDM [7]. Good agreement with the results of MC simulations allows to justify the applicability of transport level concept to disordered organic semiconductors with long-range correlated disorder.

3. Parameters of multiple trapping model

The multiple trapping model was developed originally in order to describe charge transport in semiconductors, controlled by traps (localized states) of various energies in the energy gap [8]. A carrier, which spends majority of time on these traps, transfers from a trap to another trap via delocalized states after the thermal activation to the conduction band (an electron) or to the valence band (a hole). Later on, the MT model was applied for the analysis of transport in amorphous inorganic materials like a-Si, As_2Se_3, controlled by the quasi-continuous (in energy) tail of localized states below (for electrons) and above (for holes) a critical energy – "mobility edge", which separates localized and extended states, see, for example, [9]. Thus, the total density of charge carriers, $p(x,t)$, is a sum of the density of "mobile" carriers, which occupy extended states, $p_c(x,t)$, and trapped carriers, characterized by the energy distribution of occupied traps, $\rho(x,t,E)$:

$$p(x,t) = p_c(x,t) + \int_{-\infty}^{E_C} dE \rho(x,t,E), \quad (5)$$

where E_C is the mobility edge.

Continuity equation implies that only mobile carriers contribute to the transport:

$$\partial p(x,t)/\partial t + \mu_c F \partial p_c(x,t)/\partial x = 0, \quad (6)$$

where μ_c is a mobility of mobile carriers, and F is the field strength. The 3[d] of the equations of MT-model is the balance (kinetic) equation, describing trapping and release processes (small-concentration limit is considered here:

$$\partial p(E,x,t)/\partial t = g(E)p_c(x,t)/(\tau_0 N_t) - \nu_0 \exp[-(E_C - E)/kT]\rho(E,x,t), \tag{7}$$

where N_t is a concentration of traps, τ_0 is a lifetime of carriers in conductive states. One has to note, that application of the TL concept to the calculation of mobility and diffusion coefficient and application of MT-formalism to the description of a hopping transport are not identical problems. Application of MT-formalism to the hopping transport is approved typically by the TL concept [2,10] (effective transport level E_C replaces the mobility edge, and energetically deeper states considered as traps). However, that application of MT-formalism to the disordered material with hopping transport does not require introducing the transport level explicitly. Indeed, the key feature of a disordered system of hopping sites is an exponentially large distribution of transition rates between different sites. Thus, all these sites could be characterized qualitatively as "h-centers", which contribute mostly to the hopping, and "t-centers", which serve as traps. Respectively, all carriers at any moment of time could be separated in two fractions- "mobile" and "trapped" ones, see eqs. (5), (6). If the energetic disorder prevails, and spatial and time scale of the problem exceeds the correlation radius of a percolation network [16,17], r_{corr}, and the respective time, t_c, one can write the balance equation for the functions, averaged over the mentioned scales [16],

$$r(E)^{-1}\partial \rho(E,x,t)/\partial t = \nu_0 K_0 g(E)p_c(x,t)/N_t - \nu_0 \exp(E/kT)\rho(E,x,t), \tag{8}$$

where $r(E)$ is a normalized trapping rate, which depends on the character of intersite hopping rates (in the Miller-Abrahams model, which is used here and in the work [12], $r(E)\equiv 1$); the parameter K_0 is related to the proportion of *t*-centers. Eq. (8) does not include the transport level explicitly. One can return to the form of eq. (7), providing that $K_0 = (\nu_0\tau_0)^{-1}\exp(E_C/kT)$.

One has to note, that the above consideration was carried out for the case of uncorrelated disorder. Returning to the CDM and considering a low-field transport, one has to note, that we can qualitatively accept the above reasoning, considering the potential wells around rather deep states, which control transport, as "collective traps". In the case of a weak external field, one can define the boundary of a well by the condition $U(r) = -kT$, which gives $r = r_C^* = \alpha a|E_i|/kT$, see eq. (1), in analogy with the Coulomb radius (the Onsager radius) $r_C = e^2/4\pi\varepsilon\varepsilon_0 kT$, where ε is relative permittivity. We will consider below the value $r_C^* = \alpha a(\sigma/kT)^2$, which corresponds to the characteristic energy $|E_i| = \sigma^2/kT$. A charge carrier leaves the potential well via sites located at $r \approx r_C^*$. Thus, in course of release from a deep state, a carrier covers typically a distance about $r_C^* >> a$, which is a typical spatial scale in the case of the CDM. Comparing the diffusion coefficient, which results from eq. (4) and from MT equations [7],

$$D = D_c\nu_0\tau_0 \Big/ \int_{-\infty}^{E_C} dEg(E)\exp[(E_C - E)/kT], \; D_c = (kT/e)\mu_c, \tag{9}$$

one obtains the relation $D_c\tau_0 = (1/6)(\alpha a)^2(\sigma/kT)^4$, which is specific for the CDM. The value of E_c seems to be proximal to the same for GDM, basing on the modeling of mobility [12] and MC simulations [15]. One has to note that the contribution of the temperature dependence of r_C^* to the mobility, see eq. (4), is quantitatively considerable.

4. Conclusions

Analytic modeling of mobility supports the applicability of multiple trapping model for the description

IOP Conf. Series: Materials Science and Engineering **498** (2019) 012045 doi:10.1088/1757-899X/498/1/012045

of hopping transport of charge carriers in disordered organics with long-range correlated disorder, if the spatial scale of the problem (for example, the film thickness) exceeds considerably the maximum from characteristic scales r_{corr} and r_C^*.

5. Acknowledgements

The work was supported by "improving the competitiveness" program of the National Research Nuclear University «MEPhI» and by the Volkswagen foundation, grant "Understanding the dependence of charge transport on the morphology of organic semiconductor films".

References

[1] Nikitenko V R, von Seggern H 2007 *J. Appl. Phys.* **102** 103708
[2] Nikitenko V R, Tameev A R, Vannikov A V, Lachinov A N, and Bässler H 2008 *Appl. Phys. Lett.* **92** 153307
[3] Bässler H 1993 *Phys. Stat. Sol. B* **175** 15
[4] Novikov S V, Dunlap D H, Kenkre V M, Parris P E and Vannikov A V 1998 *Phys. Rev. Lett.* **81** 4472
[5] Cottaar J, Coehoorn R and Bobbert P A 2012 *Phys. Rev. B* **85** 245205
[6] Baranovskii S D 2014 *Phys. Stat. Sol. B* **251** 487
[7] Nikitenko V R and Strikhanov M N 2014 *J. Appl. Phys.* **115** 073704
[8] Rose A 1951 *RCA Rev.* **12** 362
[9] Rudenko A I and Arkhipov V I 1982 *Phil. Mag. B* **45** 177
[10] Nikitenko V R, von Seggern H, Bässler H 2007 *J. Phys. Condens. Matter* **19** 136210
[11] Novikov S V, Malliaras G G 2006 *Phys. Stat. Sol. B* **243** 387
[12] Nikitenko V R, Saunina A Yu, Tyutnev A P and Prezhdo O V 2017 *J. Phys. Chem. C* **121** 7776
[13] Novikov S V, Vannikov A V 1995 *J. Phys. Chem.* **99** 14573
[14] Onsager L 1938 *Phys. Rev.* **54** 554
[15] Nikitenko V R, Sukharev V M 2016 *J. Phys.Conf. Ser.* **741** 012080
[16] Zvyagin I P 1984 *Kinetic Phenomena in Disordered Semiconductors* (in Russian) (Moscow: Moscow State University Press)
[17] Shklovskii V I and Efros A L 1984 *Electronic Properties of Doped Semiconductors* (Berlin: Springer)

IOP Conf. Series: Materials Science and Engineering **498** (2019) 012046 doi:10.1088/1757-899X/498/1/012046

Effect of microscopic Coulomb interactions on concentration dependence mobility of charge carrriers in organic materials

V R Nikitenko and A Yu Saunina

National Research Nuclear University MEPhI (Moscow Engineering Physics Institute), Moscow, 115409, Russia

*Corresponding author's e-mail address: vladronik@yandex.ru

Abstract. The filling of deep states is considered usually to be the reason of increase of the mobility of disordered organic materials with increasing concentration. However, at moderate concentrations the effect of microscopic Coulomb interactions (MCI) could be significant, because these interactions can reduce the activation energy of a hop. It is shown, that MCI results in significant additional increase of mobility along with concentration, in addition to the effect of the filling of deep states. The results are in qualitative agreement with existing theoretical and experimental studies.

1. Introduction

The mobility of charge carriers is one of the most important characteristics of transport. In recent years, considerable attention is paid to experimental [1] and theoretical [2-6] studying of transport in disordered organics in case, where the effect of charge carrier concentration on the mobility is significant. It was shown, that the mobility is increasing significantly with increasing concentration. This dependence is essential for organic field-effect transistors, where the relative concentrations could reach the value of 0,01 [1], but in case of ohmic contacts it should be also taken into account for organic light-emitting diodes, although the concentration values there are much lower.

The filling of deep states is considered to be the main reason of increase of the mobility with increasing concentration. In that case these states stop acting as "traps". This mechanism is studied in a couple of works, see for example [2-6]. However, at moderate concentrations the other effect could be more significant – strong Coulomb repulsion of randomly nearing charges, which accelerates the release of charges from deep states. This effect is considered in present work analytically.

2. Model for mobility calculation considering microscopic Coulomb interactions

Transport in organic materials occurs due to hopping of charge carriers between localized states, randomly distributed in energy [2,7,8]. Consider a carrier (a test charge), jumping on a typical distance Δr (in proximity to average distance between hopping centers). Energy of Coulomb interaction between the test charge and neighboring charges of the same sign, ΔU, varies randomly after a hop, since the direction of the hop is subjected to random scattering. The effect of the nearest neighbour, located at a distance r, is only significant at moderate carrier concentrations. The value of microscopic Coulomb field, $F = e/\kappa r^2$, where $\kappa = 4\pi\varepsilon\varepsilon_0$, is approximately uniform, hence it is proportional to variation of potential energy ΔU, since $\Delta r/r << 1$:

$$\Delta U = \frac{e^2}{\kappa|\vec{r}|} - \frac{e^2}{\kappa|\vec{r}+\overrightarrow{\Delta r}|} \approx eF\Delta r\cos\theta + \Delta U^{(2)} \approx eF_0\Delta r\,, \tag{1}$$

where θ - the angle between field and jump directions, F_0 is a projection of electric field on the direction of a jump. The term $\Delta U^{(2)} \cong eF\Delta r^2/r$ is negligible compared to thermal energy kT, if $c << c_{\max} \cong N_0^{-1/3}/r_c$, considering $\Delta r \cong N_0^{-1/3}$ ($r_c = e^2/\kappa kT$ is a Coulomb radius). The estimation at room temperature, $\varepsilon = 3$, $N_0 \cong 10^{21}$ cm^{-3} gives $c_{\max} \cong 0.01$. On the other hand, microscopic Coulomb interactions (MCI) could affect significantly the release rates from deep states (and, consiquently, the mobility), providing that $\Delta U > kT$. Considering eq. (1), one obtains $c > c_{\min} \cong \left(N_0^{-1/3}/r_c\right)^{3/2} << c_{\max}$. Apparently, in this approximation ΔU is nothing but interaction energy between the test charge and small-size dipole $\vec{p} = e\Delta\vec{r}$, situated on a distance r from it, see Figure. 1.

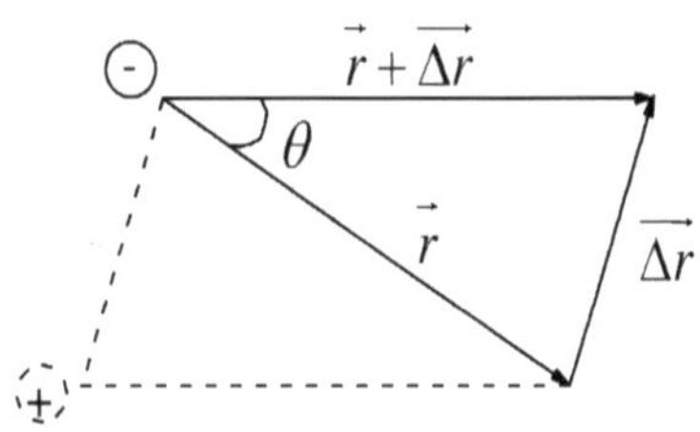

Figure 1. Scheme of test charge jump. The nearest neighboring charge and «virtual» opposite charge, forming dipole, are shown.

Since the jump directions $\overrightarrow{\Delta r}$ are equally probable, providing weak external field, the dipoles are randomly oriented. Using the Novikov's result [9], we get distribution function of ΔU :

$$P(\Delta U) = \frac{1}{2}\int_0^\infty w(r)dr\int_{-1}^{1} d(\cos\theta)\delta\left(\Delta U - \frac{e^2\Delta r\cos\theta}{\kappa r^2}\right) =$$
$$= \frac{\kappa}{2e^2\Delta r}\int_0^\infty w(r)r^2\theta\left(\frac{e^2\Delta r}{\kappa r^2} - \Delta U\right)\theta\left(\frac{e^2\Delta r}{\kappa r^2} + \Delta U\right)dr = \frac{\kappa}{2e^2\Delta r}\int_0^{\sqrt{e^2\Delta r/\kappa|\Delta U|}} w(r)r^2dr, \tag{2}$$

where $w(r)$ is the distribution function of the distance to nearest neighbour, which is selected on the base of the following reasons. Long-range asympotics corresponds to Poisson's distribution (random distribution of the nearest neighbors). At small values of r the Coulomb repulsion should be taken into account. Indeed, owing to energetic disorder, one of the carriers occupies deeper state, while the other moves adiabatically promptly. Therefore at small r the dependence $w(r)$ could be considered as quasi-equilibrium, $w(r) \propto \exp\left[-U(r)/kT\right]$, where $U(r) = e^2/\kappa r$. Thus,

$$w(r,c) = Ar^2\exp\left(-\frac{e^2}{\kappa kTr}\right)\exp\left(-\frac{4\pi}{3}N_0cr^3\right), \tag{3}$$

where A – normalization factor. Moving from potential energy to module of electric field strength in eq. (2), one gets the following distribution function with normalization factor C_F:

$$\Phi(F_0,c) = C_F \int_0^{\sqrt{\frac{e}{\kappa|F_0|}}} r^2 w(r,c)dr\ . \tag{4}$$

The function $\Phi(F_0,c)$ contains the parameter c, the relative concentration of carriers, as well as the function w. Consider uniform external electric field F_{ex}, directed along the x axis, in order to calculate mobility. Note, that mobility in materials with strong Gaussian disorder is defined by release of carriers from relatively deep states with typical energies of $-\sigma^2/kT$, $\sigma/kT >> 1$, if $c < c_*$, or from states near quasi-Fermi level, if $c > c_*$. The initial state is assumed to be deep enough in order to the jump in any direction being thermally activated. In the presence of both external and microscopic fields, the activation energy is changing by some value $\tilde{U} = \tilde{U}(F_{ex}, F_0)$. Typical release rate could be written as: $\nu(c) = \nu_0(c)\exp(\Delta\tilde{U}/kT)$, where ν_0 corresponds to zero field strength (however, it depends on relative concentration, c). The mobility is calculated as average drift velocity in single external field:

$$\mu(c,F_{ex}) = \frac{\langle \nu\Delta x\rangle}{F_{ex}} = \int_{-\infty}^{\infty} dF_0\Phi(F_0,c)\left[\frac{\nu_0(c)\langle\Delta x\rangle(F_{ex})}{F_{ex}}\int_{-\infty}^{\infty} d\tilde{U}G(\tilde{U},F_{ex},F_0)e^{\frac{\tilde{U}}{kT}}\right], \tag{5}$$

where $G(\tilde{U},F_{ex},F_0)$ is a distribution function of energy U_{ex}, which, in general, includes F_{ex} and F_0 as parameters. The analysis of this function is beyond the scope of this work. According to eq. (5), the mobility dependence on concentration could be determined by averaging of "local" mobility $\mu(F_0,c)$, see the expression in square brackets, which is calculated at a certain microscopic field F_0, with distribution function $\Phi(F_0,c)$. In the case of rather weak external field, one can neglect the dependence of mobility on that field,

$$\mu(c) = \int_0^{\infty} dF_0\Phi(F_0,c)\mu(F_0,c)\ . \tag{6}$$

The local carrier mobility $\mu(F_0,c)$ is calculated according to model, based on percolation theory [10,11]:

$$\mu(F_0,c) = \frac{\mu_0}{c}\exp\left[-\frac{E_C(F_0)-E_F(c)}{kT}\right]. \tag{7}$$

Here, the transport level E_C dependence on field strength F_0 yields well-known phenomenological, so called Pool-Frenkel field dependence of mobility, and is in accord with correlated disorder theory by Novikov et al. [12]:

$$E_C(F_0) = E_{C0} + kT\varphi\sqrt{F_0},\ \varphi = \tilde{C}\left[(\sigma/kT)^{3/2} - \Gamma\right], \tag{8}$$

where E_{C0} is an effective transport level in a low field limit [13], $\tilde{C} = 8.85\cdot10^{-5}$ (m/V)$^{1/2}$ and Γ=2.

3. Results and Discussion

Results of calculations of mobility dependence on concentration in a weak external electric field in comparison with experimental data (from well-known work [1]) and with results of Extended Gaussian Disorder Model (EGDM) [2] for two different values of disorder parameter are shown in Figures. 2 and 3. Note, that the experimental results (see empty symbols) cover only the low and the high concentrations regions. The used values of parameters $\Delta r = 1.2$ nm, $2\gamma\Delta r = 10$ - are in proximity to those used in the work [1] for analysis of experimental data. The qualitative agreement with the experimental data is achieved by the use the same values in eq. (8), as in the ref. [12].

Figure. 2 shows, that the model considering MCI gives result in qualitative agreement with experimental data (except of very high concentrations region), whereas other models do not provide sufficient mobility increase, at moderate value of disorder parameter, $\sigma / kT = 4$.

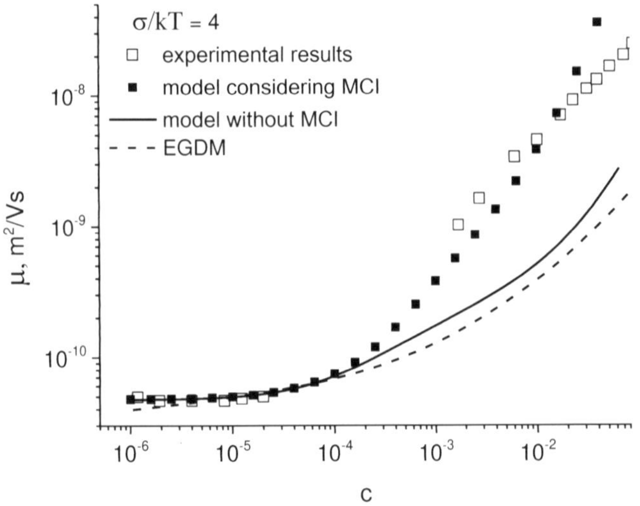

Figure 2. Concentration dependence of mobility in a weak external electric field for disorder parameter $\sigma / kT = 4$. Empty squares presents experimental data [1]. Full squares shows results of calculations according to eq. (7) considering MCI, see. eqs. (3)-(6), while the solid line is the same non considering MCI. Results of Extended Gaussian Disorder Model (EGDM) [2] are shown by line with symbols.

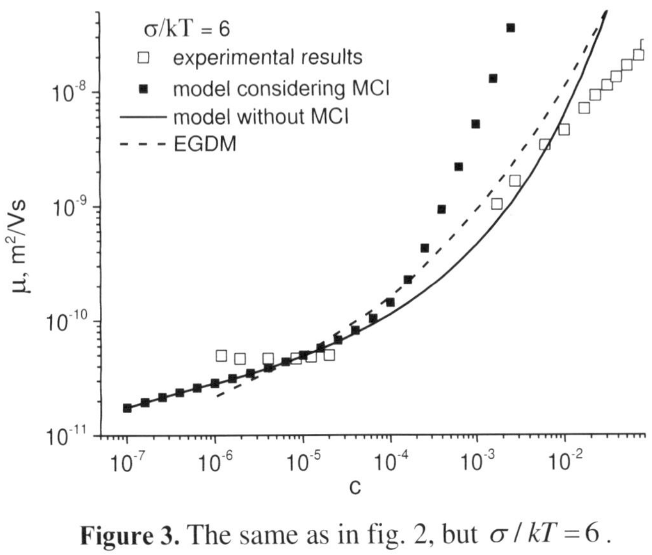

Figure 3. The same as in fig. 2, but $\sigma / kT = 6$.

The conclusion about quantitative agreement in moderate concentrations region (from 10^{-5} to 10^{-3}) could not be made due to the lack of experimental results in that region. The discrepancy between MCI model and the experimental data at high concentrations ($c > 0.01$) is not surprising, since the approximation (1) in this region is not applicable, and one need take into account not only the effect from the nearest neighbor, but also from other neighboring charges. From the Figure. 3 one can notice, that with large disorder parameter ($\sigma / kT = 6$, that gives $\sigma = 0.15$ at room temperature, proximal to $\sigma = 0.14$, used in [2]), good fit is obtained at moderate and high concentrations without considering MCI, while MCI provide too strong increase of mobility in this case. However, models not considering MCI provide too small mobility in limiting case of low concentrations, than the experiment.

4. Conclusion

Theoretical model of mobility dependence on carrier concentration in disordered organic materials at weak electric field, considering not only the filling of deep "tale" of DOS, but also Coulomb interactions between randomly nearing charges, was built. Microscopic Coulomb interactions (MCI) result in considerable increase of mobility at moderate concentrations, if the field dependence of mobility is rather strong. Previously, the effect of MCI ("short-range Coulomb interactions") on J-V characteristics (not directly on mobility) was studied by the means of Monte-Carlo modelling, and considerable *reduction* of current, at low voltages and small layer thickness, was demonstrated [14]. One has to note, that it is hard to distinguish effects of MCI on transport and injection in the case of thin (22 nm) layer and low (about 0.1 V) voltage [15]. Thus, further investigations are necessary. The analytic treatment of MCI in the present work do not claim to quantitative accuracy, being only the upper estimation, since only the interaction with the nearest charge of the same sign is considered. One has to stress, that various sets of parameters could be derived from analyses of experimental data, considering or not considering MCI. One of the aims of this work is to get the attention to a problem of independent (apart from J-V characteristics measurements) and reliable definition of disorder parameter, σ / kT, in order to validate one or another model of field dependence of the mobility, and to make conclusions about the effect of MCI on concentration dependence, using experimental concentration dependence of mobility.

References

[1] Tanase C, Blom P W M, de Leeuw D M and Meijer E J 2004 Charge carrier density dependence of the hole mobility in poly(p-phenylene vinylene) *Physica Status Solidi a-Applied Research* **201** 1236

[2] Pasveer W F, Cottaar J, Tanase C, Coehoorn R, Bobbert P A, Blom P W M, de Leeuw D M and Michels M A J 2005 Unified description of charge-carrier mobilities in disordered semiconducting polymers *Physical Review Letters* **94** 206601

[3] Oelerich J O, Huemmer D, Weseloh M and Baranovskii S D 2010 Concentration dependence of the transport energy level for charge carriers in organic semiconductors *Applied Physics Letters* **97** 143302

[4] Zvyagin I P 2008 A percolation approach to the temperature and charge carrier concentration dependence of the hopping conductivity in organic materials *Physica Status Solidi C - Current Topics in Solid State Physics* **5** 725

[5] Cottaar J, Coehoorn R and Bobbert P A 2012 Scaling theory for percolative charge transport in molecular semiconductors: Correlated versus uncorrelated energetic disorder *Physical Review B* **85** 245205

[6] Li L, Lu N, Liu M and Bassler H 2014 General Einstein relation model in disordered organic semiconductors under quasiequilibrium *Physical Review B* **90** 214107

[7] Bassler H 1993 Charge transport in disordered organic photoconductors - a Monte-Carlo simulation study *Physica Status Solidi B-Basic Research* **175** 15

[8] Baranovskii S D 2014 Theoretical description of charge transport in disordered organic semiconductors *Physica Status Solidi B-Basic Solid State Physics* **251** 487

[9] Novikov S V 2008 Thesis (Moscow: IPCE RAS)
[10] Shklovskii V I and Efros A L 1984 *Electronic Properties of Doped Semiconductors* (Berlin: Springer)
[11] Cottaar J, Koster L J A, Coehoorn R and Bobbert P A 2011 Scaling Theory for Percolative Charge Transport in Disordered Molecular Semiconductors *Physical Review Letters* **107** 136601
[12] Novikov S V, Dunlap D H, Kenkre V M, Parris P E and Vannikov A V 1998 Essential role of correlations in governing charge transport in disordered organic materials *Physical Review Letters* **81** 4472
[13] Nikitenko V R and Strikhanov M N 2014 Transport level in disordered organics: An analytic model and Monte-Carlo simulations *Journal of Applied Physics* **115** 073704
[14] van der Holst J J M, van Oost F W A, Coehoorn R and Bobbert P A 2011 Monte Carlo study of charge transport in organic sandwich-type single-carrier devices: Effects of Coulomb interactions *Physical Review B* **83** 085206
[15] Nikitenko V R, Sannikova N A and Strikhanov M N 2014 Features of the Current-Voltage Characteristics in Thin Conductive Layers of Organic Light-Emitting Diodes *Semiconductors* **48** 1494

Institute of Physics
Dirac House, Temple Back
Bristol BS1 6BE UK

ISSN: 1757-8981
ISBN 978-1-5108-8555-4

9 781510 885554

CURRAN ASSOCIATES INC.
proceedings
.com